POURING FOR PROFIT

A Guide to Bar and Beverage Management

COSTAS KATSIGRIS

MARY PORTER

JOHN WILEY & SONS

New York • Chichester • Brisbane • Toronto • Singapore

Library of Congress Cataloging in Publication Data:

Katsigris, Costas.
Pouring for profit.

Includes index.
1. Bartending. I. Porter, Mary.
II. Title.

TX950.7.K38 1983 647'.95'068 83-5887
ISBN 0-471-88900-8

Printed in the United States of America

10 9 8 7 6 5 4 3 2 1

Alcohol

It sloweth age; it strengtheneth youth; it helpeth digestion; it cutteth flegme; it abandoneth melancholie; it relisheth the heart; it lighteneth the mind; it quickeneth the spirits; it strengtheneth the hydropsie; it healeth the strangurie; it pounceth the stone; it expelleth the gravel; it puffeth away ventositie; it keepeth and preserveth the head from whirling, the eyes from dazzling, the tong from lisping, the mouth from snaffling, the teeth from chattering, and the throat from rattling; it keepeth the weason from stiffling, the stomach from wambling, and the heart from swelling; it keepeth the hands from shivering, the sinews from shrinking, the veins from crumbling, the bones from aching, and the marrow from soaking.

—An early enthusiast, quoted in J. C. Furnas, *The Life and Times of the Late Demon Rum*,

preface . . .

Here is the whole story of bar and beverage management, all in one book. It introduces you to planning a facility, equipping and staffing it, operating it, marketing it. It tells you all about the beverages—how they are made, purchased, watched over, and mixed into all of the different kinds of drinks people are asking for these days. Its focus is on the profitable enterprise as a going concern.

We have written this book to provide a usable, readable core of knowledge about beverages and beverage operations—the basics every present or future manager ought to know about every aspect of serving drinks for profit. To this end the scope is comprehensive. The level of discussion begins at the beginning, and the explanations are clear and simple. Yet the subject matter includes complexities of operation every manager must be prepared to meet.

Our orientation is practical and down to earth. This is management in practice. Nevertheless, theories and principles are still there—they are discussed in terms of choices and decisions to be made, as a manager faces them. The prospective entrepreneur can think in terms of what he or she will do, while the manager already in business will quickly identify with the questions and—we hope—find needed answers.

We have included a great deal of useful, up-to-date detail on products, procedures, practices, techniques, regulations, terms of the trade, so that the book can function as a guide and reference for anyone now in the industry or planning to enter it. We have photos and descriptions of bar equipment and how it works, useful tables and charts for easy reference, sample forms for keeping records and organizing data. Essential terms are boldfaced and defined. A Dictionary of Drinks gives today's popular recipes as well as some old standbys, well over 200 in all.

We have illustrated everything that needs to be seen to be understood. Many things are nearly impossible to describe, such as the shape of a hurricane glass or a bocksbeutel, or how to hold the bar strainer when pouring a drink, or the way to hook up beer kegs in series. We have included many how-to-do-it text/picture combinations, believing that this is the most explicit way to teach techniques short of a live demonstration.

An overall objective has been to make the reader aware of certain industry standards. We talk a lot about what it takes to make a good drink consistently—not only the structure of the drink and the liquors and the mixing technique, but the glass and ice, the care of beer lines and the pouring of the beer, and the selection and care and serving of wines. We also talk about meeting customer standards and expectations, about following health codes and government regulations, and about the point at which the profit motive yields to the customer's well-being. We discuss the age-old problem of the dual nature of alcohol and the responsibility of the seller, and we delve into history to give some perspective on government regulations and lingering attitudes toward people in the beverage industry.

This book can be used as a comprehensive guide to management for profit—from planning, equipping, and staffing to purchasing, budgeting, inventory management, marketing, and details of daily operation. Use it as a source book for ideas, as a storehouse of data and answers to specific questions, as a check on your current methods, as a mind-expanding experience in your own field. We think you will feel at home with us. One of us owns a profitable little place and the other is a discerning customer.

C. "Gus" Katsigris
Mary Porter

acknowledgments . . .

We have had a great deal of help in pulling this book together, especially since much of the material was not to be found in books and had to be assembled from people with firsthand knowledge. We are particularly indebted to the following individuals and organizations:

Ray Randuk and Norman Ackerman of HRI Consultants International for sharing their expertise on bar design, consultant services, and the many elements that go into planning an enterprise.

Doug Balthrop of The Design Market for giving us his thoughts on design trends and working with clients, and for clarifying the concept of design as packaging.

Herschel Stewart of the Bureau of Alcohol, Tobacco and Firearms for reading the chapters on beverages and regulations and for helping us thread our way through the Standards of Identity and other federal laws as well as through the corridors of the federal offices.

Harold Fager, who as beverage director of El Chico Corporation gave freely of his experience in opening operations in different states, designing a limited drink menu and developing specialty drinks, hiring and training personnel, and many other phases of beverage management.

Albert Moulin of Brookhaven and El Centro Colleges and Accent Wines and Spirits for reading two versions of the wine chapter and for repeated and continuing advice and counsel, as well as for demonstrating wine and champagne service for our camera.

Joe Kanewske of Jack Tar Hotels and Peter Heinan, Jack Morgan, and head bartender Paul Pope of the Jack Tar/San Francisco for letting us invade their inner sanctums with camera and tape recorder.

Wayne Spraggins of Hussman Refrigeration for telling us everything we always wanted to know about ice, and to Joseph E. Geary of Whirlpool Corporation for reading parts of the manuscript dealing with ice, and Sam Leake of Texas Ice Machine Company for providing plastic ice cubes for our illustrator.

Bob Watson of Food Service Industries for sharing his knowledge of bar equipment and for loaning us some items for photographs.

E. M. Wiemar of the U.S. Department of Labor Wage and Hour Division for reviewing our section on compensation and clarifying the intricacies of overtime pay, tipped employees and tip pools, and exempt personnel.

Jim Johnston of American Airlines, Jo Sinclair of Sky Chefs, and Bill Gesell of Ozark Air Lines for insights into airline beverage operations.

Stewart Ball and François Chandou of La Cave for letting us photograph their wine bistro and for adding to our knowledge of wines and our collection of wine labels.

Tony LaBarba of American Wine and Importing Company for a critical reading of the wine chapter.

Rebecca Murphy of Universal Restaurants for additional advice and counsel on wines and wine service.

Deeter Paul and Graham Compton of the Hyatt Regency/Dallas for behind-scenes glimpses of beverage operations.

Henry Africa, of Henry Africa's in San Francisco, for a mind-boggling interview on the saloon business in San Francisco and a dazzling display of the power of ambience, product value, attention to detail, and word-of-mouth advertising.

Ramona Cox of The Quiet Man for letting us capture a moment in the life of her postage-stamp domain.

Kelly Cross and Marge Alfelda of Crackers Restaurant for critiquing the mixology chapters and sharing their drink knowledge and bartending skills.

George Schepps of Julius Schepps Wholesale Liquors for reading the beer chapter and lending us his beer book.

The many many people and organizations who responded to our requests for information, including Distilled Spirits Council of the United States, United States Brewers Association, several state alcoholic beverage commissions, Jack Daniels Distillery, Hublein/Spirits Group, Anheuser Busch, Miller Brewing, Bud Nicol of American Beverage Control, Mike Tunnell of Electronic Dispensers International, Jan Senuta and Paul Richard of NCR.

The many individuals who helped us gather pictures, especially Thomas McGee of The Perlick Company, who painstakingly assembled pictures as similar as possible to the equipment Perlick installed at the Jack Tar bar; Jim MacMillan of Libbey Glass, who had prints specially made to scale for us; Jo Sinclair of Sky Chefs; Charles Nixon, Public Affairs Officer of the Bureau of Alcohol, Tobacco and Firearms, who tracked down the photo of George Washington's still even while dealing with the ramifications of the Reagan assassination attempt; Julius Schepps Co., Martin Sinkoff of Glazer's Wholesale Distributors, and Arwood H. Stowe for providing wine labels.

Dee Swope for his architectural drawings, Pat Roberts and Charlotte Speight for their painstaking camera work, and Carolyn King for providing bar-based drawings from which final illustrations were made.

All of these people have contributed immeasurably to the book. We have filtered their contributions through our own philosophy and conception of the book's needs, and any mistakes are our responsibility.

We owe a special debt to Tom Powers, not only for a particularly penetrating and sensitive final review, but for his encouragement and suggestion from the project's inception when he shepherded us into the Wiley fold. We are also indebted to the patience and persistence of Wiley editors Bob Pirtle, Judith Joseph, Maria Colligan, and Claire Thompson.

Special thanks go to Evelyn Katsigris, without whose many hours of overtime at the restaurant this book would never have been finished. Our thanks go also to Franklin Porter for his readiness to critique draft manuscript and test drink recipes.

C.K.
M.P.

contents . . .

POURING FOR PROFIT

A Guide to Bar and Beverage Management

chapter 1 . . .

THE INDUSTRY, PAST AND PRESENT

The drinking of alcoholic beverages is as old as human history, and the serving of drinks for profit is as old as profit itself. Both were, and are, taken for granted in everyday life in most cultures. In this country, however, it is as though the episode of Prohibition had somehow dropped a curtain between present and past. A look at the role of alcohol and its providers through the centuries can give a fresh and healthy perspective on the bar and beverage business today.

This chapter offers such a perspective with a look at the past and the present of beverages and the beverage industry. This sets the stage for digging deeper into specific facets of bars and beverages in chapters to come.

Fifty years ago the only bar and beverage business in the United States was carried on behind the locked doors of the speakeasy, and it was strictly illegal. Today the beverage-service industry—strictly legal—is one of the most profitable in the country. With the food-service industry (and it is difficult to separate them), the two together form the fourth-largest industry in the United States.

Fifty years ago the country was nearing the end of Prohibition—a disastrous attempt at regulating alcohol consumption by outlawing alcohol entirely. It did not work. History tells us such attempts have never worked, because people find other ways of getting what they want. And from earliest times human beings have wanted alcoholic beverages. Indeed, some historians theorize that one of the reasons our nomad forebears settled down to civilized life was to raise grain and grapes to ensure supplies of what they looked upon as sacred beverages.

What our ancestors drank and why . . .

Perhaps 8000 to 10,000 years ago someone happened to discover that fermented fruit or grain or milk or rice tasted good, made one happy, or both. One legend claims that wine was discovered by a neglected member of a Persian king's harem, who tried to end her loneliness by drinking from a jar marked *Poison* containing grapes that had fermented. She felt so much better after drinking the liquid that she gave a cup of it to the king, who named it "the delightful poison" and welcomed her back into active harem life.

Early peoples all over the world fermented anything that would ferment—honey, grapes, grains, dates, rice, sugarcane, milk, palms, peppers, berries, sesame seeds, pomegranates. When Noah settled down after the flood, he planted a vineyard "and he drank of the wine and was drunken." With all its benefits and hazards, alcohol was a universal feature of early civilizations.

These early beers, ales, and wines were considered gifts from the gods—miracle products with magical powers. People used them universally in religious rites, and they still do. The Israelites of the Old Testamant offered libations to Jehovah. Greeks and Romans honored Bacchus, god of wine. Christians used wine in their sacrament of communion. Primitive peoples used fermented beverages in their sacred rites.

Victories, weddings, and other sacred and joyous occasions were celebrated with "mellow wine" or endless supplies of ale (the word bridal comes from bride + ale). Goodfellowship was honored by sharing a loving cup passed around the table until it was emptied.

In many cultures people believed intoxicating beverages were linked with wisdom. Early Persians discussed all matters of importance twice—once when they were sober and once when they were drunk.[1] Saxons in ancient England opened their council meetings by passing around a big stone mug of beer. Greeks held their famous symposiums—philosophical discussions—during hours of after-dinner drinking. In fact, the word symposium means "drinking together." As the Roman historian Pliny summed it up, "In wine there is truth (*In vino veritas*)."

Alcoholic beverages were used for centuries as medicines and tonics, often in combination with herbs. Indeed, herbs and alcohol were among the few ways of treating or preventing disease until 100 or so years ago.

Probably the most important use of alcoholic beverages in the past was for the simple purpose of food and drink. Bread and ale or bread and wine were the backbone of any meal for the ordinary person, with the drink considered part of the food. For centuries, in fact, it *was* part of the workers' food, providing up to half the calories needed for a day's heavy labor. These beverages were considered the only ones fit to drink, and with good reason. Household water was commonly polluted. Milk caused "milk sickness" (tuberculosis). But beer, ale, and wine were disease-free, tasty, and thirst-quenching—important in societies that preserved food with salt and washed down a diet of starches.

The importance of beer in the diet is illustrated in the landing of the *Mayflower* at Plymouth. The Pilgrims were headed for Virginia, but the ship was running out of beer. So they were "hasted ashore and made to drink water that the seamen might have more beer," wrote Governor Bradford later.

About the time of the Pilgrims' troubles, new types of alcoholic drinks—distilled spirits—were coming into use in Europe. They were made by extracting the alcohol in fermented liquids, so that the resulting beverages were many times more potent. The first distilled spirits, called *aqua vitae*, or "water of life," were used as medicines, but they were quickly taken over as beverages. Highland Scots and Irish distillers made whisky. The French distilled wine to make brandy. A Dutch doctor's experiments produced gin—alcohol flavored with the juniper berry. In Russia and Poland the distilled spirit was vodka. In the West Indies rum was made from sugarcane, while in Mexico Spaniards distilled the Indians' native drink to make mexcal, great-granddaddy of today's tequila.

With increasing supplies of spirits and their high alcohol content, excessive drinking became a national problem in several European countries. In England cheap gin became the drink of the poor. They could—and did—get "drunk for a penny, dead drunk for twopence," as one gin mill advertised, adding, "Straw provided free" for sleeping it off.

Americans welcomed the new spirits, and it wasn't long before rum became the most popular drink and New England became a leading manufac-

[1]This is soberly reported by the Greek historian Herodotus in *Persian Wars*, Book I, Chapter 133.

turer. George Washington put rum to political use when he ran for the Virginia legislature, giving each voter a barrel of rum, beer, wine, or hard cider. By the end of the century whisky was challenging rum. Washington was right in the mainstream again; he made his own rye from his own grain in his own stills (Figure 1.1).

The tavern: pleasures and politics . . .

Pouring for profit developed hand in hand with civilization. The clay tablets of the Old Babylonian King Hammurabi refer to alehouses and high-priced, watered-down beer. A papyrus from ancient Egypt warns, "Do not get drunk in the taverns . . . for fear that people repeat words which may have gone

figure 1.1 George Washington's still, made in 1787 and believed to have been one of five at Mount Vernon. (Photo courtesy Bureau of Alcohol, Tobacco & Firearms)

out of your mouth without you being aware of having uttered them." Greek and Roman cities had taverns that served food as well as drink; excavations in Pompeii, a Roman city of 20,000, have uncovered 118 bars. In both Greece and Rome some taverns had lodging for the night; some had gambling and other amusements.

After the fall of the Roman Empire, life in most of Europe became much more primitive. When next the taverns reappeared they were alehouses along the trade routes, with a stable for the horses, a place to sleep, and sometimes a meal. In England the public house, or pub, developed during Saxon times as a place where people gathered for fellowship and pleasure. An evergreen bush on a pole outside meant ale was served. Each pub was identified by a sign with a picture, such as the White Swan or Red Lion, since hardly anyone could read.

As time went on the tavern became a permanent institution all over Europe. There were many versions: an inn, a pub, a cabaret, a place of refreshment with entertainment and music and dancing, a meeting place. Neighbors gathered here to exchange the latest news and gossip over a mug or a tankard. In cities men of similar interests met for a round of drinks and good talk. In London's Mermaid Tavern, Shakespeare, Ben Jonson, and other famous poets met regularly. Lawyers had their favorite taverns, students theirs. Members of Parliament formed political clubs, each meeting in its favorite tavern to discuss strategy.

Whatever its form, the tavern was a place to enjoy life, to socialize, to exchange ideas, to be stimulated. The beverages intensified the pleasure, loosened the tongue, sparked the wit, "moistened the soul," as Socrates once said.

When Europeans emigrated to America they brought the tavern with them. It was considered essential to a town's welfare to have a place providing drink, lodging, and food; in Massachusetts in the 1650s any town not having a tavern was fined. Often the tavern was built near the church so that churchgoers could warm up quickly after Sunday services in unheated meetinghouses. A new town sometimes built its tavern before its church.

As towns grew into cities and roads were built connecting them, taverns followed the roads. You find small towns in Pennsylvania today named for early taverns: Blue Bell, Red Lion, King of Prussia.

It was in the taverns that the spirit of revolution was born, fed, and translated into action. Taverns were the rendezvous of the rebels. Here the Sons of Liberty were formed and held their meetings. The Boston Tea Party was planned in Hancock Tavern, while in the Green Dragon Paul Revere and 30 companions formed a committee to watch the movements of the British soldiers. In Williamsburg the Raleigh Tavern was the meeting place of the Virginia patriots, including Patrick Henry and Thomas Jefferson. In New York's Queen's Head Tavern a New York Tea Party was planned, and many patriot meetings were held there during the war. Renamed Fraunces Tavern after the Revolution (to get rid of the reference to the Queen), it is still in

figure 1.2 Fraunces Tavern, scene of much Revolutionary history, was named for owner Samuel Fraunces, a Jamaican known as Black Sam. Portions of this reconstructed landmark building date back to 1719. (Photo courtesy Fraunces Tavern Museum, New York City)

the tavern business. Downstairs is a restaurant/bar; upstairs is the room where Washington said goodby to his fellow officers (Figure 1.2).

When Americans pushed westward, taverns sprang up along the routes west. As towns appeared, the tavern was often the first building and the town grew up around it.

By the middle 1800s the American tavern was turning into a large-scale inn for the travelers and businessmen of a nation on the move. At the same time drinking places without lodging were appearing. These kept the name tavern, while the more elaborate inns adopted the name hotel. But the hotel kept its barroom; it was often a showplace, with a handsome mahogany bar and a well-dressed bartender who might wear gold and diamonds. Certain hotel bars became famous—the Menger in San Antonio where Teddy Roosevelt recruited Rough Riders, Planter's Hotel in St. Louis, home of the Planter's Punch.

By the turn of the century the successors of the early taverns had taken many forms. There were glittering hotels that served the rich in the world's cities and resorts. There were fashionable cabarets such as Maxim's in Paris where rich and famous men consorted with rich and famous courtesans, and music halls such as the Folies Bergères. There were private clubs, cafés ranging from elegant to seedy, big-city saloons that provided free lunches with their drinks, and the corner saloons of working-class districts where many an unhappy man drowned his sorrows in drink. The restaurant industry also made its appearance in the nineteenth century, serving wines and other beverages to enhance the diner's pleasure.

Prohibition and its effects . . .

Meanwhile, in the United States, a movement was growing to curb the use of alcoholic beverages. At first the name of the game was Temperance and its target was "ardent spirits" (distilled spirits). But it soon included beer and wine and shifted its goal from temperance to total prohibition. In a century-long barrage of propaganda and moral fervor the movement succeeded in selling many Americans the idea that drink of any kind was evil and led inevitably to sin and damnation. Only outlaw Demon Rum, they believed, and sin would disappear and Utopia would naturally follow. Along with this belief went the notion that those engaged in making or selling alcoholic beverages were on the devil's side of the fight.

The fervor was fed by the proliferation of saloons set up by competing breweries to push their products, many of them financed by money from abroad. By the late 1800s there was a swinging-door saloon on every corner all over small-town America as well as in the cities. It was often an unsavory place. There were far too many of these saloons to survive on sales of beer and whisky alone, and many became places of prostitution and other illegal goings-on.

The prohibition movement was also an expression of religious and ethnic antagonisms, of nativist, fundamentalist middle Americans against the new German and Irish Catholic immigrants. The brewers were German and the bartenders were Irish, and both brought with them a culture in which alcohol was an everyday fact of life. The movement also pitted small-town and rural America against what was perceived as big-city licentiousness.

During World War I the drys won their battle. The Eighteenth Amendment, passed during the wartime fever of patriotism and self-denial, prohibited the manufacture, sale, transportation, and importation of intoxicating liquors in the United States and its territories. Ratified by all but two states, Connecticut and Rhode Island, it went into effect in 1920.

Prohibition had a short and unhappy life—not quite 14 years. There was simply no way to enforce it. While legal establishments closed their doors, illegal speakeasies opened theirs to whispered passwords. Legal breweries and distilleries closed down, and illegal stills made illegal liquor—"moonshine"—by the light of the moon in secret hideouts. Illegal spirits were smuggled into the country from Canada and Mexico and from "Rum Rows" offshore—bootleg supply ships that sold to small fast boats that made the run to shore. Illegal beer and wine and gin were made at home.

Far from decreasing drinking, Prohibition almost seemed to invite it, as though flouting the law was the "In" thing to do. After nine years of Prohibition, New York City had 32,000 speakeasies, about twice as many as the number of pre-Prohibition saloons. To add to the problems, organized crime took over the bootleg business in many cities. Gangsters quickly became rich, powerful, and practically immune to the law. Adding racketeering and gang warfare to bootlegging, they became a national problem. Everyone agreed Prohibition was not working, including the drys. In 1933 Congress passed the Twenty-First Amendment, repealing the Eighteenth.

The wet-dry controversy was far from ended, however. States, counties, towns, and precincts, taking over the controls, made a wet-dry checkerboard of America. Much of it persists today.

After Repeal the beverage manufacturing industry, which had been the fifth-largest in the country when Prohibition shut it down, made a quick comeback in spite of stiff taxes and a lingering cloud of hostility. Today's successors to the early taverns are doing well too, though there still may be some feeling that it is not quite respectable to sell drinks to people who want to buy them.

These attitudes are part of what social chronicler J. C. Furnas has aptly called a "cultural hangover" from the long crusade against alcohol. Somewhere along the way, history books dropped from sight the positive contributions of alcoholic beverages and even the part they played in ordinary daily life. You probably never knew before that beer was once safer to drink than water, that Columbus carried half a gallon of red wine per sailor per day on the *Santa Maria*, that the Pilgrims never reached their destination because their beer was running out, that George Washington had five stills at Mount Vernon, that Paul Revere stopped to fortify himself with a couple of mugs of rum before he went off on his midnight ride, that founding father Samuel Adams was a brewer.

Here is another one: In 1833 Abe Lincoln and a business partner took out a tavern license. It allowed them to sell whisky, rum, wine, brandy, and Holland and domestic gins at retail, and it specified the prices. The price of whisky was 12½ cents a pint.

Today's beverage service industry . . .

Tavern prices are a little higher these days—$2 or more for a 1-ounce shot of whisky compared to less than a penny an ounce in Lincoln's day. Many other things have changed too, even in the years since Prohibition ended. New beverages have become part of the scene—vodka, tequila, and a host of liqueurs—and Americans are becoming wine drinkers. New equipment has speeded service—automated soda systems dispense carbonated mixers at the touch of a button and liquor-dispensing systems measure and count the drinks they pour. Computerized cash registers not only take care of money and keep records but provide instant sales analysis and can even pour drinks and adjust the inventory record.

One of the biggest changes is the extent to which alcoholic beverages have become an accepted part of the American scene. In 1934 only 28 states allowed the sale of distilled spirits. Today all 50 states and the District of Columbia allow their sale, and only 3% of the population lives in dry areas where local laws prohibit liquor sales. In 1934, right after Repeal, per capita consumption of distilled spirits in states allowing their sale was 0.65 gallons; in 1980 it was 1.99 gallons per capita and 2.76 gallons per adult. In 1935 less than half the adults in the country drank alcoholic beverages; today more than two-thirds of us do. More women drink than they used to; they drink almost as much as men, but they drink more wine than men, somewhat less spirits, and much less beer.

Not only are Americans drinking more themselves, they are accepting the serving of liquor in restaurants and bars as a normal part of the culture. Though in the Thirties most of the drinking was done in bars, taverns, and restaurants, drinking in a public place wasn't always considered socially acceptable, and people in the bar business were often looked down on. Women seldom went to bars, and never alone, and men often didn't want them there—a bar was considered male territory. (It is only recently that McSorley's Ale House in New York City consented to let women cross its threshold, and then only because of equal-rights laws.)

Today people drink both at home and in bars and cocktail lounges and wherever liquor is served. Restaurant patrons expect to be able to buy drinks and wine with their food, and restaurants that don't serve liquor often have a hard time competing. Although there may still be some traces of snobbery toward pouring for profit, more and more people on all social levels are getting a piece of the profits as owner-investors. (This seems to be particularly attractive to doctors and lawyers.) In any case, beverage service is a flourishing business: in 1980, 213,953 licenses were issued for on-premise sale of distilled spirits. A survey by *Institutions* magazine (October 1, 1980) reported that about 40% of all alcoholic beverage sales were on premise—that is, in a bar or restaurant.

Of these beverages, said the *Institutions* survey, 50 to 60% were distilled spirits, 10 to 20% were wine, and less than 20% were beer. (The figures don't quite add up, but the relationships are clear.) Wine and light beer sales were on the rise, and there was an increasing interest in premium beers. "White goods"—vodka, rum, tequila, gin—led in popularity. Amaretto was popular everywhere. Specialty drinks were the newest enthusiasm.

A year later the same magazine reported lower percentages for on-premise sales of spirits and gains for beers and wines, with wine percentages increasing dramatically at higher income levels. In total sales (both on- and off-premise) light beers and wines showed the greatest increase, with wine leading spirits for the first time.[2]

Today's beverage service is a lively business, full of growth and competition and change, reflective of the society in which it operates. It has many manifestations. It has left behind the old-time saloon, yet there are nostalgic echoes of the best of that species in today's neighborhood bars and taverns where regulars congregate for fellowship and good cheer. Gone too are yesterday's Grand Hotels; most of them died in the early years of Prohibition, unable to make it financially without beverage service and the clientele that went with it. But today's hotels are flourishing successors, oriented to the business traveler, the vacationer, the conventioneer, all of whom keep up the demand for alcoholic beverages. Alcohol is even appearing at fast-food counters.

Let us look at a few different types of beverage service. It is really impossible to divide bars into a few categories; there are almost as many variations as there are bars. But certain kinds have distinct characteristics and types of service, and it may be revealing to see how they differ and what they have in common.

The beverage-only bar. The simplest kind of beverage enterprise is the bar that serves beverages alone, with no food service beyond thirst-provoking snacks such as peanuts, crackers, and pretzels and maybe some cheese at happy hour. It serves beer, or wine, or mixed drinks, or any combination of the three. It may be a neighborhood gathering place, or a way station on the homeward trek, or a bar at an airport or bus terminal or bowling alley, or perhaps a stop along the highway.

Business at such places typically has a predictable flow—a daily pattern of peaks and valleys, a weekly pattern of slow days and heavy days, with the heavy days related to paydays and days off. There may also be seasonal patterns. In airports and bus terminals, business is geared to travel patterns, daily, weekly, and seasonally. Because there is only one product—alcohol—and because business is generally predictable, the operation of such enterprises is relatively simple, from production to staffing to purchasing to budgeting to keeping track of the beverages and the money and the profits.

This type of bar usually has a specific reason for success: it may be location, reputation as a friendly place, extra-good drinks, lack of competition, or maybe it has always been *the* place where everybody goes. Whatever its special reason, it fills the needs and desires of enough customers to make it go.

An example of a beverage-only bar is The Quiet Man in Dallas (Figure 1.3), a tiny off-beat place frequented by college professors, writers, and individualists of every income level who shun the typical bar scene. On a site

[2] *Restaurants & Institutions*, October 1, 1981.

figure 1.3 The Quiet Man, an oasis on a busy city street. (Photo by Pat Kovach Roberts)

that has housed a bar for 56 years (during Prohibition it was a gas station with a speakeasy in back), it serves beer, wine, mixed drinks, snacks, and conversation. It has a loyal coterie of regulars who cherish it for its lack of nearly everything other bars have—music, noise, food, decorator decor. They cherish it too for its other interesting customers, with whom they play darts, bridge, chess, and other games of skill, or just drink and talk, or just drink. Although it looks too small to pay its way, its present owners have netted a comfortable profit for 15 years.

Beverage-only bars are definitely a minority today. Although some are highly profitable, most bars find that operating on the basis of liquor alone is not enough. So the majority of bars have something else—either entertainment or food.

Bar/entertainment combinations. Bars offering entertainment range from the neighborhood bar with pool, pinball, or shuffleboard to nightclubs with big-name entertainers and ballrooms with big bands. Along the way are cocktail lounges and night clubs with live entertainment—piano bars, country-and-western entertainment, jazz groups, pop and folk music, rock groups. Then there are the disco bars with canned music and stylized dancing—great moneymakers while they last, but in many areas country-and-western dancing is pushing them out. Enterprises whose lure is invested in trendy types of entertainment must always be thinking ahead to the day when the fickle public moves on to the next fad.

In many such enterprises the beverage operation is subordinate to the entertainment in bringing in the crowds, but it is the drinks that provide the profits. If there is a cover charge it is likely to go to the entertainers. The fortunes of the bar rise and fall with the popularity of the entertainment unless the place has something else going for it.

Probably the most stable of the bar/entertainment enterprises is the smaller place with an attractive ambience, good drinks, and local entertainment with a regular following. Such places have much the same predictability as the bar-only enterprises. Large operations featuring out-of-town entertainers have a higher but riskier profit potential. It is likely to be either a feast or a famine. The bar gears up for each crowd with temporary extra help, a large investment in liquor inventory, and probably extra security personnel. A snowstorm bringing traffic to a halt can wipe out everything but the unused liquor.

Billy Bob's Texas in Fort Worth is the ultimate example of a liquor-plus-entertainment enterprise. With 177,000 square feet of building (3 acres) and 40 pouring stations, it is the world's largest night club. Entertainment includes top country-and-western singers and bands, professional boxing matches, and a live rodeo. Customers can ride live bulls too. In its first month (April 1981) Billy Bob's grossed more than $600,000 in liquor alone, paying the largest gross-receipts tax in Texas history. Whether Fort Worth and its tourist trade are enough to support this huge place remains to be seen.

Food and beverage combinations. The most common form of beverage operation is one that is linked with some form of food service. One type is the restaurant bar whose products are simply additions to the food service. Drinks and wine are served at the dining table by the waiters and waitresses who serve the meal. The drinks may be poured at a service bar out of public view, or at a pickup station in a bar that serves customers while they are waiting for a table. In such an enterprise the major portion of the sales comes from the food service. However, the beverage sales often turn the profit for the enterprise. The only added costs are the wine and liquor, the bartender, and a minimum investment in equipment. The other necessities—the service personnel and the facility—are already built into the restaurant operation.

Another type of food-beverage combination is the bar that serves light food in addition to its drinks. It is the beverages and the bar atmosphere that dominate in this case, with the food just an added service offered to attract customers. In this type of enterprise the major sales volume comes from the bar.

A special variation of the drink-plus-food combination is the wine bar, emerging in the past three or four years. Here the customer can choose from a selection of wines by the glass or by the bottle, beginning with inexpensive house wines and going up in quality and price as far as the entrepreneur cares to go. Typical food offerings are fruit and cheese trays and hors d'oeuvre specialties.

Wine bars are part of the great American discovery of wine, and they face certain problems. One is that serving only wine limits the clientele to wine lovers plus those who like to follow the latest fads and fashions. Another is that wine by the glass can be a chancy business: the opened bottle cannot be saved for another day since it deteriorates quickly. If the last glass or two in the bottle is not sold, the profit for that bottle disappears. Finally, purchasing appropriate wines takes an expertise few people have and can require an investment few are willing to make.

As a result many wine bars have a full-service bar as well, in order to maintain a broad appeal and realize the necessary profit margin. In effect they are simply bars that have a wine-oriented ambience and serve wines as a specialty. Other wine bars may broaden their offerings by serving meals and, in effect, are restaurants with an emphasis on wines.

Still another variation of the wine bar is the one developed by La Cave in Dallas (Figure 1.4). It is a cozy, casual, candlelit place that serves a small selection of quality wines by the glass and a large, expertly chosen selection of wines by the bottle at retail price plus $1.50. Its special twist is that it doubles as a wineshop, selling wine to take home by the bottle or the case. About 60% of its sales come from the wineshop, but the wine bar is the major drawing card. Customers come in to linger over the wine and the paté

figure 1.4 La Cave ("The Wine Cellar"), a Dallas wine bar. (Photo by Pat Kovach Roberts)

maison or the French onion soup, and discover the wineshop while beguiled by the atmosphere and the wine. Food is French and is limited mainly to soups and salads, patés and cheeses, with a daily hot entrée made on a hot plate behind the bar, such as cassoulet or boeuf bourguignon.

La Cave's clientele is mostly over 30, professional, affluent, and wine-oriented. The wine bar and shop are the nucleus of other wine-related activities such as wine classes, tastings, and an annual wine tour in France.

This combination of on-premise service and take-home sales is not an option everywhere. Beverage laws in many areas do not allow it.

A popular type of food-beverage combination links a bar and a restaurant on an equal, semi-independent basis, with a common roof, a common theme, mutual services, and the same management. This type of enterprise may have a bar and restaurant in separate areas, or even in separate rooms, serving both the drop-in bar customer and the mealtime patron. The food/drink sales ratio is likely to reflect the equal status of food and drink, with bar and restaurant each doing better than it would without the other. In many cases neither side of the house would make a go of it alone, but together the customer attraction is doubled, the income is doubled, and the overhead costs are split between them.

Hotel beverage operations. In hotels the beverage operation is different in many ways from that of the bar or the bar-restaurant combination. There may be three or four bars under one roof, each with a different purpose and a different ambience—say a lobby bar, a cocktail lounge, a restaurant bar, a disco club. In addition there is room service, including mixed drinks, wine, and champagne. Above all, there is banquet service, including conference and convention service, which is pretailored to the client's needs and served from portable bars (Figure 1.5) by extra personnel hired for the occasion.

Added to these different facets of hotel beverage service is the up-and-down nature of demand. Since a hotel bar's primary clientele is its overnight guests, demand rises and falls according to the occupancy rate. However, even this is unpredictable: a hotel can be 100% full for a convention and yet have very little bar trade, depending on what kind of convention it is. On the other hand, a very low occupancy rate can be accompanied by a lot of bar business, again depending on who the hotel guests are.

An example of a typical commercial hotel is San Francisco's Jack Tar Hotel. Its 403 rooms and 17 meeting rooms serve an extensive travel and convention business producing annual beverage sales of $600,000. A single bar (Figure 1.6) with a regular staff of three bartenders serves the cocktail lounge, coffee shop, dining room, room service, and a thriving happy-hour business from an adjacent office building complex. Service for conventions, business meetings, and special events is provided from portable bars with temporary help as the occasion demands. Since San Francisco is a union town, there are always trained bartenders and serving personnel

figure 1.5 One type of portable bar. (Photo courtesy Carter-Hoffmann Corp.)

figure 1.6 Jack Tar Hotel bar, San Francisco. (Photo by Charlotte F. Speight)

available—at union prices, of course. A food and beverage manager oversees all aspects of beverage service. Purchasing and storage coordinate with beverage service but report to other departments.

As hotels go, this is a small operation. But the basic elements are typical of the industry as a whole. Resort hotels and luxury hotels often have several bars and restaurants, with a variety of entertainment, food, and drink, to keep the hotel guests on the premises and to attract an outside clientele as well. In San Francisco, however, most hotel guests go out to explore the city.

Airline beverage service. Another type of beverage service that must adapt to special conditions is the serving of drinks on airline flights. The restrictions of space, time, weight, and equipment are formidable. Of necessity the drink menu is limited. The liquors, beers, wines, and a few premixed cocktails are handed out in individual bottles (50 milliliters for spirits, 200 milliliters for wine, and 12-ounce cans of beer). The glasses are nesting plastic disposables, and the mixes are poured from small cans. Flight attendants push a double-ended cart down the aisle and, working from both ends, ice and garnish the glasses, pour the mixes, hand out the little bottles, and collect the money. It is all a marvel of organization.

For first-class passengers, drinks are free and service typically includes real glassware, unlimited refills, a choice of wines, champagne for breakfast, and sometimes specialty drinks—American Airlines, for example serves Cappucchino. Ozark Air Lines features a "wine cellar in the sky," in which passengers may sample three wines-of-the-day. Thus beverage service can act as merchandising for the airline itself.

Purchasing for airlines is done at base cities. The federal government and most states have special liquor permits that meet the needs of beverage service on the move. Tight control systems follow the little bottles everywhere, since they are extraordinarily tempting.

Similarities and differences. Grouping types of beverage service into these rather arbitrary categories does not really say much about the character of individual enterprises. Many establishments do not fit into any category completely, and those within categories can be as different as day and night.

Yet all categories have certain similarities. They sell the same merchandise. They all have similar staff structures, similar patterns of purchasing and inventory, similar ways of controlling the merchandise, similar items in the budget, and similar methods of accounting. They must all meet certain government requirements and operate within certain government regulations. Even the prices charged for the same drinks are not wildly different from one enterprise to another.

Yet no two enterprises are alike, unless they are part of a chain. The successful enterprise is one that meets the needs and desires of a certain clientele and, since people's needs and desires differ from group to group, enterprises differ. Beyond that, the successful enterprise is deliberately different from others serving the same type of clientele in order to compete in attracting customers. A third reason for the variety of bar operations is the special circumstances of each operation and each entrepreneur. This contributes to variety but not necessarily to success. The entrepreneur must put clientele above all else in shaping an enterprise.

Summing up . . .

Today's flourishing beverage service industry is witness to the growing acceptance of alcohol as a normal and desirable part of life. Even so, there are still lingering attitudes stemming from the prohibition movement, and beverage sellers must still operate within the restraints of numerous regulations. Indeed, there may always be regulations of some kind because there will always be the potential for alcohol abuse. This side of alcohol, too, is as old as history.

No one is more aware of this potential than the responsible bar manager. It is a cardinal principle of good bar management never to pour that extra drink for the customer who has had too much, no matter what the profit involved. Only through such responsible attitudes can an enterprise stay healthy and maintain a respected place in the community.

Today's industry is building a reputation for responsibility, except for a few operators who make it hard for the rest of us. When you contrast today's successful bars and taverns with the typical bars and taverns of the late-nineteenth century, you can see how far we have all come.

chapter 2 . . .

THE BAR

In order for you to get the most out of this book, assume that you are going to open a bar or renovate an existing one. If you have no such plan, why not dream one up? Then, as you read, you can apply the text material in your mind's eye to an enterprise of your own design.

This chapter deals with the physical facility in which drinks are served. The physical surroundings set the stage for the patrons' pleasure and even make the drinks taste better. Many ingredients go into creating just the right environment—atmosphere, decor, the efficient use of space, the bar itself. Since every enterprise is different, the discussion centers on basic questions to be answered and principles and guidelines to be followed in finding answers.

If you were to pick up a 10-year-old entertainment guide to almost any city, you would be astonished at the number of bars and restaurants listed in it that no longer exist, and the number of establishments flourishing today that did not exist 10 years ago. The food and beverage industry is famous for its volatility, for meteoric rises and falls of fortune, and often for the magnitude of individual failure while across the street some little shoe-string operation becomes a year-after-year gold mine.

The reasons for successes and failures are often complex. Like the entertainment business, food and beverage facilities are vulnerable to bad luck, the fickleness of fashion, and the weather. But more often success/failure is a consequence of good/poor management or good/poor planning, or both.

The elements that make up good planning and good management are extraordinarily interdependent, and it is somewhat arbitrary to isolate one at a time, since in practice they are all mixed together and interact like ingredients in a cake batter. Nevertheless we can't consider everything at once. We'll take one ingredient of success at a time, and by the end of the book you can bake your own cake. This chapter's ingredient is the physical facility.

In real life the bar is often part of a larger operation that includes food service. Some of the discussion that follows, therefore, will apply to a total facility—you could not, for instance, plan the decor of a bar in a restaurant without considering the restaurant too. But we will focus only on those aspects of the total facility that affect the physical setting of the beverage service.

Clientele . . .

The starting point for designing a successful bar is its target clientele—the people you want to attract and serve, who are going to pay your bills and provide your profit and give you the pleasure of making them happy. They are the focal point around which everything else revolves: the atmosphere you create, the drinks you serve, your decor, entertainment, sound, lighting, dress—the total impact. They influence your location, your floor plan, the equipment you put in your bar, the kind of staff you hire, everything. So it's good to know as much as you can about them.

Many people inexperienced in the bar and restaurant business dream of opening a little place that will be their idea of perfection—such perfection that the world will recognize its excellence and flock through its doors. Well, it doesn't work that way. Everybody's idea of perfection is different, and it's impossible to please everyone. Disco music drives some people crazy. Elegance may frighten the couple worried about paying the baby-sitter. Mom and Dad are not going to take the kids out to dinner at a bar/grill that has overtones of an X-rated movie.

So it's best to choose a certain type of customer, or a few similar types, and appeal directly to the tastes and habits and needs of those people. Then if other types come along too, that's fine.

Researching customer tastes. Who do you want your customers to be? Jet set? Singles? Professionals—lawyers, doctors, politicians, newspersons, industry tycoons? Regulars—jazz buffs, sports fans, disco dancers, campus crowd, Mom and Dad? Transients—tourists, vacationers, business executives, conventioneers, expense-accounters? No one group is inherently a better bet than another. There's a profit to be made with any type of clientele, if their needs are met and satisfied.

Think about them. How old are they? Where do they come from? How much money do they like to spend? Why do they go out and buy drinks and what kinds of drinks do they buy? What kind of atmosphere do they respond to? What turns them on?

Do some homework: visit the favorite places your target clientele is flocking to now. Talk to those customers. Watch their reactions. What do they like and dislike about the place (*not* what do *you* like and dislike)? Study everything about the operation—the decor, drinks, layout, ambience, food, entertainment or absence of it. Talk to the bartender; study the bottles behind the bar, the wine list, the menu.

Keep your research local: tastes and interests vary widely from place to place even among the same age-groups and income levels. Look at ads in local papers; read the local restaurant reviews. Talk to people who go places. Ask local beverage wholesalers; they are some of the best-informed people around. They know who is buying what kind of liquor and what kinds of customers are drinking it and what places are raking in the money and which ones are having trouble paying their bills on time.

Planning your services. Once you have decided which customers you are going to attract to your facility, you must decide what you are going to do for them: what services will you perform for them, and how? Will it be drinks only, drinks and dinner, drinks and entertainment? Music? Dancing? If you serve food as well as drinks, which will have the major emphasis, and how will they relate?

What kind of bar will it be? A stand-up bar for people in a crowd or in a hurry? Table service in a cocktail lounge? Dining-room service from a service bar? A holding area for people waiting to eat in your restaurant?

What kind of drinks will you serve? Beer? On tap or bottled? Wines? By the glass or by the bottle? For the casual light drinker or the connoisseur? Mixed drinks? Fancy drinks? Frozen drinks? Flaming drinks? House specialties? Soft drinks? Coffee drinks? As in everything else, you need to work out your answers in terms of what appeals to your intended clientele, not necessarily what you yourself would buy as a customer.

Defining your identity. What will be your own special character that will entice people into coming to you instead of to your competitors? ***Favorable uniqueness***, it is sometimes called, or ***image***, or ***identity***. It's what springs to mind when the name of your place is mentioned, what people look forward to before they come, what lingers in memory afterwards as a special pleasure, so that they will come back again and tell their friends about you. This is the most intriguing, most elusive, most worth-striving-for element of success.

In today's most successful bars and restaurants, that uniqueness is likely to be a carefully worked out ***total concept*** rather than one thing such as special house drinks or a mahogany bar with a brass rail from the 1890s. The name of the place, the drinks, decor, layout, service, uniforms, and menu all fit into the total concept and reinforce it. The customer may remember only one thing, like the crazy menu full of puns, or the 2-ounce drinks, or the houseful of Playboy bunnies, or the revolving panorama of the city at night from atop a tall building, but these things are only symbols of a total ambience, crystallizations of the concept.

To assure success, then, you need an overall concept that can tie everything together, unify it. A concept should be an idea that can be stated simply—for example, a neighborhood bar/restaurant for families, or a wine/cheese/fruit bar, or a natural foods restaurant where all the drinks are made with fresh fruit juices. Decor, layout, lighting, menu, and service will all be planned in reference to the concept.

Your overall concept will grow out of your understanding of your chosen clientele and the careful planning you do to serve them. It will be shaped by your observations of the kinds of places your clientele is responding to. But it will not copy; it will have its own personality, its own identity, as the magnet to your front door. It is a real challenge.

Atmosphere and decor . . .

The atmosphere of your place will determine whether people come to buy drinks from you, what kind of people come, how long they stay, how much they spend, and whether they come back and bring their friends.

After all, why do people go to bars? They go for pleasure. They don't go just to drink; they can buy a bottle at a package store for much less money and drink at home. They go to relax, to socialize, to rendezvous with old friends or meet new ones or perhaps to be alone with a special person; they go to escape the everyday mood and scene. If you can transport them from a world of problems and frustrations to a world of pleasure, you have the first ingredient of success. Just what will transport your chosen clientele will determine how you create this other world.

Whatever the specific attractions you choose, they should be inviting from the very first moment. The atmosphere should convey a message of welcome, of festivity, of concern for the customers' well-being. Some of this will come from you and your personnel, and some will come from your other customers—most people enjoy being among others like themselves who are having a good time. But the physical surroundings are equally important. They create the first impression, set the stage, strike the keynote.

Creating atmosphere through decor. The kind of decor you choose for your facility will be the visual expression of its mood. Decor includes the furniture and its placement; the walls, floor, ceiling, lighting, and window treatment; plants and other accessories; special displays; and the front and back of the bar itself. Each element should be planned in relation to the total concept. In effect it is the packaging of your concept; not only does it help to create mood, but it merchandises your product.

What kind of mood are you after—what does your research show your target clientele responds to?

Do you want a sense of spaciousness, relaxation, restfulness, a place where people come to talk to each other without shouting, or are you after a noisy, crowded, stimulating, hyped-up atmosphere? Do you want to convey elegance, opulence, luxury? Or modest comfort and terrific value for the price of your food and drink?

Soft colors and rounded shapes are restful; bright colors and bold patterns are stimulating. Noise is muted by carpets, drapes, upholstered chairs, fabric-covered walls; it bounces off tile and concrete floors, plaster walls and ceilings, glass. High ceilings give a sense of space; low ceilings make a room seem smaller and more intimate. Ceilings that are *too* low and rooms that are *too* small can give your customers claustrophobia. Soft light and candlelight send messages of intimacy, romance, intrigue. Bright lights and flashing lights go with noise, crowds, action, excitement. Firelight is restful, dreamy, romantic—and may also be against fire regulations.

Luxury and opulence can be conveyed by lavish use of expensive fabrics, furniture, and accessories; by museum pieces and art objects; by dramatic effects such as waterfalls and magnificent views and murals by famous artists; by gleaming silver and crystal, masses of fresh flowers, ice sculptures, an elaborate list of vintage wines, tuxedoed waiters, expensive food and drinks, valet parking, and perfect service. Terrific value can be conveyed by simple inexpensive but imaginative decor, good drinks at moderate prices, and quick friendly service.

Investors sometimes spend a fortune on decor; it is often necessary in order to compete for certain types of customer or to build a certain image in a national or international market. If you are going to spend even $30,000 or $40,000 on decor, you will want a professional designer.

But not all decor involves a lot of money. It does involve a lot of thought and good taste. Paint, plants, prints, judiciously chosen and placed, plus the look of the furnishings and the way they are grouped, the right lighting, and the right sound (or its absence) can create a mood. A few inexpensive conversation pieces can add to the fun. The trick is to keep mood and clientele constantly in mind, and pick colors, textures, shapes, furniture, fixtures to mesh together for a total, finished look.

Carry your decor and mood right into your service, drinks, uniforms, small details. A decor at variance with the other aspects of an operation sends a mixed message: crystal chandeliers say one thing; shabby restrooms with cold water and no soap say something else.

Work toward a decor that doesn't simply copy a competitor, that isn't built on a passing fad. Yesterday's dimly lit red and black lounges are already out of date, says Doug Balthrop of Design Market in Dallas; today's plastic theme recreations of other times and places are heading that way too. The new trend, he feels, is toward honesty and authenticity in decor, with more light, more windows, more informality, more warmth and neighborly feeling. In fact, he sees the small, informal neighborhood bar with inexpensive food as the trend of the immediate future. (Tomorrow, when you read this, it may be something else.) Joseph Giovannini of Los Angeles suggests using people themselves as a design element, to create a warm and human environment.[1]

Layout . . .

Whether you are starting from scratch or remodeling, you'll want to think through your layout carefully with your basics—customers, services, atmosphere—in mind. Now add to those a fourth one: efficiency.

Even if you are planning to use a professional designer, there are many decisions to be made before a designer is called in. And if you are considering buying or leasing a certain facility, understanding the elements and necessities of layout may help you to make up your mind. It will prompt you to investigate things you might otherwise overlook, and it will test whether this particular setting can be made to function according to your needs.

The basic elements of layout are the space available, the activities going on in that space, the furniture and fixtures required to carry out the services and other activities, the volume of business or number of people to be served, and the relationship of the bar area to other aspects of a larger facility such as a restaurant, hotel, or club. Other practical concerns and restrictions come into play: plumbing, refrigeration, lighting and other electrical requirements, heating and air conditioning, health and fire regulations, and state laws and regulations. Last but not least comes profit—the most profitable use of the space available.

Again, the final layout will come from a mix of all the factors. But let's discuss them one at a time in order to appreciate the impact of each.

[1] In *Restaurant Design*, Winter 1980.

Available space. Available space as an element of layout includes not only square footage but also the shape of the area, the position of entry and exit, and the sharing of space with other activities such as dining, dancing, or live entertainment.

The square footage will set an outer limit on the number of customers you can serve at a time. It may also determine whether you have a stand-up bar that can serve crowds of standees, or whether you will have seating at the bar or at tables, lounge style.

The shape of the room is critical to the arrangement of the furniture and fixtures. Think of three rooms of the same square footage—one long and narrow, one square, and one L-shaped (Figure 2.1). Shape will affect the number and arrangement of tables, the position of the bar itself for the best visual and psychological impact, and its size and shape. It will influence the channeling of traffic for entry and exit and for service. And it will certainly affect the sharing of space with activities such as dining or dancing.

Entry and exit deserve special thought. The relationship between entrance and bar will influence the movement of customers into the room and the way the room fills up. Do you want your customers to move immediately into noise and conviviality? Do you want them to stop at the bar before going on to dine and dance? Or will a crowd around a bar near the entrance block the way both visually and physically, leaving the room beyond it empty?

If the bar is associated with a restaurant, should it have a separate entrance, or should it be part of the restaurant? Will it be only a holding area for the restaurant or do you want it to have its own patrons?

Sharing space with other activities takes careful planning. Think about the amount of space each element requires—the bar, the activities, and the furniture and fixtures. You may even want to measure everything and plot it to scale on a floor plan. It is very difficult to estimate the sizes of things just by looking at them. An empty room usually looks smaller than it is. A room with furniture already in it is very difficult to picture accurately with a different arrangement. Most empty dance floors don't look big enough for more than three or four couples, but in action they may accommodate many more. (People may bump into each other, but that's part of the fun.) Musicians and their instruments take up a surprising amount of space.

If you plan to have live entertainment, do you need a stage? Is the space available big enough to accommodate the sound, the physical space the entertainers need, and enough customers to support the entire activity? Is the space *too* big? Can you fill it? Compare in your mind the feeling of a half-filled room with the impact of a small place that is crowded with people having fun.

You need to be sure that the space as a whole will do what you want it to do, while there is still time to change your plans. You need to set space priorities and guidelines. Then, whether you do the layout yourself or have a designer do it, there will be a minimum of problems and surprises.

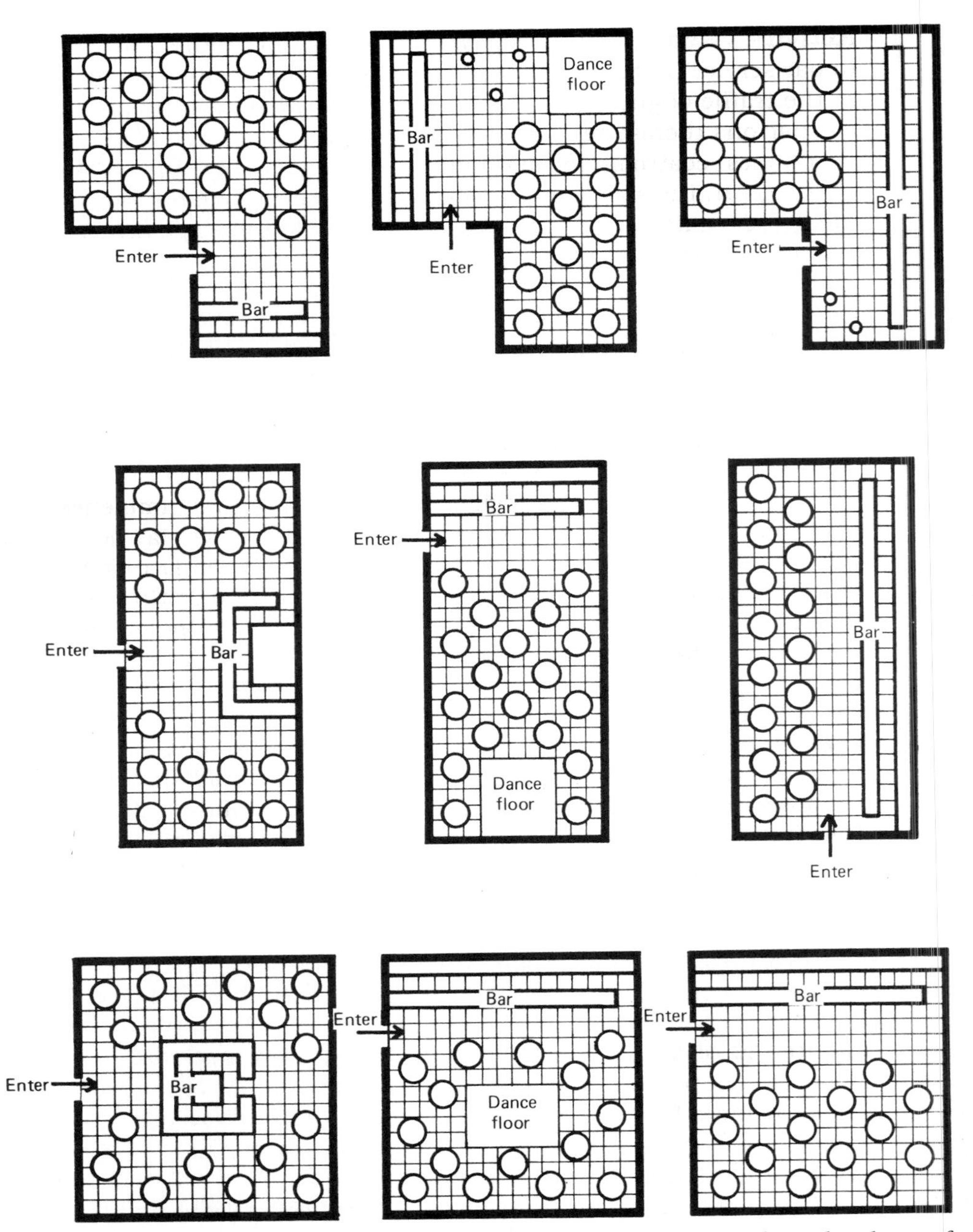

figure 2.1 These rooms all contain the same square footage. Notice how the shape of the room affects the room arrangement.

Activities and traffic patterns. What goes on in the room? In addition to bar service perhaps you have dining, or dancing, or live entertainment, or the traffic of guests in a hotel lobby. You also have the movements of the bar patrons themselves—entry and exit, comings and goings to restrooms, telephones, coatroom, or just milling around among themselves. For an efficient layout, the space and directional needs of each activity must be defined and located and the doors, furniture, and fixtures so placed that these activities move efficiently.

In particular, a good layout will establish efficient traffic patterns to and from the bar for table service, to and from the kitchen and service areas for dining service, as well as customer entry and exit, so that there is an orderly flow instead of chaos. The bar must also have easy access to storage areas; you don't want the bartender trundling a tub of ice through a throng of customers. Figure 2.2 illustrates good and poor traffic patterns for the same space.

A good layout will also consider clientele, mood, and ambience as well as efficiency. The position of the bar itself can play up or play down liquor service. In a family restaurant you might place a service bar discreetly in the background. In a singles cocktail lounge you might position an island bar in the middle of the room, center stage, striking the keynote of fun and festivity, like a merry-go-round at a carnival.

Furniture. All furniture should be chosen and laid out in relation to the total bar concept. Bar stools and lounge chairs should look inviting and feel comfortable. Chair designers claim they can control the rate of customer turnover simply by the degree of comfort of the seat cushion! Lounge chairs should arrange well around cocktail tables. The tables themselves can be small if they are only for drinks, and in a busy bar you can crowd them somewhat, adding to the conviviality. The sizes and shapes of tables and chairs are important elements in layout.

Influence of utilities, codes, and licensing restrictions. Plumbing is important in positioning the bar. Supply pipes and drains should not travel long distances because they are expensive to install and because there is more to go wrong. If you have a kitchen, it is most efficient to coordinate plumbing for both.

The same thought must be given to electrical requirements in relation to layout. In addition to lighting designed for mood and decor, there are all the special electrical needs of the underbar equipment. If you have live entertainment, you have to plan an electrical supply for electronic gear and a public address system. This can affect your general layout: you may think twice about putting the bar or the musicians in the center of the room.

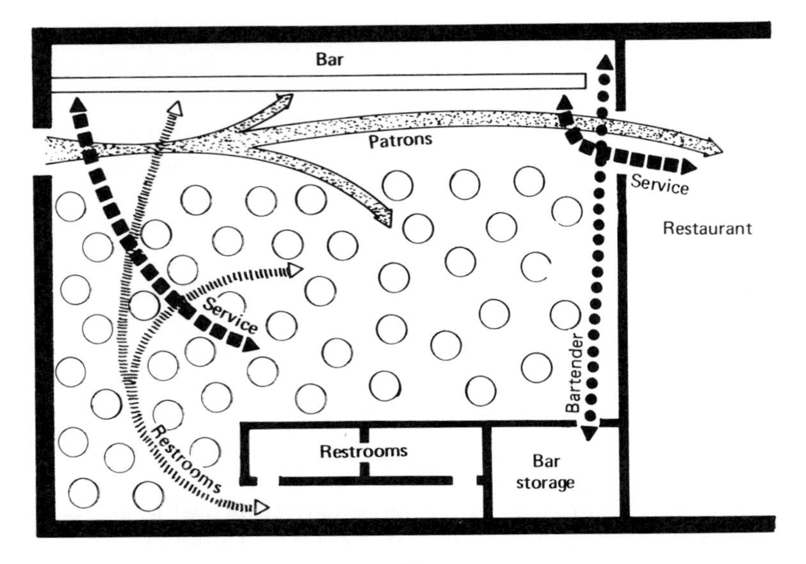

(a)

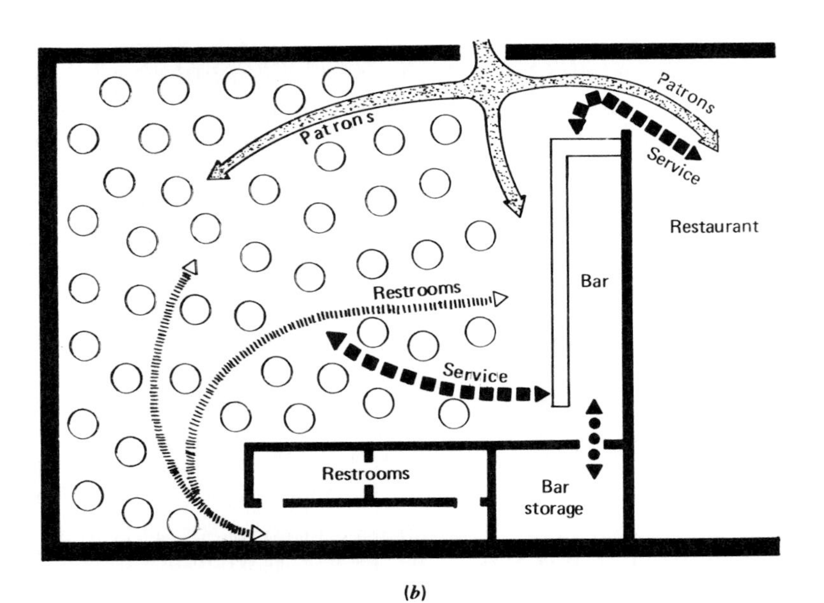

(b)

figure 2.2 (a) A tangle of conflicting traffic patterns in a bar lounge. (b) The flow of traffic smoothed out by repositioning the entry and the bar.

Heating/air-conditioning ducts and vents, smoke eaters, and air circulators can also affect layout by their output, the space they occupy, and their visual effect. A ceiling fan, for example, needs to have a certain visual relationship to the furniture and fixtures below it.

Local health and fire regulations often impinge on layout, especially in respect to exits and traffic aisles. Health department requirements about washing glasses can affect the space requirements inside the bar, and the plumbing and electrical requirements too.

Some state and local requirements for liquor licensing can affect layout. In some places an open bar is not allowed, so that a restaurant must have a private club in a separate room in order to serve liquor. In other places food must be served in the same room if liquor is to be served.

The bar itself . . .

Determining the size, shape, and placement of the bar itself is a design problem with two facets—the element of decor and the element of function. The size and shape of the bar, its appearance, and its position in the room are typically planned by the owner, architect, or interior designer, whose primary concerns are layout and decor. The working areas, where the drinks are poured, are planned by a professional bar-design consultant or by an equipment dealer. Sometimes they all work together from the beginning. More often a consultant or dealer is called in after the bar has been positioned and its dimensions set, and does the best job possible within the allotted space.

All too often a space in the room is simply assigned to the bar without considering the drinks to be served, the projected volume of business, and the space and equipment needed to serve those drinks in those numbers. Only after money has been spent in building the bar and buying the equipment does the owner discover its inadequacies. A poorly thought-out bar can cost more initially, can limit profits, and can cause daily frustration to those who work it. As we examine the bar in detail you will see why.

Parts of the bar. A bar is made up of three parts: the ***front bar***, the ***backbar***, and the ***underbar***. Each has its special functions. Figure 2.3 shows you these parts in profile, as though we had sliced through the middle from front to back. The dimensions given are those of a good workable design for a typical bar. The length of the bar will vary according to need.

The front bar. The front bar is the customers' area, where they order their drinks and where the drinks are served. The bar is typically 16 to 18

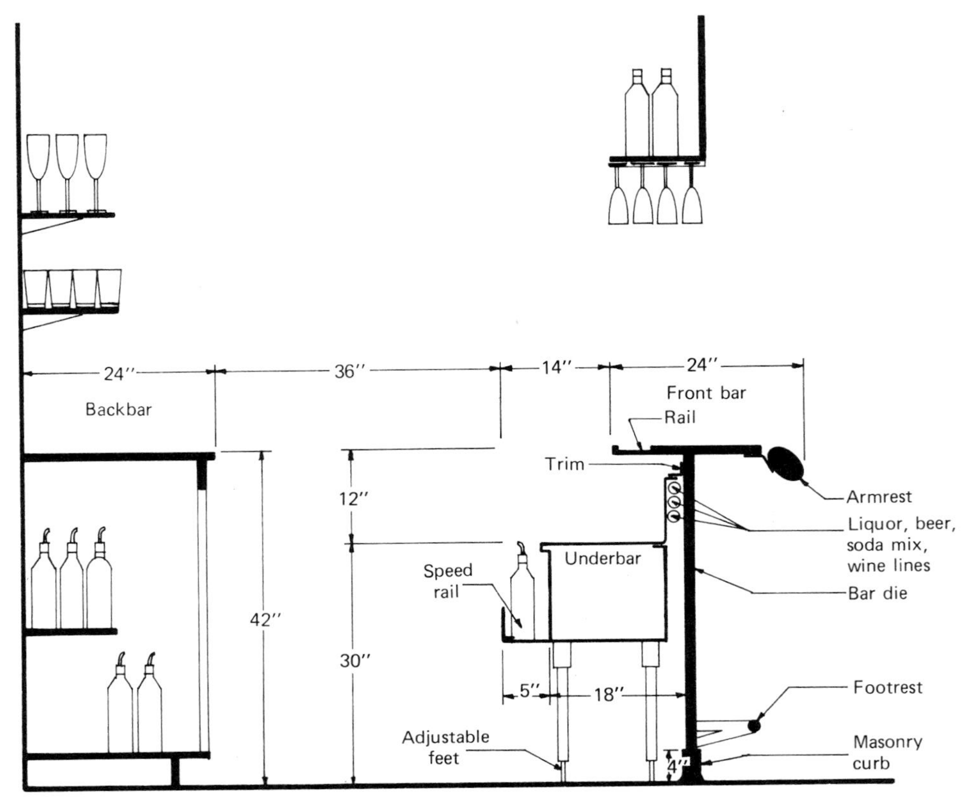

figure 2.3 The bar in profile, showing dimensions.

inches wide, with a surface that is alcohol-proof and waterproof, usually of laminated plastic. An armrest along the front edge, often padded, adds another 8 inches to its width. The last few inches of the back edge are usually recessed, and it is here that the bartender pours the drinks, to demonstrate liquor brand and pouring skill. It is known variously as the ***rail***, ***glass rail***, ***drip rail***, ***spill trough***.

The vertical structure supporting the front bar, known as the ***bar die***, is like a wall separating the customer from the working area. Seen in profile, it forms a T with the bar, making a kind of table on the customer side, with the other side shielding the underbar from public view. There is usually a footrest running the length of the die on the customer side, about a foot off the ground. On elegant mahogany bars of the 1800s the footrest was a brass rail, and underneath it were brass spittoons every few feet. The Prohibitionists made the brass rail a symbol of the wickedness of drink, along with swinging doors and Demon Rum. Figure 1.6 shows a typical modern front bar.

The height of the front bar, 42 to 48 inches, is a good working height for the bartender. It also makes the front bar just right for leaning against with one foot on the footrest, in the time-honored tradition of the nineteenth-century barroom. All underbar equipment is designed to fit under a 42-inch bar.

If it is a sit-down bar, it will have stools tall enough to turn the bar into a table (Figure 1.6). Each stool is allotted a 2-foot length of bar. The stools should look and feel comfortable; often they have upholstered backs and seats. Since the seats are high off the ground, the stools typically have rungs for footrests, or else the footrest of the bar is within reach of the feet. Even numbers of stools make it convenient for couples.

All of the front bar—the surface, die, armrest, footrest, and stools—must be planned as visual elements in the total decor.

The backbar. The backbar has a double function, the decorative function of display and the work function of storage. Traditionally it is the area where bottles of liquor and rows of sparkling glassware are displayed, their splendor doubled by a mirror behind them. In the Old West—or at least in Old West movies—the mirror had another function: it showed the man at the bar whether anyone was coming up behind him, gun in hand.

The typical modern bar still follows the same tradition of bottles, glassware, and mirror. Some people feel it is just not a bar without them. There are functional reasons too: the liquor and glassware are part of the bartender's working supplies, and the backbar is a good place to display call brands as a subtle form of merchandising. The mirror adds depth to the room; it also gives customers a view of others at the bar and of the action going on behind them. Bartenders sometimes use it too, to observe customers without being noticed.

New fashions in backbar decor are branching out to include stained glass, paneled or textured walls, murals, posters, wine racks, mood pieces and conversation starters. Stemware hanging from slotted racks overhead is popular as a design element as well as for functional glass storage.

The base of the backbar is likely to be storage space, refrigerated or otherwise. Or it may house special equipment such as a glass froster, an ice machine, or a mechanical dishwasher. If you are featuring specialty drinks, your frozen-drink or espresso machine will probably be on top of the backbar. The cash register is usually on the backbar too, in a recessed space.

Whatever its uses, the backbar must be visually pleasing from top to bottom, since customers look at it, and it must coordinate visually with the decor of the room as a whole.

The underbar. The underbar is the heart of the entire beverage operation and deserves the most careful attention to its design. In its space the equipment and supplies for the products you are selling must be arranged compactly and efficiently, with speed the overriding concern.

Each bartender must have an individual supply of pouring liquor, ice, mixes, glasses, blender, and garnishes, all within arm's reach. Each pouring station has an ice bin and one or more bottle racks for the most-used liquors and mixes. (We will discuss equipment in detail in the next chapter.) The supply of glasses may be upside down on the glass rail, or on drainboards near the ice bin, or on special glass shelves, or in glass racks stacked beside the station, or on the backbar, or in overhead racks, or in all these places, grouped according to type and size. The blender, and probably a mixer, may be on a recessed shelf beside the ice bin, while the garnishes are typically on the bar top in a special condiment tray. Figure 2.4 shows the underbar of a well-designed pouring station.

Most operations use an automatic dispensing system for carbonated mixes. Such a system has lines running from bulk supplies to a dispensing head at each station, which you can see in Figure 2.4. If the bar has an automated liquor-dispensing system the setup is similar. Note the lines behind the backsplash in Figure 2.3.

figure 2.4 Underbar of Jack Tar pouring station in process of setting up. Garnishes are on bar top (above picture); glasses are on counter to left of station. (Photo by Charlotte F. Speight)

The number of stations at a given bar will depend on the volume and flow of business. The bar should be designed with enough stations to handle the peak periods, with the equipment it takes to do it. Figure 2.5 shows the plan of the bar in Figure 1.6 and Figures 3.2 and 3.3. Notice the three stations located to serve different areas—dining room, coffee shop, and bar lounge.

Where drinks are served from the main bar for table service, the bar must always have a ***pickup station***—that is, a section of the front bar set off from the customers' bar area, where serving personnel turn in and receive orders and return empty glasses. It doesn't work to have them elbow their way through among the customers; confusion reigns and spills occur, and your profits may end up on the jacket of a celebrity who has just dropped in for a drink, or an ice cube may find its way down someone's neck. The pickup station should be near a pouring station and the cash register. On Figure 2.5 you can identify two of the pickup stations (top right and bottom left) by the railings (M) setting them off.

Another area of the underbar contains equipment for washing glasses—a three- or four-compartment sink (I on plan) with drainboards on both sides, or in some cases a mechanical dishwasher. The underbar must also have provision for waste disposal and a hand sink (G on plan). These are typical health department requirements.

Underbar and backbar together must provide enough storage for the day's reserve supplies of everything—liquor, mixes, wines, beers, ice, garnishes, and such nonbeverage supplies as bar towels, cocktail napkins, picks, and stir sticks. All these must be arranged so that they require a minimum of movement: movement is time, and time can be money, as well as good will and good cheer.

Three feet is the customary distance between the backbar and the underbar (see Figure 2.3), to accommodate the bartenders' movements and the opening of doors to storage cabinets. The doors must not be so wide that they block passage when open. Storage areas must be available to each bartender without interfering with another's movements.

Special drinks require special planning for the equipment they need. If you plan to have beer on tap, you must place the standards (faucets)—Q on Figure 2.5—so that they are easily accessible to the bartender (but not to the customer!), and there must be refrigerated storage space for each keg either at the bar or in a nearby storage area with lines bringing the beer to the bar. (The latter arrangement is not nearly so good as having the kegs at the bar, and often the beer isn't so good either.) Frozen-drink dispensers, ice cream equipment, and glass frosters are all bulky and must be designed into the overall scheme.

Hidden but essential factors in underbar and backbar design are the plumbing and electrical needs of the equipment. Faucets, icemakers, soda guns, and dishwashers need a water supply. Sinks, refrigerators, glass frosters, ice bins, icemakers, dishwashers, and waste disposal need proper

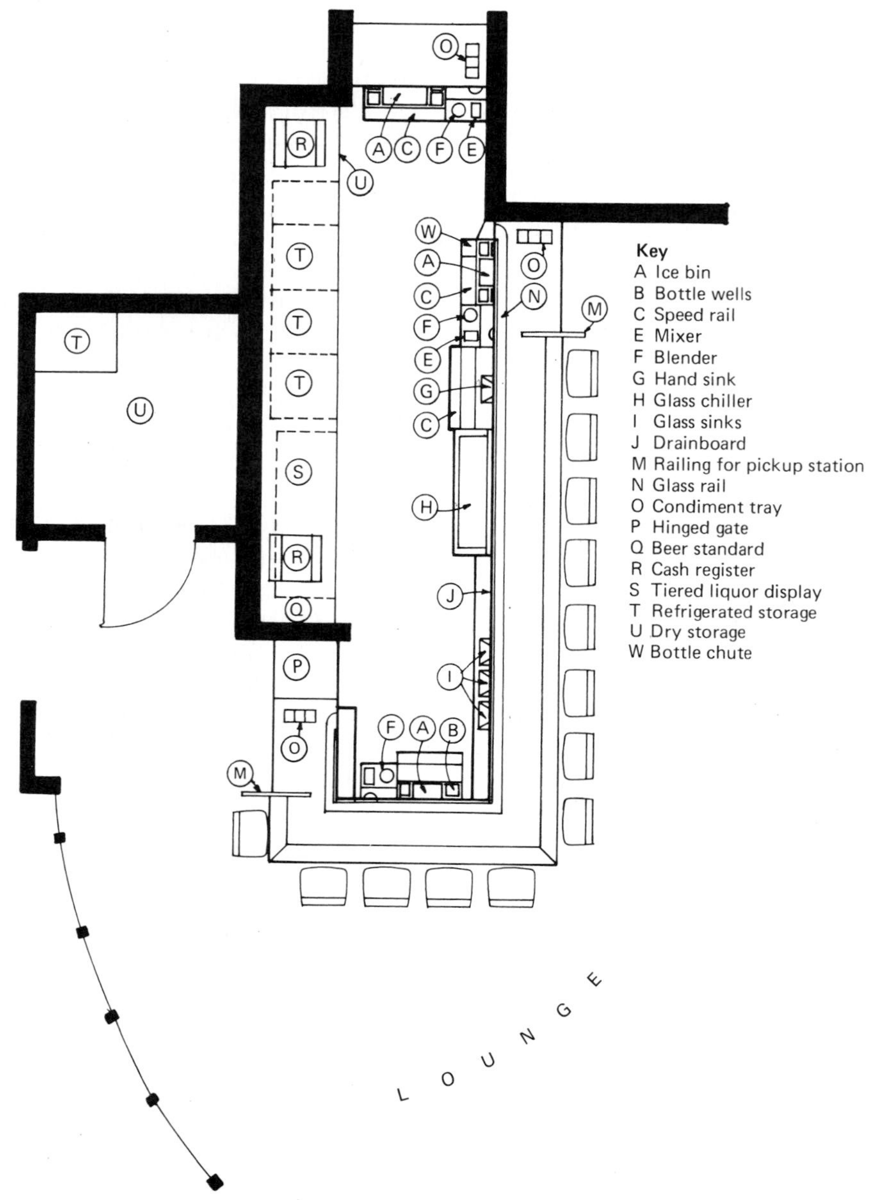

figure 2.5 Plan of Jack Tar bar. Station at top (A, C, O) serves coffee shop across the corridor. Right-hand station (A, C, M, O) serves cocktail lounge and bar. Station at bottom (A, B) serves bar and pickup station (M, O) for dining room beyond storage room to left (U). Rear portions of underbar equipment are not visible in this view. (Plan used with permission of Norman Ackerman, designer, HRI Consultants International, Dallas)

drainage. Much of the equipment needs special electrical wiring. All this must have ready access for repairs.

The entrance to the bar must be large enough to accommodate the largest piece of movable equipment, in case it must be sent out for repairs. Sometimes the entrance is a hinged piece of the counter that lifts up (P on plan). Sometimes one end of the bar is open, though this makes the liquor supply more vulnerable to tampering and makes control more difficult. Sometimes there is a doorway in the backbar.

The bar floor. Think about the bartenders' comfort and safety in planning the bar floor. They are on their feet for hours, and you ask them to look fresh and smiling. The floor under their feet must have a nonporous surface, such as tile or sealed concrete, to meet sanitary code requirements. Wood and carpeting are not acceptable. A tile or concrete surface is cold, hard, and slippery when wet. As the evening wears on, ice cubes, beer foam, soapy water, debris from empty glasses, and broken glass may accumulate.

There are ways to improve comfort and safety—none of them ideal. Slotted plastic panels allow spills to go down between the slats, to minimize hazards of slipping. They must be taken up for cleaning, which is a nuisance, and if it isn't done often they become stale and unsanitary. They are also hard on the feet. Rubber or plastic mats minimize slippage and are easy on the feet, but they must also be cleaned frequently.

Size, shape, and position in the room. From the front of the front bar to the back of the backbar, the overall depth should usually be about 8 feet (Figure 2.3). The minimum length of the bar should be determined from the inside, according to equipment needs. Additional length and shape will be determined from the outside, according to the number of seats (if there are seats), the size of the room, and the overall design requirements.

The inside factors are determined mainly by the kinds of drinks served and the number of stations needed to meet peak volume. The outside factors have to do with your overall concept, your clientele, your decor, and the available space.

Unfortunately the last consideration—the space available—is usually the tail that wags the dog. Often the space available is what is left over after everything else has been planned. Many times an inadequate bar space will limit what you can serve and how much, thereby decreasing your drawing cards and your profits. Or it may require expensive and complicated equipment solutions to problems that would be simple to solve in a larger space.

In sum, the best way to proceed is to plan your drink menu first, with your clientele in mind. Figure carefully the volume you can expect at peak periods. Size your bar to accommodate space and equipment needs for those drinks in that volume, or have a specialist do it. Don't box yourself into a bar that is too small.

If your facility is already built and space is predetermined, it becomes even more important to think through your bar design and equipment to make the most profitable use of the space you have. Again your clientele and your drink menu are your starting points. You may, for example, have to choose between beer on tap and ice cream drinks in frosted glasses, but if you know your clientele you can make the most profitable choice.

Bars can be seen in many different shapes—straight, curved, angled, horseshoe, round, square, or free-form. Shape, too, is a decision involving many factors—room size and shape, mood, decor, function. Again, the functioning area of the bar is often the stepchild of the design. Unusual shapes are tricky. Most underbar equipment is factory-made in standard sizes that may not fit an irregular shape. Custom work increases cost and sometimes does not work as well in action. It can also cause problems of maintenance and repair.

Usually a bar has its back to the wall, but in a large room it may be the centerpiece or focal point, a free-standing square, round, oval, or irregular island, with stations facing in several directions and a backbar in the middle. Obviously an island bar will have special design considerations. The backbar will be smaller and the front bar larger, and the underbar will be visible to the patrons. There may be special plumbing and electrical problems.

Whatever its shape, the bar's position in the room deserves as much thought as its shape and size—and may affect both. Do you want it to be seen from the street? Do you want it to set the tone of your establishment or take second place to your food service, your bowling alley, or your disco dancing?

Consider the customers' reactions as they enter the room. Crossing the room to get to the bar is inhibiting. Some people may turn around and go out rather than cross an empty room at 4 P.M. or, a couple of hours later, plow through crowds of standees or thread their way through the lounge amid staring eyes. The best place for the bar is near the door where customers can head straight for it.

Make these kinds of decisions before you draw up any plans.

The bar as a control center. While its major function is the dispensing of drinks, the bar is also a control center in which record is kept of the stock on hand, the drinks poured, and their sales value. After each serving period the sales record is checked against money received, to verify that one equals the other. To this end we have the most important piece of equipment in the bar—the cash register. It is the core of the system of controls by which management assures that its liquor is sold to the customer with little or no "evaporation" en route. In some operations the bartender also takes in the money; in others this is done by a cashier at a separate register.

In a large and busy bar each bartender may have a separate register, or bartenders may share a register with a separate drawer for each one. In any case, a register must be close to each station so that a minimum of time and motion is lost, and close to the pickup station. Since the register is usually within full view of customers, its placement also becomes a design element to be reckoned with.

Service bars. The term ***service bar*** refers to a bar that pours for table service only, usually in conjunction with food service. It does not serve customers directly but deals only with filling drink orders brought by waiters and waitresses. Usually a single station is enough to handle the volume except in very large restaurants.

Sometimes a service bar is part of the dining room, but more often it is out of sight, in which case it is smaller and simpler in design. Instead of backbar display there is room for bulk supplies of beer, mixes, liquor stock, and no need to camouflage or hide ugly or noisy equipment. Mechanical dispensing systems are often used here in preference to hand pouring, to increase speed and reduce liquor loss, whereas at a public bar there may be customer resistance.

In its basics, however, a service bar is like any other bar. It has the same functions, uses the same kinds of equipment, and performs the same tasks of recording and controlling the pouring and selling of drinks. It needs the same forethought in its planning as any other bar.

Working with a designer or consultant . . .

A whole range of expertise is available to you in planning the physical environment of your facility. The problem is sorting out what you need.

Consultants. At one end of the range is the consultant firm that will do the entire job for you, from surveying the market and developing a concept that will sell to your clientele, through completion of the job, right down to choosing your glassware and designing your matchbook covers. Many consultant firms offer a full spectrum of design services, including architects, interior designers, graphics artists, and foodservice facilities designers who are specialists in kitchens and bars. Other consultant firms have only specialized pieces of the package within their firms but can put together a whole design team for you nevertheless by subcontracting with other specialists. The term ***consultant***, then, can mean many things, and the only way to find out which consultants have what you need is to know what you need and then ask.

For bar design, the most specialized service a consultant has to offer is the designing of the space inside the bar by the ***facilities design consultant***. These specialists know code requirements countrywide, all the needs, all the equipment available, all the typical problems and the ways to solve them, in a way no interior designer or architect does. They usually work closely with an interior designer or architect and should be brought into the picture before the size and shape of the bar structure are frozen.

Consultants are typically used in projects where a sizable investment is planned. For such projects the right consultant will make you money in the long run by doing the most appropriate, efficient, and profitable job. Many foodservice consultants also do small jobs. Since their professional organization (Foodservice Consultants Society International) does not look kindly on advertising, you may not know about them unless you ask around. If a facilities design consultant designs your bar and puts it out for bid, you have a choice of equipment from several companies and you may get a better bar for less money than you'd pay if you tried to put it together yourself or have a restaurant supply house do it using the equipment lines it carries.

Interior designers. Another kind of professional designer is the interior designer who specializes in restaurants. Here is another term with many meanings. An ***interior designer*** is trained in the aesthetics of design, and a good interior designer experienced in doing restaurants is familiar with restaurant design trends and with the commercial furniture, fabrics, and other materials available. An interior designer will typically do your space planning (layout), select your furnishings, design your floors, ceilings, walls, window treatments, and plan your lighting, coordinating the entire plan to fit your overall concept and supervising all installations.

There are other interior designers who have done only residential work or other kinds of commercial establishments; they are of little help in designing a bar or restaurant because of the latter's specialized needs. In many states there is nothing to stop almost anyone from appropriating the title of interior designer, with no specific skill or knowledge to back it up.

Working arrangements. Consultants and designers work on contract—an advantage to you both. Usually there is a retainer to start off the job. Some interior designers will work on the basis of a design fee plus a commission on the furniture and materials you buy. Some will do a design on an hourly fee basis and let you carry out parts or all of it yourself.

Whatever your investment is to be, the contract should spell out clearly the scope of the job your investment is to cover and the various stages of the project at which payment is to be made. It may also be wise to have an open-ended contract providing that either party may terminate the contract at the completion of certain stages of the work.

Choosing design assistance. How do you choose a designer or consultant? Shop around. Find a bar or restaurant you think has been particularly well designed—it doesn't matter whether it's your kind of concept or even your kind of clientele; the question is whether it is well done in terms of its own purpose and concept. Find out who designed it. Then find out what else that person has designed and visit as many of those places as possible. If they are all on target, go ahead. If not, do some more exploring until you find someone who does the terrific job you are looking for.

The small end of the scale. What if you have only a limited budget, or what if you want to open in a small town or remote area where specialists are not available to you? (Specialists will go anywhere for the right money, but you will pay their travel expenses.) You can get design help from dealers—from restaurant supply houses, commercial furniture dealers, and the like. Investigate as many as you can. Some of the help will be good (many foodservice design consultants started designing in restaurant supply houses). Some of it will be no better than you could do alone if you studied things out. Some if not all of it will be limited to the products people are selling.

Sometimes a local interior designer can work with you on an hourly basis. Sometimes an art teacher can help with design and color coordination. Sometimes help is available through manufacturers' representatives. Maybe you yourself have a talent for design.

More often, though, people who have the operational know-how are concerned with the product, the staff, and getting the money together, and they neglect the physical ambience. Yet in today's market, the bar/restaurant business is becoming part of the entertainment business, and the physical setting is half the story. Where competition is keen, it may be the whole story.

If in the end you go it alone, give your place a simple, clean look. Don't clutter it up; don't try for pseudosophistication. Today the clean line is sophisticated; it is currently one of the major design trends.

Checklist of bar design essentials. Whether you hire a designer or do it yourself, you should have your needs and wants clearly worked out in your own mind. The more information you can give a designer and the clearer your own goals, the better the result will be and the easier the collaboration.

Here is a checklist of basics you ought to be on top of before you meet with a consultant or designer or go ahead on your own:

- Target clientele.
- Services to be offered.
- Overall concept.
- Local bars with which you expect to compete.
- Projected volume of business (number of bartenders or drinks per hour or customers or tables to be served).
- Types of beverage you will serve.
- Size and shape of bar area (architect's plans for an existing structure if you have them).
- Activities going on in bar area.
- Entry/exit and relationship of bar to dining and service areas (kitchen, storage).
- Existing decor, equipment, furniture, and fixtures you expect to keep, including those in adjacent areas.
- Licensing, zoning, health, fire, and building code requirements.
- Your time limits.
- Your budget limits. This is a must before you consult anyone else, unless you are hiring a consultant just to estimate what a given job would cost.

Summing up . . .

The starting point for turning a building, room, or space into a popular and profitable bar is the clientele to be served. The next decision involves the products and services to be offered to this clientele. The third is to define a unifying concept with a special uniqueness or identity. Only then is it time to deal with the physical facility. The goal is to make it an environment that attracts the desired clientele, gives them pleasure, and makes them want to come again.

Decor is a large factor in creating this environment. Color, light, arrangement of furniture and fixtures, efficient use of space, and the appearance of the bar itself all contribute. The bar should be designed from the inside out, beginning with the drink menu; its decor should develop in partnership with functional needs.

In sum, the bar should be the result of total planning with the market in mind—a basic step in merchandising the enterprise.

YOUR CHECKLIST OF BAR DESIGN ESSENTIALS

Target clientele

Age group ________ Income level ________ Life-style ________
Reason for coming ________________________

Services to be offered

Standup bar ____ Table service (lounge) ____ Table service (dining) ____
Live entertainment (type) ________ Dancing ________ Other ________

Overall concept, ambience ________________________

Competition (local bars) ________________________

Peak volume of business

Most customers at one time ______ Number of tables to be served ______
Most bartenders needed at one time ______ Most drinks per hour ______

Types of beverage

Cocktails/highballs ____ Hand-pour ____ Automatic ____ Frozen drinks ____
Ice cream drinks ____ Draft beer ____ No of brands ____ Bottled beer ____
Wine/glass ____ Wine/bottle ____ Soda system ____ Bottled sodas ____

Size/shape of bar area

Dimensions: Length ________ Width ________ Square feet ________
Sketch of area (separate sheet): Show shape ____ Outside entry/exit ____
Plot traffic to dining room ____ Storage ____ Kitchen ____ Rest rooms ____

Size/shape of existing bar

Dimensions ________ Shape ________ Number of stations ________

Activities in bar area

Table service (drinks) ____ Dining ____ Entertainment ____ Dancing ____

Existing decor, equipment, furniture, fixtures to be retained (include those in adjacent areas) ________________________

Code requirements

Permit/license requirements ________________ Zoning ______
Special restrictions ________________________
Special health department requirements ________________

Target date for opening ____________ Budget limits ____________

chapter 3 . . .

THE EQUIPMENT

A thoroughgoing familiarity with the bartender's bailiwick—the underbar and backbar and their equipment—is essential to profitable bar management. To produce the drinks wanted, of the quality desired, with the speed and efficiency needed to satisfy the customers and meet the profit goal, the right equipment must be in place. Few managers really comprehend equipment needs, the effects of equipment on drink quality, and the limitations that equipment and space can impose on a drink menu.

This chapter deals with the basic equipment essential to any bar and the alternatives most commonly used to meet varying needs.

When Abe Lincoln was selling whisky back in the 1830s he didn't use much equipment at all. Most likely a customer would bring a Mason jar into Abe's general store in New Salem, and Abe would pour whisky out of the barrel tap into the jar.

Earlier, in colonial taverns, drinks were poured from heavy glass decanters into tumblers, mugs, and tankards. Mixed drinks were stirred with a toddy stick—an early form of muddler—or a loggerhead—a metal bar on a long handle that was heated in the open fire and then thrust into the mug or punch bowl to stir up a hot toddy or flip. Along with the beverage equipment there might be a dice box, to see who would pay for the drinks.

In the saloons of the Old West, customers poured their own whisky, neat, from the bottle set on the counter by the bartender. By the mid-1800s hotel bars were serving mixed drinks, often with ice that was scraped, chipped, or pounded from large cakes transported from frozen lakes or rivers. By 1890 a hotel might have its own ice machine, which could make a large block of ice in 15 hours. Mixed drinks spawned tools for mixing—barspoons, measures, shakers, seltzer bottles. The cash register was born in a tavern in 1879.

Today's bar equipment stems mostly from post-Prohibition days. And marvelous equipment it is, fitted into minimum space, geared to high-speed individual service, and manipulated by that master of dexterity and showmanship, the bartender. Let's take a look at it.

Underbar and backbar equipment . . .

All underbar equipment must meet local health department requirements. These typically follow the standards set by the National Sanitation Foundation (NSF). Equipment meeting these standards carries the NSF seal (Figure 3.1). The major pieces of underbar equipment have surfaces of stainless steel, which is durable, cleans easily, and is unaffected by chemical cleaners

figure 3.1 NSF seal of approval. (Courtesy National Sanitation Foundation)

needed to kill bacteria. It also looks nice and easily takes a high polish.

Work surfaces of underbar equipment are a standard 30 inches high with a depth of 16 inches to the backsplash at the rear. Units from the same manufacturer fit side by side and give the appearance of being continuous.

Each piece of equipment is either on legs six or more inches high, for access to plumbing and ease of cleaning, or else it is flush with the floor. The legs have bullet feet (feet tapered like bullets) for ease of cleaning. All these features are NSF standards. The feet are adjustable to accommodate uneven flooring.

In Figures 3.2 and 3.3 you see the underbar and backbar of the Jack Tar Hotel in San Francisco. This is the bar shown in earlier chapters, Figures 1.6 and 2.5. It serves the hotel's cocktail lounge, dining room, and coffee shop. For this it has three stations. The station at center top in Figure 3.2 is the pouring station for serving customers seated or standing at the front bar, as well as the pickup station for cocktail waitresses serving the lounge. At the bottom right in Figure 3.2 is the pickup station for the dining room, seen from the waiter's side. You see this station from the bartender's side in Figure 3.3 on the left. This station also serves customers seated or standing at the bar. The third station, at top left in Figure 3.2, serves the coffee shop. You see it from the waiter's side at bottom right in Figure 3.3.

All three stations are set up in the morning, and a single bartender works from all three, according to where the calls for drinks come in. Two bartenders are on duty for the busy happy-hour and evening periods. Serving personnel garnish the drinks at the pickup stations; notice the condiment trays on the bar top. A shelf below on the server's side holds ashtrays, napkins, and other server supplies. Railings (M) set these pickup stations off from customer use. They are also return stations for used glasses; notice the waste dumps below.

Equipment for mixing drinks. Each of the stations in Figures 3.2 and 3.3 is outfitted with the following equipment:

- Ice chest, ice bin (A).
- Containers for bottles—bottle wells (B) and speed rails (C).
- Handgun for dispensing soft-drink mixes (D).
- Mixer (E) and blender (F) on recessed shelf.
- Glasses, overhead, on the backbar, on drainboards, almost anywhere there is room.

The centerpiece of any pouring station is an ***ice chest (ice bin)*** (A), with or without ***bottle wells*** (B), usually having a ***speed rail*** (C) attached to the front. This piece of equipment is variously known as a ***cocktail station, cocktail unit, beverage center***, or, colloquially, ***jockey box***. Figure 3.4 shows an ice chest similar to the Jack Tar's but with double rows of bottle

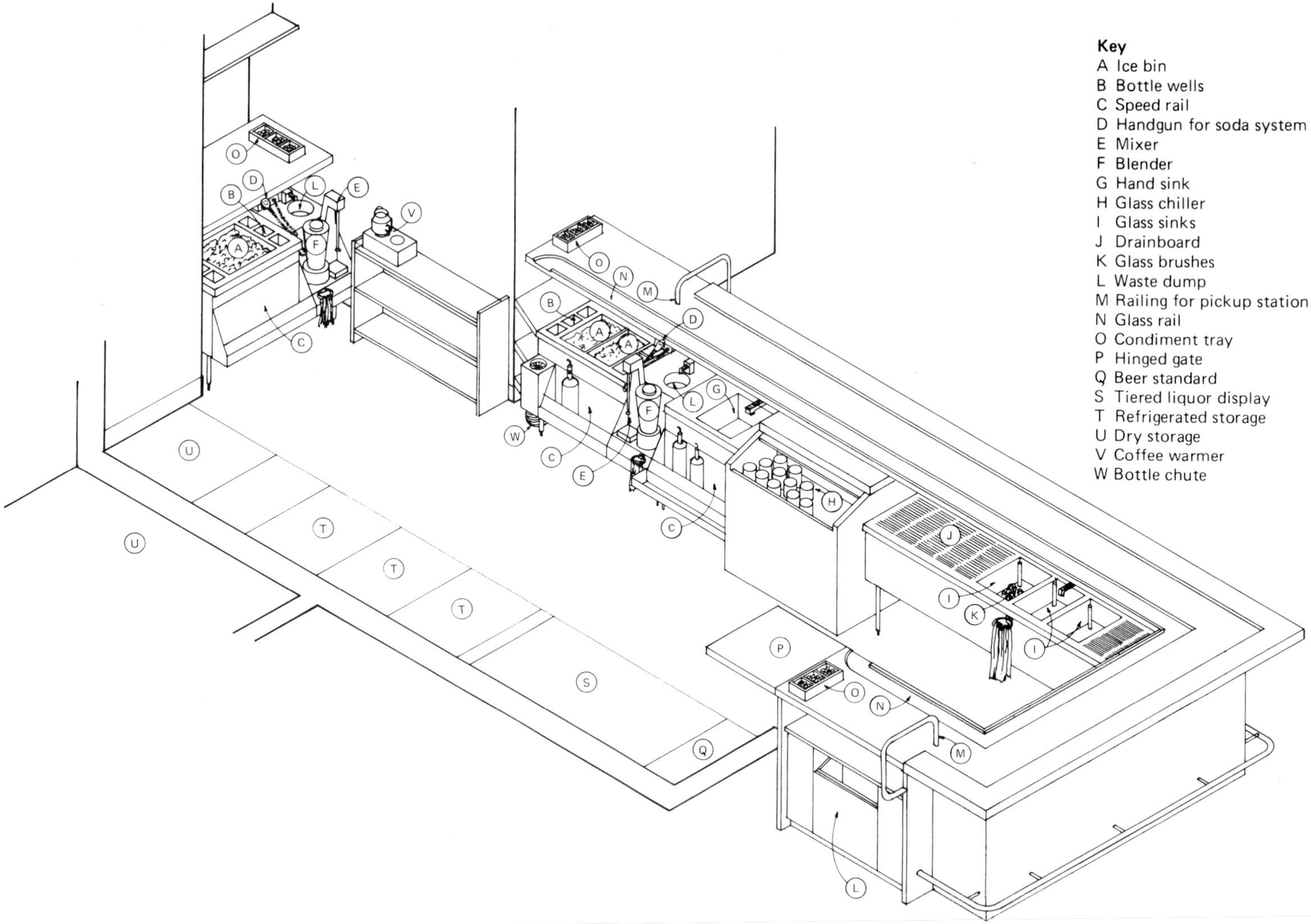

Key
A Ice bin
B Bottle wells
C Speed rail
D Handgun for soda system
E Mixer
F Blender
G Hand sink
H Glass chiller
I Glass sinks
J Drainboard
K Glass brushes
L Waste dump
M Railing for pickup station
N Glass rail
O Condiment tray
P Hinged gate
Q Beer standard
S Tiered liquor display
T Refrigerated storage
U Dry storage
V Coffee warmer
W Bottle chute

figure 3.2 Underbar, Jack Tar Hotel, San Francisco.

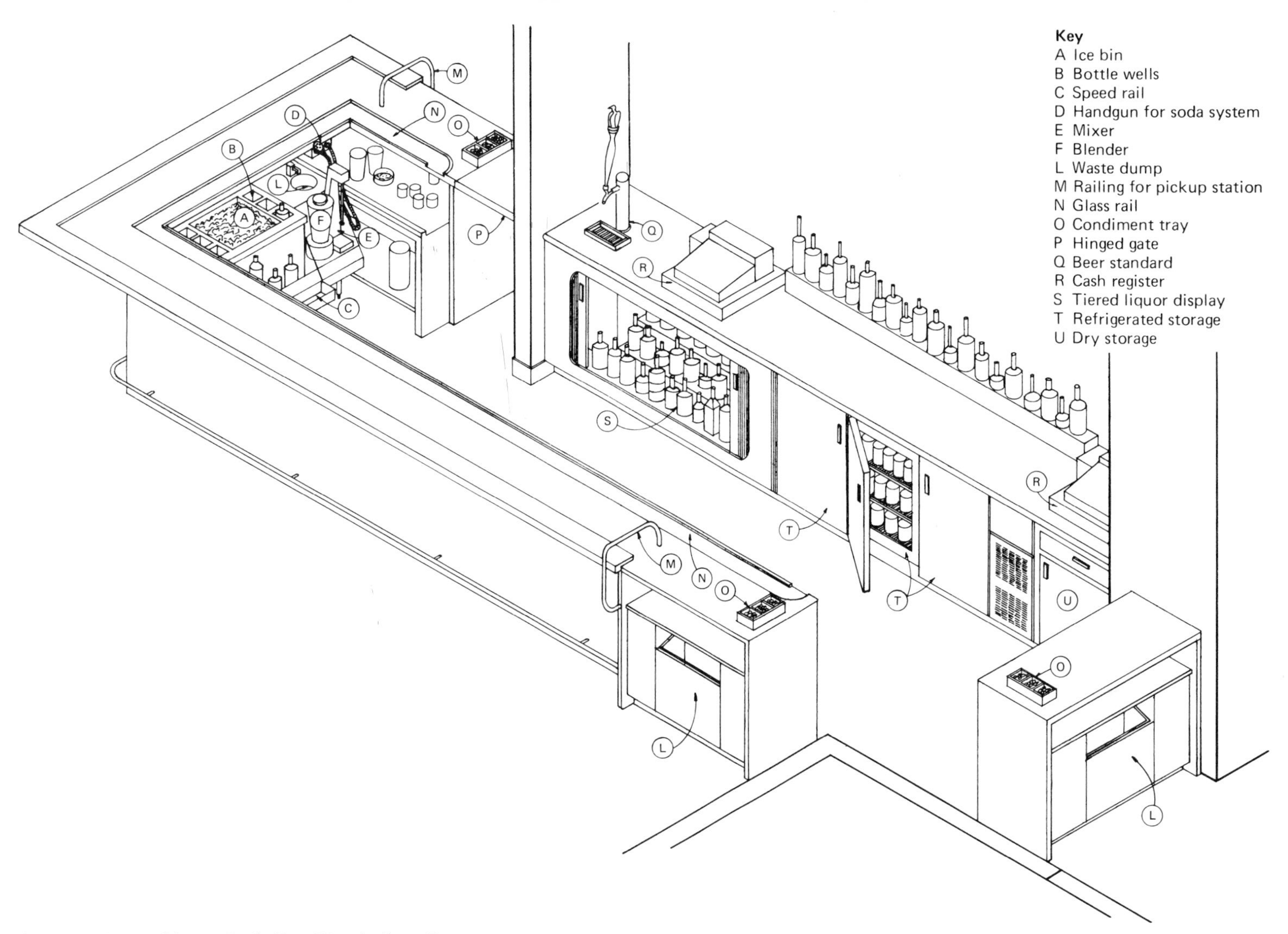

figure 3.3 Backbar, Jack Tar Hotel, San Francisco.

figure 3.4 Ice chest with bottle wells. The front cover fits under the back, or the entire cover may be removed. (Photo courtesy The Perlick Company, Milwaukee)

wells. Wells adjacent to the ice chest are chilled by the ice and are used for juices, preprepared mixes, milk, and cream. An ice chest may have a divider, enabling a station to have both cube and crushed ice; one of the Jack Tar ice bins is so divided.

Attached to each cocktail unit in Figures 3.2 and 3.3 is a unit called a ***blender station***, with a recessed shelf for the station's mixer and blender and a special dump sink (L) and faucet. In addition to speed rails on the cocktail stations there is a double rail on the hand sink.

A speed rail typically contains the most frequently poured liquors (usually scotch, bourbon or a blended whisky, gin, vodka, rum, brandy in Wisconsin, tequila in the Southwest). The liquor supply at a bartender's station is known collectively as the ***well***, and the brands used there are called ***well brands*** (or ***house brands*** or ***pouring brands***)—brands the house pours when a drink is ordered by type rather than by brand name. Popular ***call brands*** (brands customers ask for by name), vermouths, white wines, and the current favorites in liqueurs are set up within easy reach. Additional liquors—more call brands, liqueurs, premium brandies—are typically displayed on the backbar. Many bars have tiered liquor displays containing reserve supplies as part of the backbar itself, such as (S) on Figure 3.3.

At each station of the Jack Tar bar is the ***handgun*** (D) that dispenses the carbonated mixes (Figure 3.5). Nicknamed the cobra or the six-shooter, this instrument consists of a head having a nozzle and seven pushbuttons that deliver plain water and carbonated mixes (one per button) such as soda, tonic, cola, 7-Up, collins mix, ginger ale, Dr. Pepper, Sprite—whatever half-dozen you choose. Behind scenes are bulk supplies of concentrated syrups and a tank of carbon dioxide under pressure.

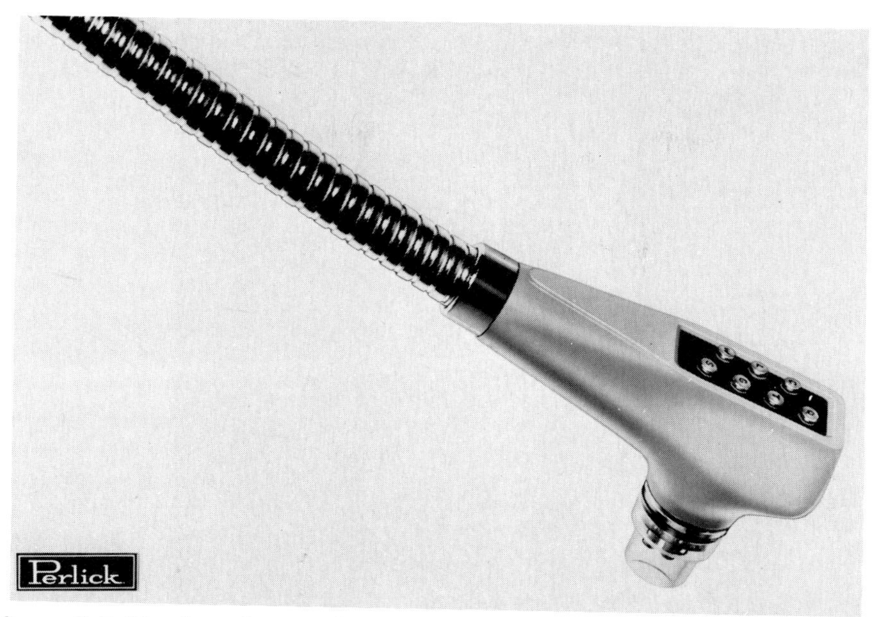

figure 3.5 Handgun for a soda system. (Photo courtesy The Perlick Company)

At the Jack Tar these supplies are on the garage level directly below the bar (Figure 3.6). Syrup lines run up from the tanks to the underbar and through an ice-cold plate on the bottom of each ice chest, made especially to quick-chill them. The CO_2 line goes to a motor-driven carbonator under the ice chest where the CO_2 is mixed with filtered water. A carbonated-water line then runs from the carbonator through the cold plate; so does a line with plain filtered water. Finally all the syrup and water lines run through a flexible metal hose (***flexhose***) to the head of the gun. There the syrup mixes with carbonated water in a 5:1 ratio at the touch of the proper button, or plain chilled filtered water is dispensed.

All this together is known as a ***postmix*** dispensing system because the soda is mixed at the time of service, as opposed to a ***premix*** system in which the complete beverage (except for the bubbles) is supplied in bulk containers that have been mixed at the manufacturing plant. In a premix system the bubbles are supplied from a CO_2 container directly to the dispensing head. The premix lines from the bulk supplies are run through ice to cool the product—not always effectively. A good postmix or premix drink should be cooled to between 37° to 42°F in order for it to maintain good carbonation. Premix systems are seldom used in today's bars, except for portable bars for special-occasion use. Postmix systems are far cheaper per drink (about two-fifths the cost) and far more compact (about one-fifth the size).

Neither a premix nor a postmix system comes ready-made. To assemble a

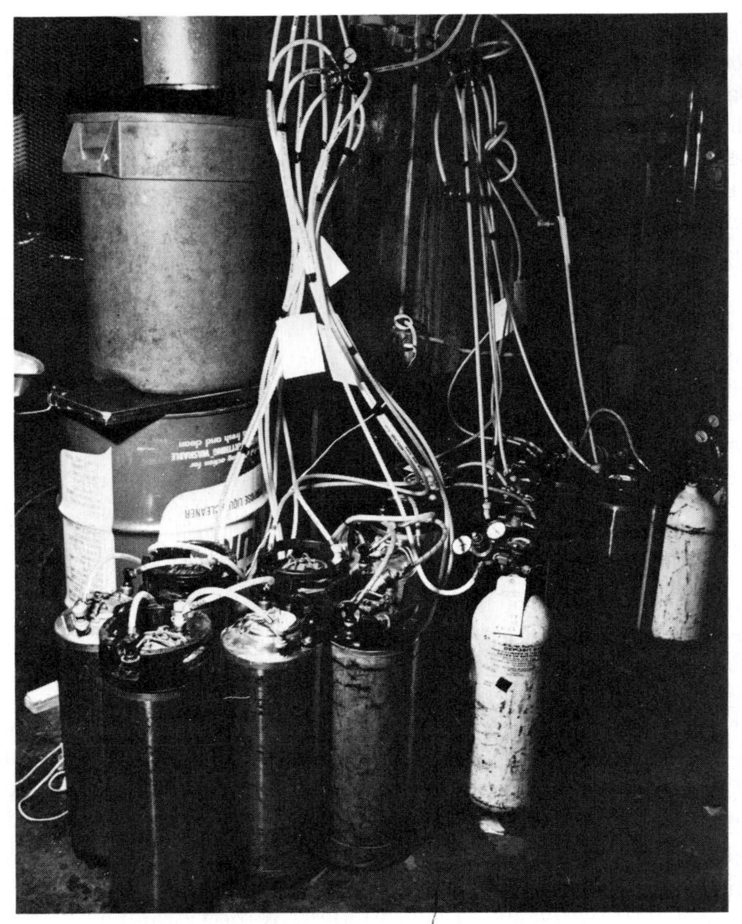

figure 3.6 Postmix syrup tanks at Jack Tar Hotel. White cylinder is carbon dioxide. Bottle chute from bar empties into trash can behind syrup lines. (Photo by Charlotte F. Speight)

postmix system the gun and carbonator are purchased from one manufacturer, the cold plate is part of the cocktail station, the syrups come from the individual distributors and the CO_2 from a CO_2 supplier, and these components are assembled on site. Usually the purchase and assembly of supplies and refills can be arranged through a single soda distributor.

A third type of soda system is the use of bottles, purchased and stored by the case. This is far more expensive than a postmix system—in portion cost, in labor, in time, and in storage space both at the bar and in the storeroom. Yet there are bars that use bottles by choice. Why? Because chilled bottled soda tastes better and keeps its carbonation longer (after all, the bubbles were put in at the factory), and discriminating customers know the difference. Bottled mixes are a specialty item used by such bars as Henry Africa's Cocktail Lounge in San Francisco and the Drake Hotel in Chicago, where a top-quality drink is part of the total concept.

At many bars today not only the soda system but also the liquor pouring is automated. A number of electronic dispensing systems are on the market, pouring preset amounts of liquor and counting each drink. Different systems measure and pour anywhere from half a dozen well brands to a complete spectrum of mixed drinks.

Some systems use a handgun mounted on a flexhose that is similar to the gun for a soda system (Figure 3.7*a*). Buttons on the gun activate the flow from half-gallon or 750-milliliter bottles mounted upside down on the walls of a remote storeroom (Figure 3.7*b*). A preset amount of liquor is delivered, and counters keep track of each drink. A "long shot" and a "short shot" can also be poured. Another kind of dispenser is a series of faucets that are activated by touching a glass to a button under the faucet. Systems like these can dispense and control perhaps three-fourths of the volume of liquor poured. Beyond the 6 or 12 brands a system controls, the call brands, liqueurs, juices, cream, and so on are dispensed by hand in the usual way.

In still another type of system electronic dispensing equipment is integrated or interfaced with a computerized cash register. To operate, the bartender or server must ring up a guest check before drinks can be poured. This gives high control, but the precheck system can slow down operations in a fast, high-volume operation.

The sophisticated system shown in Figure 3.8 is an integrated, computerized system that will pour not only liquor but mixed drinks, including drinks containing cream and juices—up to 1200 different combinations. The bartender touches the desired squares on the register face (Figure 3.8*a*), ices the glass, and raises it to the pouring tower of the cocktail station (Figure 3.8*b*). This act triggers the pouring of all the drink ingredients simultaneously, in 3 seconds or less.

A remote storage rack holds the liquor supply (46 brands, 112 bottles), capable of serving up to four stations. A refrigerated cabinet below the register holds fresh cream, fruit juices, and special mixes that also feed into the pouring system. In addition, a postmix soda system, housed below the pouring tower, is integrated with the pouring system.

Drinks are recorded on a guest-check printer as they are poured. When the bartender touches the "Total" square, the check is totaled and the cash drawer opens. The data accumulated by the register are available through a separate printer for instant sales and inventory analysis.

Automatic liquor-dispensing systems have caused considerable controversy since their introduction a dozen or so years ago. In addition to the sizable investment required, there is some resistance to the whole concept. Some bartenders are unhappy with mechanical pouring. Many operators fear customer resistance—the service seems less personal and customers

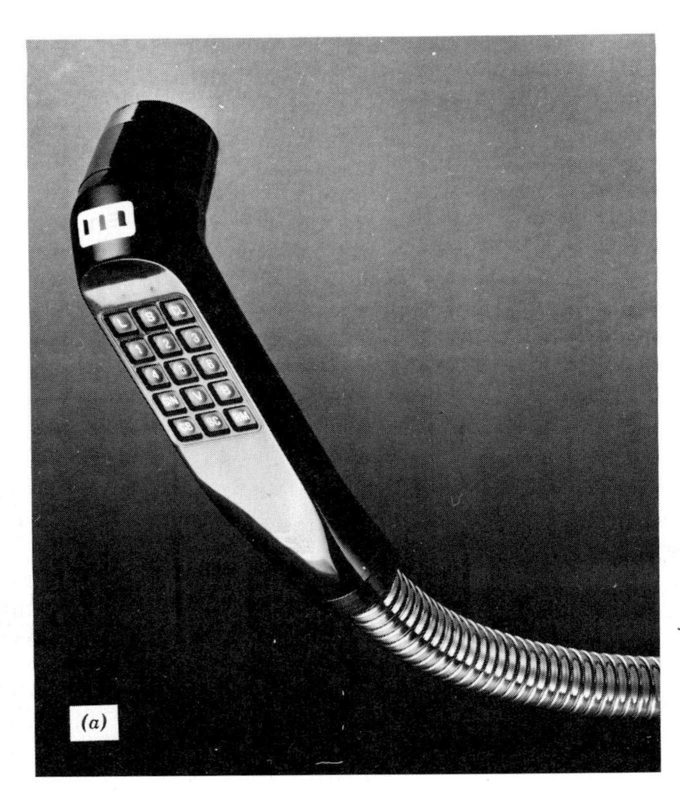

figure 3.7 An electronic dispensing system. (*a*) Handgun. (Photo courtesy Electronic Dispensers International) (*b*) Remote supply in storeroom at Hyatt Regency, Dallas. (Photo by Pat Kovach Roberts)

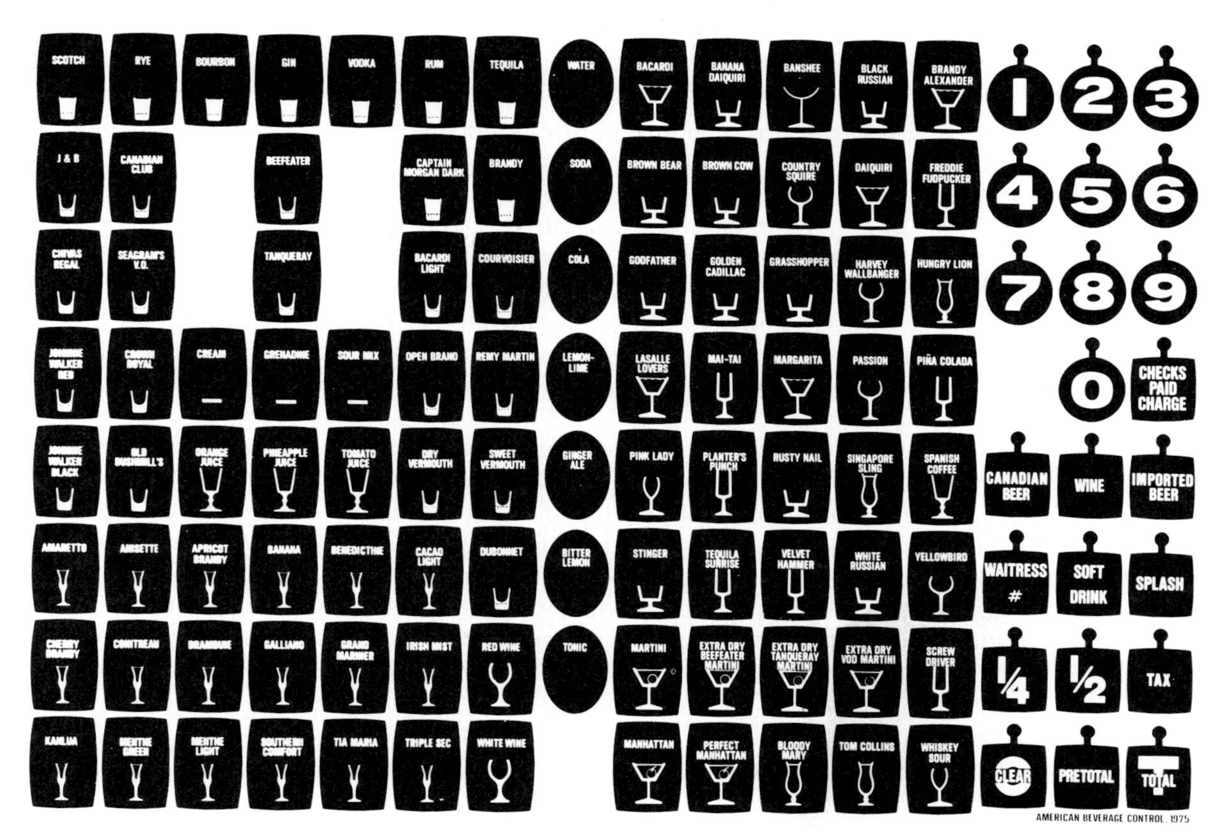

(a)

(b)

figure 3.8 ABC Computer Bar™. (*a*) Register keyboard. (*b*) Cocktail station. (Photos courtesy American Beverage Control)

may suspect that liquor poured out of sight is inferior to the brands they can see being poured at the bar rail.

But for the most part customers have accepted the system—if they notice it at all. In Crazy Ed's bar/restaurant in Phoenix they can't help noticing: the remote system is on public display—a brightly lit showpiece advertised as "Crazy Ed's Incredible Whiskey Machine." Having spent $25,000 on it, the owner wasn't going to hide it in the back of the house. In Stratford, Connecticut, the Mermaid Tavern has its bottles arranged in a handsome backbar display of premium labels; a green light shows which bottle is being poured.

On the plus side, the automatic systems cut pouring costs in several ways. There is the savings of using the larger bottles rather than liters and quarts; the savings of getting that last ounce out of the bottle that in hand-pouring clings to the bottle's sides; the savings in labor because of faster pour—often eliminating the need for an extra bartender at peak periods. Norman Hannay, in *The Cornell Hotel and Restaurant Administration Quarterly* for November 1977, estimates a savings of 14% fewer man-hours over hand-pouring the same volume—if the system is not tied to the cash register. Preringing on the cash register slows things down, his study reveals. At the Hyatt Regency in Dallas the one bar out of four that does not have automated pouring has a slightly higher bar cost—1 to 2%—than the other three.

One of the great advantages of an automatic system is the consistency of the drink served—a very desirable goal that is hard to achieve with hand-pouring, even measured hand-pouring. The gun also makes for a smooth, swift operation, with less handling of glasses.

Automatic dispensing systems can also provide tighter liquor controls, and this is one of their major selling points. They do not, however, automatically eliminate all losses through spills, mistakes, and pilferage. Drinks are measured and counted, yes, but the count must still be checked daily against sales and inventory. It is far easier to spot a discrepancy with an automatic system, but the cause of it must still be removed before this kind of savings takes place.

Each enterprise must weigh these and other savings against the cost of the equipment in relation to the volume of business being done. In a small operation an automatic system might not be worth its high price tag. In a high-volume bar it may be the thing that makes that volume possible, and it can pay off its own high price tag in a couple of years or less.

Going back to the pouring stations in Figures 3.2 and 3.3, you see a mixer (E) on the shelf next to the cocktail unit. It is like the machine that makes your shake at McDonald's, and in this book we refer to it as a ***shake mixer***. It has a shaft coming down from the top that agitates the contents of its cup. It is used for cocktails made with fruit juices, egg, sugar, cream, or any other ingredient that does not blend readily with spirits. It is one of today's mechanical substitutes for the hand shaker.

figure 3.9 Frozen drink machine, usually positioned on backbar near cocktail station. (Photo courtesy Sani-Serv Division of Catalox Corp.)

The machine on the shelf beside the shake mixer is a ***blender*** (F), which can take the mixing process one step further. In the bottom of its cup are blades that can grind, puree, and otherwise refine ingredients put into it. Some drinks require a blender, such as those that incorporate food or ice—a Banana Daiquiri or a Frozen Margarita, for example. Many bars have both mixer and blender; you see them at the Jack Tar in Figure 2.4 (page 32).

Bars making a specialty of a particular frozen drink may have a ***frozen drink dispenser*** (Figure 3.9). Similar to the soft-ice-cream machine at the Dairy Queen, it soft-freezes a large quantity of a premixed drink. You pour your 5 gallons of Strawberry Daiquiris or Margaritas or whatever into the top of the machine, and in a few moments it is frozen to a slush. To serve an individual drink you hold a glass under the tap and move a lever. At the end of the day you drain off what is left, store it in the refrigerator, and clean the machine so that it won't get gummy.

To complete the equipment for drink mixing, there are the glasses in which the drinks will be served. In a typical bar there are glasses everywhere. Stemware may be stored in overhead racks, arranged according to type (see Figure 1.6). This is a decorative as well as sanitary feature of many of today's bars. Other glasses are upside down on drainboards, ridged shelves, or heavy plastic netting, to allow air to reach the inside of the glass. This is a typical health code requirement, to keep bacteria from growing in a damp enclosed space. There is sometimes the temptation to put towels under netting to catch the runoff from wet glasses, but this cuts off the air, canceling the effect and making the problem worse.

A bar needs a way to chill glasses for straight-up cocktails, frozen drinks, and ice cream drinks. Some bars make a special promotional point of serving drinks in frosted glasses or beer in frosted mugs. Both needs may be served by a ***glass chiller*** (***glass froster***), a top-opening freezer that chills glasses at temperatures around zero degrees Fahrenheit (Figure 3.10 and H on Figure 3.2). When the glass is removed from the freezer it immediately acquires a sensuous coat of frost.

There are two schools of thought about glass frosters. One says that the

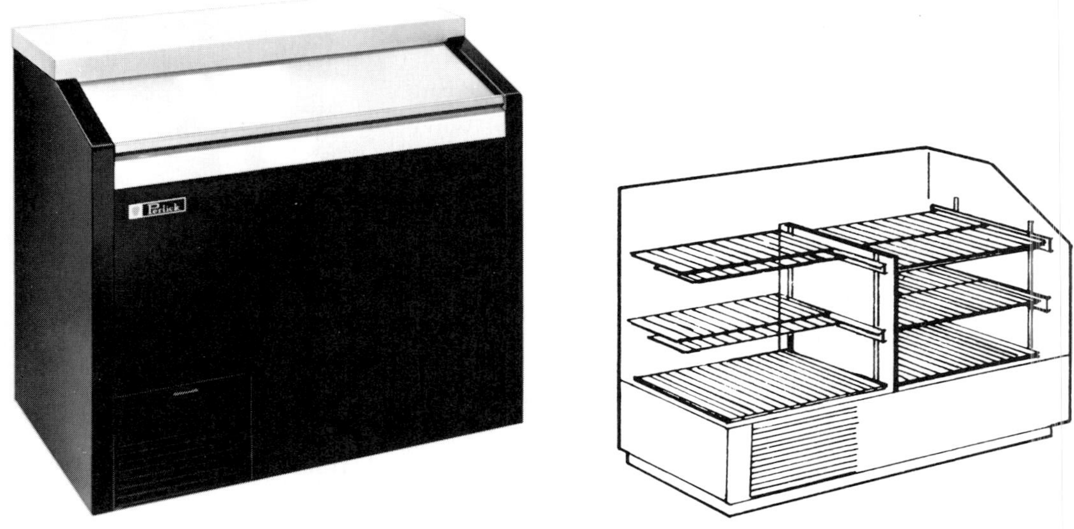

figure 3.10 Glass chiller. Drawing shows how interior racks can be slid one under another to reach glasses below. (Photo and drawing courtesy The Perlick Company)

frosted look and feel and taste makes a great merchandising tool. The other says that freezer chilling is hardly necessary: ordinary refrigerator temperatures will frost a wet glass, while a beer served in a frosted mug actually tastes hot to some people and the glass can stick to your lip. In any case there must be refrigerator space somewhere for chilling glasses, or the glasses must be iced by hand when the drink requires a chilled glass.

Equipment for washing. Equipment for washing both glasses and hands is usually specified in detail in local health codes. It typically includes these items shown on Figure 3.2:

- A three- or four-compartment sink (I).
- Drainboards (J).
- Special glass-washing brushes (K).
- Hand sink (G) with towel rack (attached to blender station).
- Waste dump (L).

The sink with drainboards is usually a single piece of equipment (Figure 3.11) placed near a bartender station or between two stations. One compartment is for washing, one for rinsing, and one for sanitizing (killing bacteria with a chemical solution). In a four-compartment sink the fourth com-

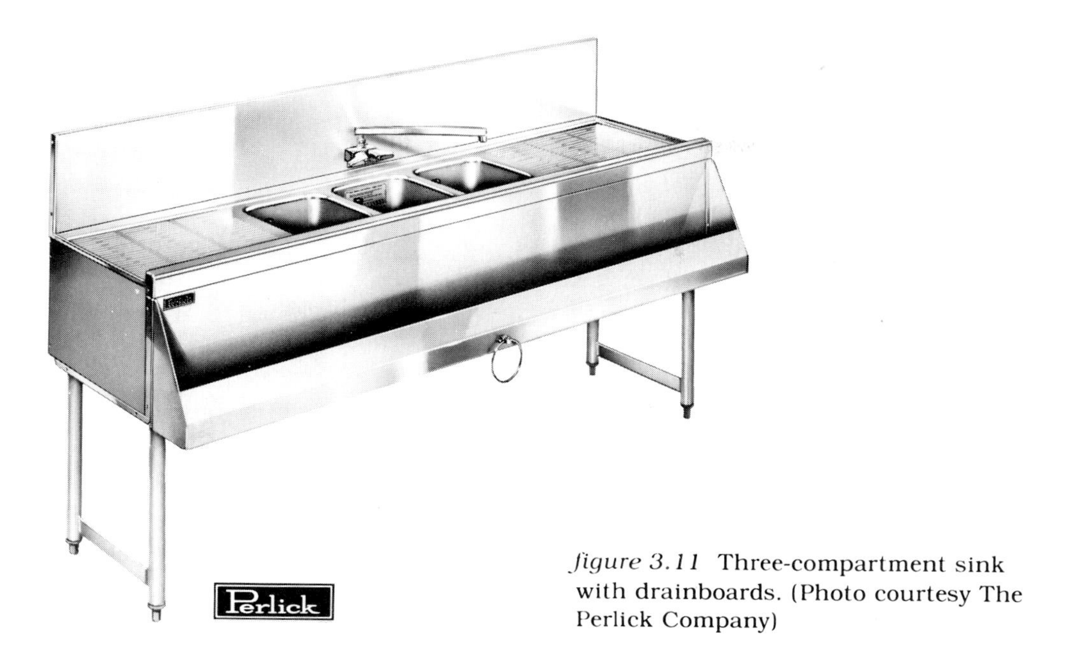

figure 3.11 Three-compartment sink with drainboards. (Photo courtesy The Perlick Company)

figure 3.12 Hand sink. (Photo courtesy The Perlick Company)

partment is usually used as a waste dump by placing netting in the bottom to catch the debris from used glasses. Sometimes it can function as the hand sink. A separate hand sink is shown in Figure 3.12.

Glass brushes stand up in the soapy water of the wash sink. Figure 3.13 is a motorized model: the bartender places a glass over a brush and presses a button to make the bristles spin. With hand models the bartender twists

figure 3.13 Motorized glass brushes. (Photo courtesy Bar Maid Corporation)

the glass around and between the brushes to clean the inside and rim. Then the glass goes into the rinse sink, and then the sanitizing solution, and finally onto the drainboard, upside down, to air-dry.

An alternative method of glass washing is a mechanical dishwasher that delivers water hot enough to kill bacteria. A few local health codes require this method, and a few do not allow it. There are small stationary under-counter models. There are also conveyor types on which servers can place dirty glassware, which is then conveyed through the machine and delivered clean to the bartender in the underbar area. This type would be used only in high-volume operations.

In some restaurants bar glasses are washed in the kitchen dishmachine. This has several drawbacks. They have to be carried back and forth. They are much more likely to get broken. They must be run separately from dirty dishes so as not to be exposed to grease. The smallest trace of grease on a glass can spoil a drink.

At the Jack Tar bar each pickup station has a waste dump on the server's side of the bar, which you can see in Figures 3.2 and 3.3 (L). Cocktail servers returning with dirty glasses dump the debris here. Behind the bar are removable trash cans.

A bottle chute (W on Figure 3.2) conveys empty beer and soda bottles to a trash can in the garage below (Figure 3.6). Empty liquor bottles are accumulated at the bar to be turned in to the storeroom in exchange for full bottles. In a bar that has a storeroom below it, a bottle chute can convey the empties directly to the storeroom.

Ice and ice machines. A cocktail bar could not run at all without ice, so a plentiful supply is essential. Every bar operation has an ***icemaker*** (***ice machine***); some have more than one. If the bar is large enough the icemaker can be part of the underbar or backbar (Figure 3.14); if not, it must be behind scenes and ice must be bused in. At the Jack Tar it is bused in from the kitchen.

What kind of ice will you use? Figure 3.15 shows different types and sizes of cubes. Large cubes melt more slowly. Smaller sizes stack better in the glass. Round cubes fit the glass better at the edges. Rectangular cubes stack better than round ones, leaving fewer voids. In a fast pour, liquor hitting a round cube can splash out of the glass. Which one will fill your special needs? Your clientele, your drink menu, your glassware, your type of pour, and your service all make a difference.

It is important to choose your ice before you buy an ice machine, because each machine makes only one type and size. In some machines you can adjust the cube size; the machine in Figure 3.14, for example, will make a ¾″ × ¾″ cube varying from ¼ to ½ inch in thickness according to a simple

figure 3.14 Icemaker. Used without legs, this model will fit under the bar. (Photo and capacity table courtesy Whirlpool Corporation)

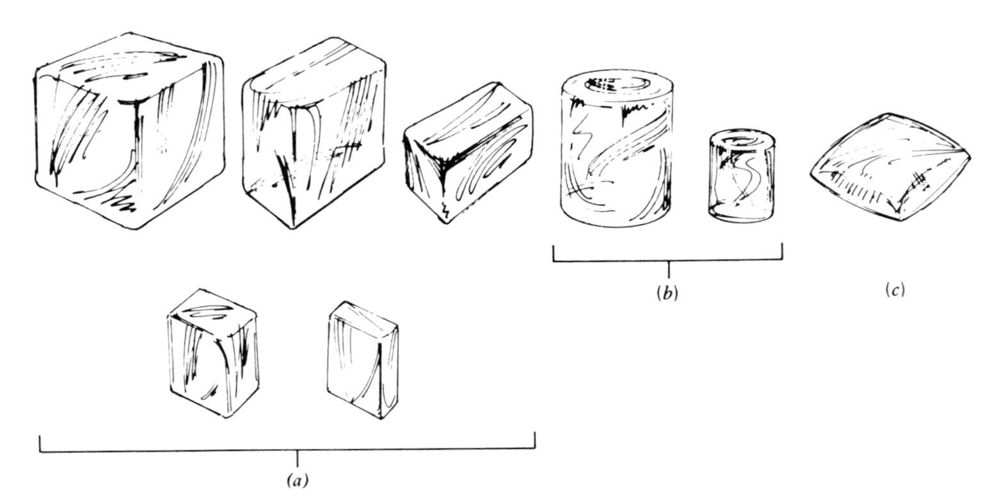

figure 3.15 Types of ice. (*a*) Cubed cubes of various sizes and shapes. (*b*) Round cubes. (*c*) Contour Cube™ (pillow cube).

adjustment you can make yourself. A different grid for the same machine will give you a big 1¼″ × 1½″ cube ½ to 1 inch thick. But no machine makes more than one size cube at one time.

In addition to cube ice there is crushed ice, cracked ice, flake ice, and shaved ice. Crushed ice can be made by running cube ice through an ***ice crusher***. Another type of machine produces crushed or cracked ice from scratch instead of from ice cubes. Both machines make small random-size pieces of hard clear ice. Some drinks call for crushed or cracked ice, and it is used in making frozen drinks.

A ***flake-ice machine***, or ***flaker***, produces a soft beady ice like snow, which is used mostly for keeping things cold. In a wine bucket, for example, flake ice will assume the shape of the bottle. If you use cubes in a bucket they will slide to the bottom when you take the bottle out and you can't get the bottle back in, which more or less defeats the purpose of the bucket, except for show.

Flake ice is not suitable as a bar ice, except for frozen drinks. In an ordinary mixed drink it melts fast, dilutes the drink, and tends to create a water cap on the surface, which makes the drink taste weaker than it is. Shaved ice, made by a machine that scrapes ice from large blocks, is similar to flake ice, soft and opaque, and has the same uses.

The size icemaker you need (expressed in pounds of ice produced in 24 hours) will depend on four things: the size glass you use (or an average of several sizes), how much of the glass you fill with ice, the size of the cube, and the maximum number of drinks you expect to sell in your busiest 24-hour period. For example, an 8-ounce highball glass filled three-quarters full with small cubes will add up to about 4 ounces of ice. Multiply that by, say,

600 drinks in 24 hours, and you find you need 150 pounds of ice per day. Making allowances for spillage, staff use, and growth in your business, you need a 200-pound machine to service your bar alone.

Where you put the machine, however, can drastically affect its production. The warmer the air around it and the warmer the water you feed it, the less ice it will produce. The table alongside the machine in Figure 3.14 shows this effect.

Other things affecting location are the available space and, in many cases, sharing it with a kitchen. Wherever you put it, the place must be well ventilated in order for it to function properly, and the area must meet the same sanitation standards as the bar itself.

Essentials for draft-beer service. A draft-beer serving system (Figure 3.16) consists of a ***keg*** or ***half-keg*** of beer, the ***beer box*** containing it, the ***standard*** or ***tap*** (faucet), the ***line*** between the keg and the standard, and a CO_2 tank connected to the keg with another line. The beer box, also called a ***tap box***, is a refrigerator designed especially to hold a keg or half-keg of beer at the proper serving temperature of 36° to 38° F. Generally it is located

figure 3.16 Draft-beer system. Keg in middle is a reserve. (Photo courtesy The Perlick Company)

right below the standard, which is mounted on the bar top, so that the line between keg and standard is as short as possible. If more than one brand of draft beer is served, each brand has its own system—keg, line, and standard, either in its own beer box or sharing a box with another brand.

The supply of beer at the bar should be sufficient to last the serving period, since bringing in a new keg of beer and tapping it is a major operation. It sometimes takes several kegs to provide enough beer for a high-volume bar.

A beer system may be designed into either the front bar or the backbar. If there is not room at the bar, the beer boxes may be located in an adjoining storage area with lines running into the standards. In remote systems, temperature and pressure are more difficult to control.

On the backbar of the Jack Tar (Figure 3.3) you see the beer standard (Q), but the rest of the system is on the garage level below. The beer box consists of a large walk-in cooler, and a beer line runs behind the liquor display up to the standard.

Storage equipment. You need enough storage space at the bar to take care of all your needs for one serving period—liquor, mixes, bottled drinks, wines, beers, garnishes, and miscellaneous supplies—cocktail napkins and the like. Generally this storage forms the major part of the backbar.

The day's reserve supplies of liquor—all the unopened bottles backing up those in use—are stored in dry (i.e., unrefrigerated) storage cabinets with locks, or they are displayed on the backbar. Also in dry storage are jug red wines for pouring by the glass or carafe, and reserve supplies of napkins, bar towels, matches, picks, straws, stir sticks, and other nonfood, nonbeverage items.

Undercounter and backbar refrigerators hold supplies of special mixes and juices, bottled beer, bottled mixes if used, white wines, fruits and condiments for garnishing, cream, eggs, and other perishables. They may also be used to chill glasses. Backbar refrigerators usually look from the outside just like the dry-storage cabinets. Some backbar units are half refrigerator and half dry storage.

Turning back to Figure 3.3, you can see how this works out at the backbar of the Jack Tar. Backbar refrigerator cabinets (T) store supplies of perishables in one section, while the other sections store bottled beers and soft drinks (Figure 3.17). To the right at the backbar is a dry-storage cabinet (U) for bar towels, napkins, and other supplies. Reserve supplies of liquor are stored in locked overhead cabinets above the front bar.

The Jack Tar keeps its daily supply of wines by the bottle for dining-room service in a storeroom around the corner (Figure 2.5, page 34), which includes a large cooler set at 40°F for the white wines. Other restaurants may store their daily wine stock at the bar, or they may have it displayed in wine racks as part of the room decor, or the wines may all be stored under lock and key in a behind-scenes wine cellar.

figure 3.17 Backbar refrigerator. (Photo courtesy The Perlick Company)

A recent development is an automatic wine-dispensing system for wines that are served by the glass or carafe. In purpose and operation it is similar to an automatic liquor-dispensing system, with a wine barrel or plastic container and a measuring and counting system either at the bar or in a remote location. Bulk wine is, of course, cheaper than bottled wine, but unless volume is high it is not practical, since wine deteriorates quickly in a half-empty container. Some states limit the size container from which wine may be poured.

A newly invented machine pours wine from individual bottles by displacing the wine with nitrogen under pressure. Since nitrogen is an inert gas, it does not interact with the wine remaining in the bottle, which therefore retains its quality until the bottle is emptied.

Bar tools and small equipment . . .

Stainless steel is the metal of choice for small equipment and utensils, just as it is for large underbar pieces, and for the same reasons. It looks good and stays that way; it is durable and easy to clean.

Most of the small bar equipment is used for mixing and pouring. A second group of utensils is used in preparing condiments to garnish drinks. A third group is involved in serving.

Small wares for mixing and pouring. The indispensable tools for mixing and pouring by hand are these:

- Jiggers.
- Pourers.
- Mixing glass.
- Hand shaker.
- Bar strainer.
- Barspoon.
- Ice scoop.
- Ice tongs.
- Muddler.
- Fruit squeezer.
- Funnel.

A ***jigger*** (Figure 3.18) is a measure of ounces or fractions of ounces. Jiggers are used to measure out liquors for cocktails, highballs, and other mixed drinks.

There are two types of jigger. The double-ended stainless-steel jigger has a small cup on one end and a large cup on the other. It comes in several combinations of sizes, such as ½ ounce/1 ounce, ¾ ounce/1 ounce, 1 ounce/1½ ounces. The most-used combinations are probably the ¾/1½ and the 1/1½ ounce sizes, but what you need depends on the size drink you serve.

The second jigger type is made of heavy glass with a plain or elevated base. It comes in several sizes, from ⅞ ounce to 3 ounces, either with or without a line marking off another measure, as, for example, a 1-ounce glass with a line at ½ or ⅝ ounce, or a 1½-ounce glass with a line at ½, ⅝, ¾, ⅞, or 1 ounce.

figure 3.18 Jiggers: two types, two sizes, seven measures altogether. (Photo by Pat Kovach Roberts)

To measure using the steel jigger the bartender fills the cup to the brim. To measure in the glass jigger the bartender fills to the line. After pouring the drink the bartender turns the jigger upside down on the drainboard so that any residual liquor drains out and one drink's flavor will not be carried over to the next. If a jigger is used for something clinging, like cream or a liqueur, it is rinsed out.

A glass jigger may also be used as a ***shot glass*** when a customer orders a straight shot.

A ***pourer*** (Figure 3.19) is a device fitting into the neck of a beverage bottle, so constructed that it reduces the rate of flow to a predictable and controllable amount. A pourer is used on every opened liquor bottle at the bar. There are three categories—slow, semifast, and fast.

Pourers are available in either stainless steel or plastic. The plastics come in different colors and can be used to color-code different types of liquor. The stainless-steel pourers are better-looking and last longer, except for the corks that fit them into the bottle necks. These wear out and must be replaced from time to time.

There are also pourers that measure the liquor poured and cut off automatically when a preset amount is reached. They are expensive and bartenders don't like them, but they are a form of control not to be overlooked if they will save more money and aggravation than they cost.

A ***mixing glass*** (Figure 3.20*a*) is a heavy glass container in which drink ingredients are stirred together with ice. A typical mixing glass has a capacity of 16 to 17 ounces. It is used to make Martinis and Manhattans and other drinks whose ingredients blend together readily. It is rinsed after each use. Mixing glasses should be heat-treated and chip-proof.

A ***hand shaker*** (Figure 3.20*b*) is a combination of a mixing glass and a stainless-steel container that fits on top of it, in which drink ingredients are shaken together with ice. The stainless-steel container is known variously

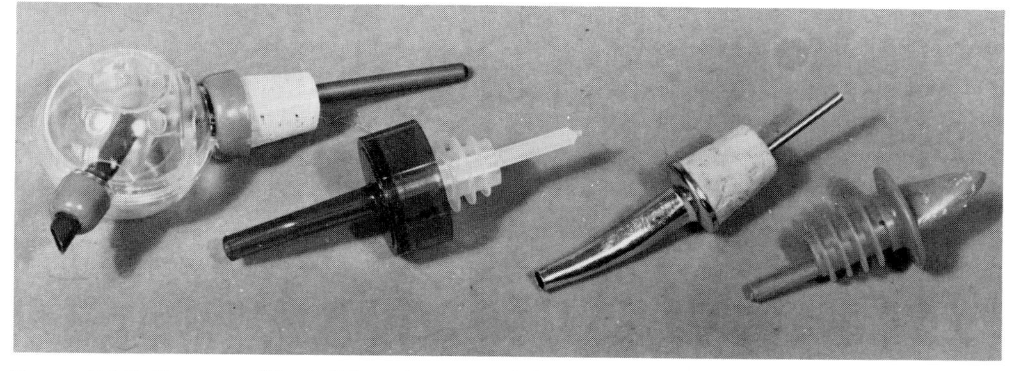

figure 3.19 Pourers. The left-hand pourer delivers a measured portion. Second from right is a stainless-steel pourer. The other two are plastic. (Photo by Pat Kovach Roberts)

figure 3.20 (*a*) Mixing glass. (*b*) Shaker cup fits tightly over mixing glass for hand shaking. (Photos by Pat Kovach Roberts)

as a ***mixing cup, mixing steel, mix can***. Ingredients and ice are measured into the mixing glass, and the cup is placed firmly on top, angled so that one edge is flush with the side of the glass. The two are held tightly together and shaken. The cup must be of heavy-gauge, high-quality stainless steel; if it loses its shape it will not fit tightly over the glass. Usually a shaker comes in a set with its own strainer.

A shaker is used for cocktails made with fruit juices, egg, sugar, cream, or any other ingredient that does not mix readily with spirits. It is rinsed after each use.

The mixing container of the shake mixer mentioned earlier is also called a mixing cup, steel, or can. This machine has largely supplanted the hand shakers at today's bars. It is faster and more efficient. It can even make ice cream drinks, which is something the hand shakers can't do.

A ***bar strainer*** (Figure 3.21, left) is basically a round wire spring on a handle, which fits the top of a shaker or mixing glass; it has ears that fit over the rim to keep it in position. The strainer keeps ice and fruit pulp

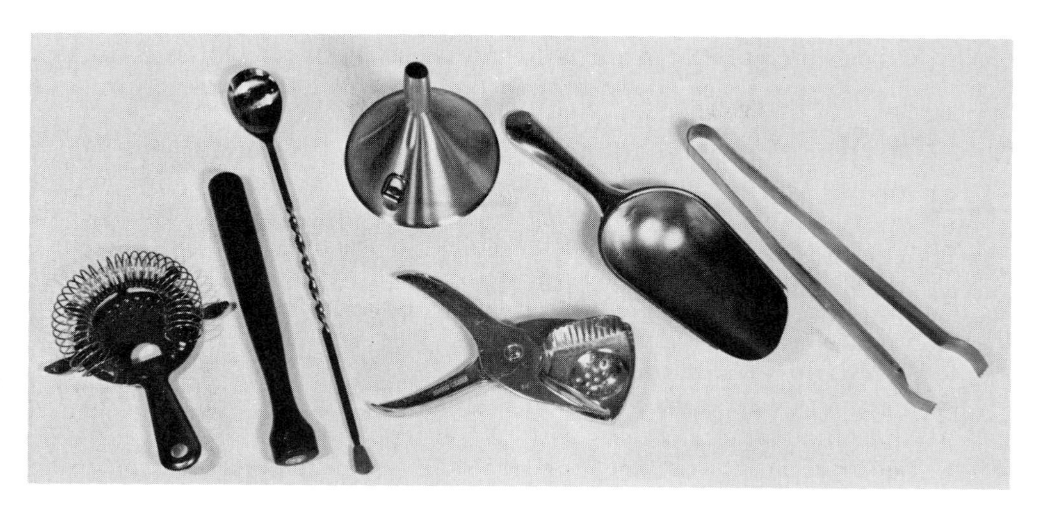

figure 3.21 Tools for mixing. Left to right: strainer (fits over glass or mixing cup), muddler, barspoon, funnel, hand squeezer, ice scoop, tongs. (Photo by Pat Kovach Roberts)

from going into the glass when the drink is poured. It is used with mixing glasses and shaker and blender cups.

A barspoon (Figure 3.21, third from left) is a shallow spoon having a long handle, often with a bead on the end. Spoon and handle are stainless steel, typically 10 or 11 inches long. The bowl equals 1 teaspoon. Barspoons are used for stirring drinks, either in a drink glass or in a mixing glass or cup.

Humorist George Ade, writing nostalgically during Prohibition in *The Old-Time Saloon*, describes the use of the barspoon in pre-Prohibition days when, he says, a good bartender would have died of shame if compelled to use a shaker: "The supreme art of the mixing process was to place the thumb lightly on top of the long spoon and then revolve the spoon at incredible speed by twiddling the fingers . . . a knack acquired by the maestros only." Perhaps this mixing method explains the traditional bead on the end of the handle.

An ***ice scoop*** (Figure 3.21, second from right) is, as its name implies, an implement for scooping up ice from the ice bin. It usually has a 6- or 8-ounce capacity. A standard size makes it easy to get just the right amount of ice with one swoop of the scoop. Actually, bartenders often scoop with the glass, a *very dangerous* practice! It is too easy to break or chip the glass, leaving broken glass in the ice bin and a cutting rim on the glass. It shouldn't be done.

Ice tongs (Figure 3.21, right) are designed to handle one cube of ice at a time. One of the lesser bar tools, tongs are a relic from the days when all ice cubes were great big things. Nevertheless they are still used—in airline service, for example. They are important, since ice that will go into a drink should not be touched by the hands.

A ***muddler***, or ***muddling stick*** (Figure 3.21, second from left), is a wooden tool one end of which is flat for muddling or crushing one substance into another, such as sugar into bitters in an Old-Fashioned. The other end is rounded and can be used to crack ice. It looks like a little baseball bat. The muddler, too, is a relic from another day; now we use simple syrup instead of lump sugar in our Old-Fashioneds, and we don't need to crack ice very often.

A bar-type ***fruit squeezer*** (Figure 3.21, lower middle) is a hand-size gadget that squeezes half a lemon or lime for a single drink, straining out pits and pulp as it squeezes.

Funnels (Figure 3.21, upper middle) are needed in several sizes for pouring from large containers into small ones, such as transferring special mixes from bulk containers into plastic bottles for bar use. Some funnels have a screen at the wide end to strain out pulp and other such things.

Tools and equipment for garnishing. It is usually part of the bartender's job to set up the fruits and other foods used to enhance or garnish a drink. These are typically lined up ready to go in a multicompartment ***condiment tray***. Often this is mounted on some part of the underbar at the serving station. It should not, however, be directly above the ice bin; many health codes define it as a health hazard because of the likelihood of dropping foods into the ice.

An alternative to the installed condiment tray is a glass or plastic tray on the bar top or glass rail. Such a tray may be moved about at the bartender's convenience and is cleaned more easily than one fixed to the underbar. If the servers garnish the drinks, which is often the case, the garnishes must be on the bar top at the pickup station, as in Figure 3.2 and 3.3. A plastic condiment tray is shown in Figure 8.9 (page 209).

The tools for the task are few but important:

- Cutting board.
- Bar knife.
- Relish fork.
- Zester, router, or stripper.

They are pictured in Figure 3.22.

A ***cutting board*** for the bar can be any board having a surface that will not dull the knife. Rubber or plastic is the best material. Wood is the pleasantest surface to work on, but many health codes rate it a health hazard

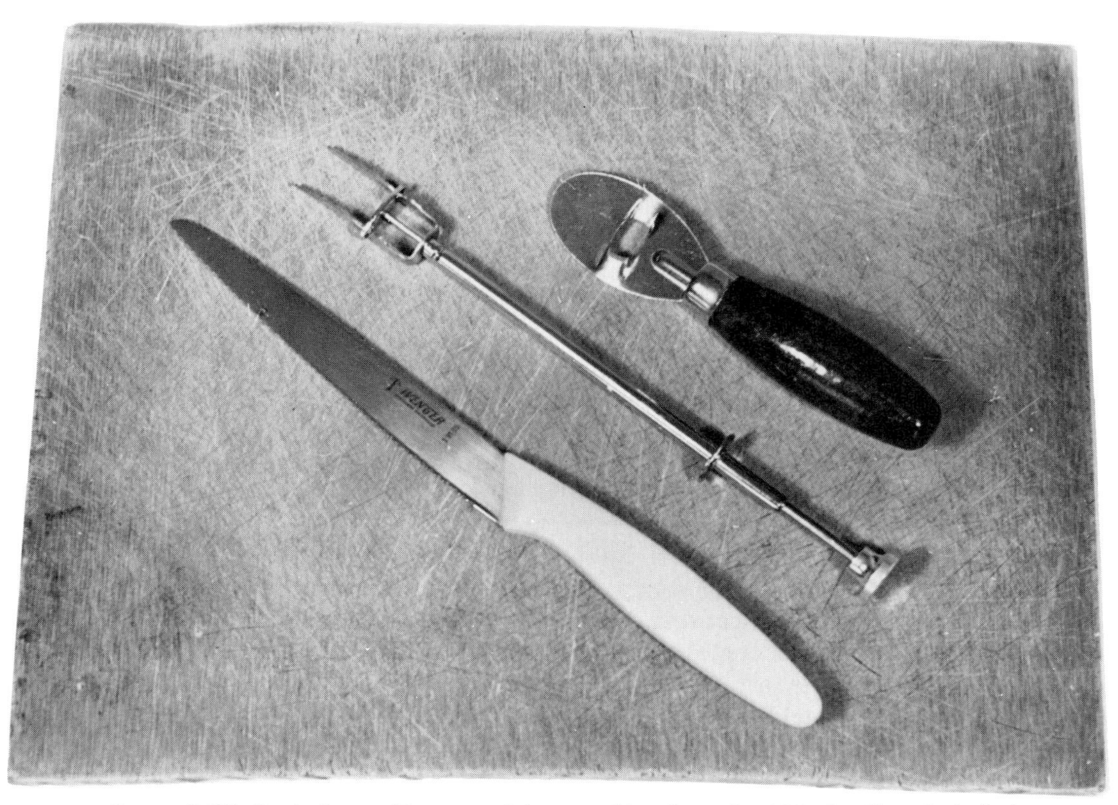

figure 3.22 Tools for cutting garnishes: cutting board, utility knife, relish fork, zester. (Photo by Pat Kovach Roberts)

because it is hard to keep bacteria-free. A small one is all you need and all you will have room for.

A ***bar knife*** can be any small to medium-size stainless-steel knife such as a paring or utility knife. Stainless steel is essential; carbon steel will discolor and pass the color to the fruit being cut. The blade must be kept sharp; not only does a sharp knife do a neater, quicker job; it is safer because it will not slip. The one in the picture has a serrated blade, which is especially good for cutting fruit.

A ***relish fork*** is a long (10-inch) thin two-tined stainless-steel fork designed for reaching into narrow-necked bottles for onions and olives. The one pictured has a spring device that helps to secure the olive or onion firmly.

The ***zester, router,*** and ***stripper*** are special cutting tools for making that twist of lemon some drinks call for; they peel away the yellow part of the lemon skin, which contains the zesty oil, without including the white underskin, which is bitter. The one pictured in Figure 3.22 (top) is a zester or router; the stripper is similar but cuts a broader, shallower swath.

Tools and equipment used in serving. This is a short and rather miscellaneous list, but no bar could get along without these items:

- Bottle and can openers.
- Corkscrews.
- Round serving trays.
- Tip trays.

Any type of ***bottle*** or ***can opener*** that is good quality and does the job is acceptable. Stainless steel is best; it is rust-free and easily cleaned. And these openers *must* be kept clean; it is easy to forget this.

There are many different kinds of ***corkscrews***, a few of which are pictured in Figure 3.23. Each one is designed to extract corks from wine bottles.

The business part of a corkscrew, the ***screw*** or ***worm***, should be 2¼ to 2½ inches long and about ⅜ inch in diameter with a hollow core in the middle like the two on the right. A solid core will chew up the cork. There should be enough spirals to take the screw clear through the cork. Edges should be rounded, not sharp. It should be made of stainless steel.

The ***waiter's corkscrew***, top right, is specially designed for opening wines tableside. It includes the corkscrew itself, a small knife for cutting the seal of the bottle, and a lever for easing out the cork. Made of stainless steel, it folds like a jackknife.

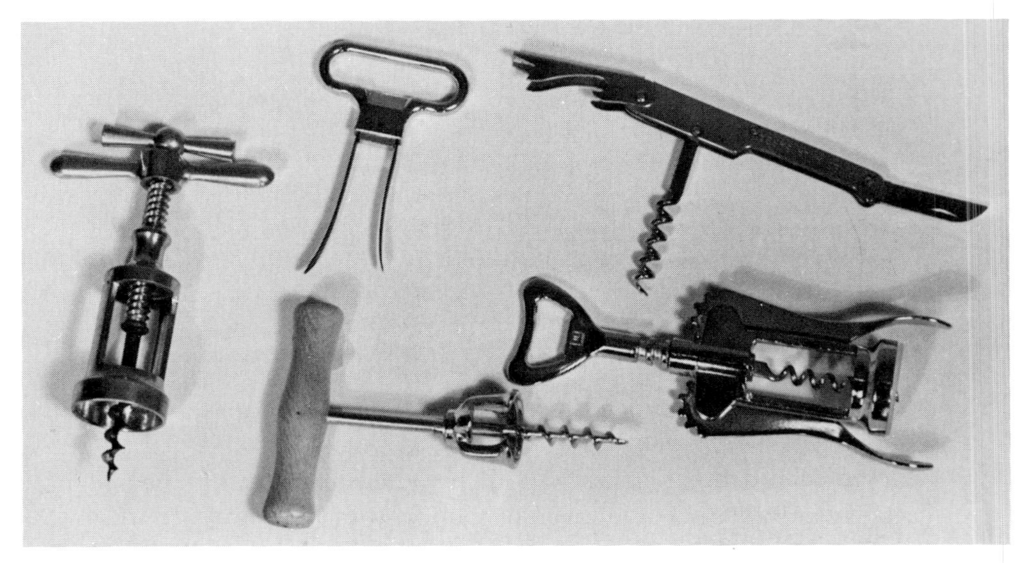

figure 3.23 Corkscrews. Waiter's corkscrew, top right, is best adapted for restaurant service. (Photo by Pat Kovach Roberts)

figure 3.24 Cork-lined serving trays keep glasses from slipping. Tip tray is in foreground. (Photo by Pat Kovach Roberts)

The one below it in the picture is efficient. Its wings rise as the screw is twisted in; when they are pushed down again, this pulls out the cork. It is fine at the bar but too bulky for the table server's pocket.

The device pictured at top middle is a pair of prongs that straddle the cork. You maneuver them about and bring the cork out whole without a puncture in the middle. It is slow and frustrating to most people and not really suitable for commercial use.

The other two corkscrews pictured have solid screws that may chew up corks. They are not recommended.

You will need ***round serving trays*** in two sizes, 10-inch and 14-inch (Figure 3.24). Bar trays should have cork surfaces to keep the glasses from slipping.

A ***tip tray*** (Figure 3.24) is used to present the check, return the change or the credit card, and remind the customer to leave a tip.

Glassware . . .

The glassware you use in serving drinks plays several roles. It is part of your overall concept; its style, quality, and sparkle express the personality of your bar. As functional equipment it has a part in measuring the drinks you serve, and it conveys them to your customers. It is a message carrier: glass size and style tell your guests that you know what you are doing—you have served each drink ordered in an appropriate glass. It can be a merchandising tool: subtle or flamboyant variations of custom in glassware excite interest and stimulate sales—oversize cocktails in wine glasses or beer mugs, coffee drinks in brandy snifters, special glassware for your own specialty drinks.

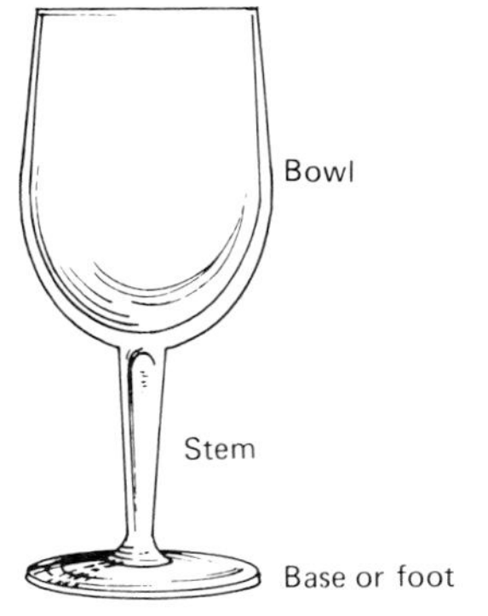

figure 3.25 Parts of a glass.

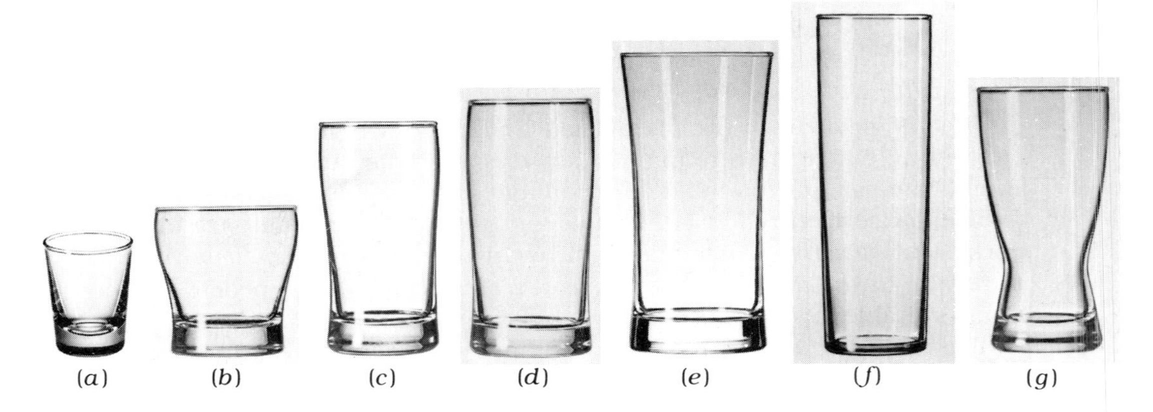

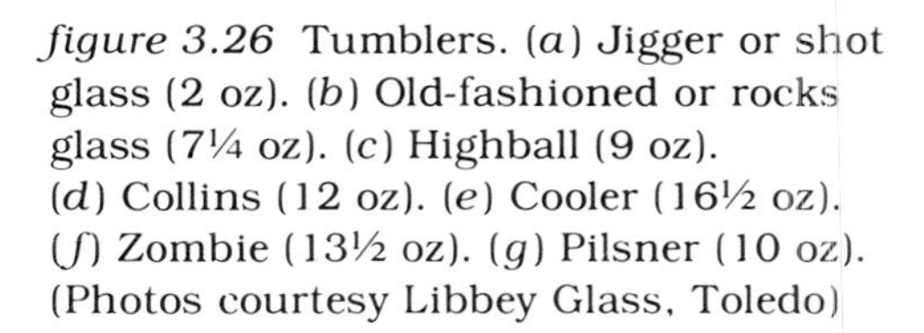

figure 3.26 Tumblers. (*a*) Jigger or shot glass (2 oz). (*b*) Old-fashioned or rocks glass (7¼ oz). (*c*) Highball (9 oz). (*d*) Collins (12 oz). (*e*) Cooler (16½ oz). (*f*) Zombie (13½ oz). (*g*) Pilsner (10 oz). (Photos courtesy Libbey Glass, Toledo)

Glass terms and types. There are three characteristic features of glasses: ***bowl, base*** or ***foot***, and ***stem*** (Figure 3.25). A glass may have one, two, or all three of these features. The three major types of glassware—tumblers, footed ware, and stemware—are classified according to which of these features they have.

A ***tumbler*** is a flat-bottomed glass that is basically a bowl without stem or foot. Its sides may be straight, flared, or curved. Various sizes and shapes of tumbler are known by the names of the drinks they are commonly used for: old-fashioned, rocks glass (for cocktails served on the rocks), highball, collins, cooler, zombie, pilsner (Figure 3.26). Glass jiggers and shot glasses are minitumblers.

Footed ware refers to a style of glass in which the bowl sits directly on a base or foot. Bowl and base may have a variety of shapes. Traditional footed glasses include the brandy snifter and certain styles of beer glass (Figure 3.27). Today footed ware is also popular for on-the-rocks drinks and highballs. In fact, any type of drink can be served in a footed glass of the right size.

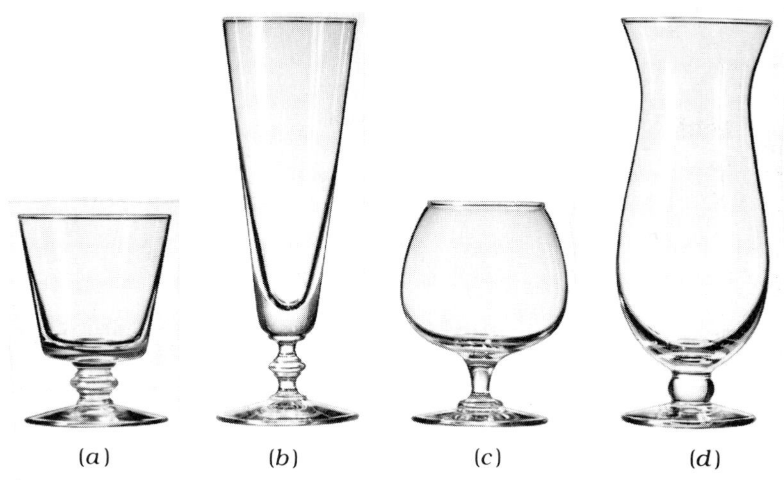

figure 3.27 Footed glasses. (*a*) Rocks (7 oz). (*b*) Beer (10 oz). (*c*) Brandy snifter (12 oz). (*d*) Hurricane (22 oz). (Photos courtesy Libbey Glass, Toledo)

Stemware includes any glass having all three features—bowl, foot, and stem. Again, there is a variety of shapes. Wine is always served in a stemmed glass, as is a straight-up cocktail or a straight liqueur. Certain shapes and sizes of stemware are typical of specific drinks, such as wine, sour, Margarita, champagne (Figure 3.28). Stemware is the most easily broken glass type.

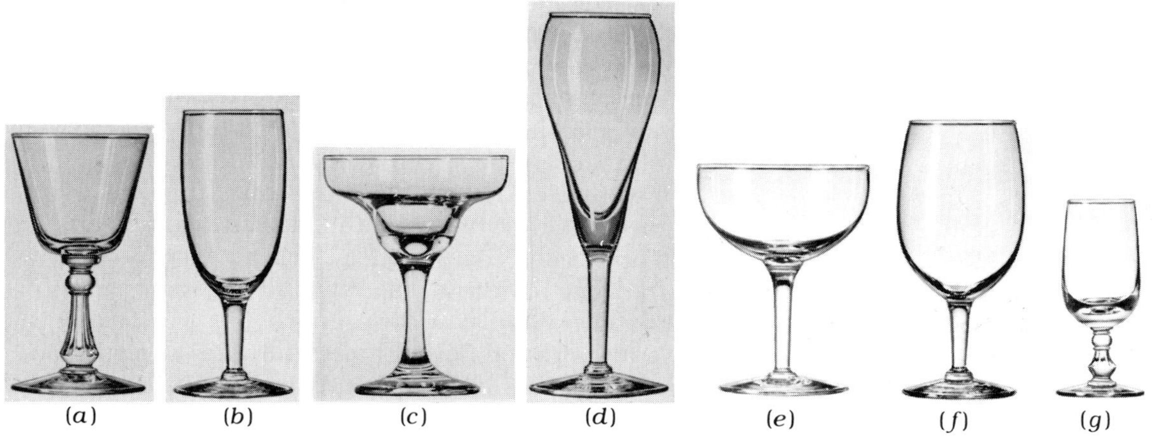

figure 3.28 Stemware. (*a*) Cocktail (5¼ oz). (*b*) Sour (5½ oz). (*c*) Margarita (5 oz). (*d*) Flute champagne (6 oz). (*e*) Champagne saucer (6½ oz). (*f*) All-purpose wine (8¼ oz). (*g*) Liqueur or brandy (2 oz). (Photos courtesy Libbey Glass, Toledo)

figure 3.29 Beer mug (10 oz.). (Photo courtesy Libbey Glass, Toledo)

A fourth type of glass is the ***mug*** (Figure 3.29). You can think of it as a tumbler with a handle or as a tall glass cup. It is usually used for serving beer.

Glass names and sizes. The name by which a glass is called in bar terminology comes from the drink usually served in it, and that drink in turn is related to glass size. Thus a highball glass is typically 8 to 10 ounces; a collins glass is typically 10 to 12 ounces. In mixing a drink the bartender uses the glass size as both guide and control. In the making of a highball, for example, the glass is a measure of the amount of ice to be used and the amount of mix to be added. If the bartender uses the wrong size glass, the drink may be too weak or too strong.

It follows, then, that before you purchase glassware for your facility you must decide how strong a drink you will serve—that is, how much spirit you will use as your standard drink—1 ounce, 1¼, 1½, or whatever. Then you will select glass sizes that will produce the right-tasting and right-looking drinks, the drinks your customers expect.

Tips on glass purchase. In selecting glasses, size is a better guide than the name of the glass, since a glass with a specific name will come in many sizes. Table 3.1 gives, for example, the range of sizes offered by one manufacturer for various types of glasses in various styles. In addition, nearly all glass types come in giant sizes for promotional drinks.

Buy glass sizes that you will never have to fill to the brim; they will surely spill. A glass for dinner wine should be only half full, so the drinker can swirl the wine around and appreciate the bouquet. A brandy snifter serves the same purpose; no matter how big the glass, only one to two ounces of brandy is served, so the customer can savor the aroma.

Most facilities buy only a few of the different types and sizes of glass. One type and size can work for Old-Fashioneds, rocks drinks, highballs; one cocktail glass for straight-up cocktails including sours, for sherry and other fortified wines, and maybe even for champagne. On the other hand, if you are building a connoisseur's image, you will probably want the traditionally correct glass for each type of drink, and you will want several different

table 3.1 **SELECTION OF GLASS SIZES**

glass	*available sizes (oz)*	*recommended size (oz)*
Beer	6 to 23	10 to 12
Brandy, snifter	5½ to 34	Your choice from middle range
Brandy, straight up	2	2
Champagne	3½ to 8½	4½ or larger[a]
Cocktail	2½ to 6	4½ for 3-oz drink
Collins	10 to 12	10 to 12
Cooler	15 to 16½	15 to 16½
Cordial	¾ to 1¾	1¾, or use 2-oz brandy glass
Highball	7 to 10½	8 to 10
Hurricane	8 to 23½	Your choice for specialty drinks
Margarita	5 to 6	5 to 6
Old-fashioned	5 to 15 (double)	7
Rocks	5 to 12	5 to 7
Sherry	2 to 3	3 for 2-oz serving
Sour	4½ to 6	4½ for 3-oz drink
Whisky, straight shot	⅝ to 3	1½ to 3 depending on shot size
Wine	3 to 17½	8 to 9 for 4-oz serving; larger sizes OK
Zombie	12 to 13½	Your choice

[a]The tall thin tulip bowl or flute is recommended over the broad shallow bowl.

wine glasses if you serve a different wine with each course of an elaborate dinner.

How many glasses should you buy? For each type of drink you may want two to four times as many glasses as the number of drinks you expect to serve in a rush period. Four times would be ample; two times would probably be enough for restaurant table service only. There are so many variables you may have to be your own judge, since you know your own clientele best.

In making your glass selection, remember that glassware is about the most fragile equipment you will be using. Consider weight and durability. Consider heat-treated glass if you use a mechanical dishwasher. Consider design, and buy glasses that do not need special handling; flared rims, for example, break easily. Then consider the breakage factor in figuring the numbers you need.

Care of glassware. Handle with care—though it's difficult to do in rush periods. Do not stack glasses or nest them one inside the other; this is bound to cause breakage. Do not wash glasses mixed with plates and silverware in dishmachines. Never use them for scooping ice.

A chipped or cracked glass is a broken glass. It may cut a lip or spill a drink. Throw it out.

Do not pour hot liquid into a cold glass. If you are making Irish Coffee, for example, pour the coffee over a spoon held in the glass.

Cash registers . . .

Since its invention 100 years ago by a tavern owner, the cash register has been a rock of Gibraltar at the bar. It still is, though the slimmer silhouette of the electronic register is fast replacing the old pot-bellied model with the ringing bell. These days, too, the bar's cash register does not always handle cash. In many operations the customer pays a cashier rather than the bartender or cocktail waitress, and credit cards are as common as cash. But the function of recording each bar sale remains at the bar, no matter how or where payment is made.

There are many kinds of cash registers. They all perform two basic functions: they record sales and add them up. The simplest mechanical cash register—probably an antique—does that and only that. Today's most sophisticated cash register not only records and totals but is part of a computer system that can do almost anything a restaurateur might want, from analyzing sales and figuring payroll to dispensing premeasured drinks and inventorying the liquor poured.

Each register has sets of keys that record sales in dollars and cents. Another set of keys represents categories of sales, such as spirits, wine, beer, food, tax, or whatever the operation needs or wants to have figured separately. There may be a third set of keys representing different bartenders or serving personnel. On some registers each of these keys controls a cash drawer, so that each person is responsible for his or her own take.

Each register prints an item-by-item record on a ***detail tape***, plus the totals, category by category. This becomes your daily sales record. It also forms the source document for all your reports to your state control board. The kind of information required varies from state to state. What your state board requires will determine the basic key categories you need on your register.

Some registers are designed to prerecord each order on a guest check. This is known as a ***precheck*** system. It acts as a double control against losses when the order is checked against the sale and both are checked against the money in the register at the end of the serving period.

A more elaborate electronic register may have single keys representing specific drinks with their prices. These keys are known as ***presets***. Each key will print both item and price and make the correct extensions both on the tape and on the guest check. When you touch the "Total" key the register will add up the check, figure the tax, print the total on the guest check, and record everything on the detail tape. For a picture of a preset keyboard, turn back to the register face of the ABC system on page 53.

In a fully computerized system the register not only records sales and totals checks but also functions as an input terminal, gathering data at the point of sale and feeding it into the computer's central processing unit. This type of system is known as a ***POS/ECR*** (point-of-sale electronic cash register). The register data provide material for instant, detailed sales analysis as well as for the bar's accounting records. This register system is commonly used in hotels and other large operations. In some chains the register at every bar feeds into a system-wide computer that can collect and analyze sales and inventory data from all over the country at the touch of a few computer keys.

A computerized system can transmit a drink order from a station in the dining room to a screen at the bar, so the drink can be ready by the time the server reaches the bar. It also prints the guest check at the bar to be picked up with the drink.

Computerized register systems have many advantages. They are more accurate and more efficient than human labor, and humans need far less training to operate them than to write guest checks and ring them up. A POS/ECR system also provides tighter controls over losses, and it supplies data for watching and analyzing operations almost as things happen instead of days or weeks later.

On the other hand, the more sophisticated the register system, the larger its initial cost. For some operations the cost savings and other advantages make a large investment well worth the money. For others the simplest cash register is quite satisfactory. Each operation is unique in its needs. It is well worth your while to analyze your own needs carefully before you buy a register.

Some general considerations . . .

Look for quality. It makes very good business sense to invest in high-quality equipment for your bar. This is true across the board, from the large underbar units right down to the jiggers and pourers and the wine and cocktail glasses. There are a number of reasons why:

- ***Survival.*** Quality equipment will last longer and will withstand better the wear and tear of a high-speed operation. Heavy-gauge surfaces will resist dents, scratches, and warp. Heavy-duty blenders will better survive the demands of mixing frozen drinks. Quality glasses will break less easily than thin brittle ones.

- ***Function.*** High-quality products are less likely to break down. Breakdowns of any kind hamper service and give a poor impression of your operation. If your pourer sticks, you've got to stop and change it. If your corkscrew bends, you may crumble the cork and lose your cool as you present the wine, and the customer may refuse it. If your icemaker quits, you are in real trouble. Repairs or replacements can be frustrating, time-consuming, and costly. Quality products, moreover, usually come with guarantees.
- ***Appearance.*** Quality products are usually more pleasing to the eye, and are likely to maintain their good looks longer. Cheap glassware becomes scratched and loses its gleam. Cheap blender containers get dingy-looking. So do work surfaces. Since much of your equipment is seen by your customers, it is important to have it project an image of quality, cleanliness, and care.
- ***Ease of care.*** High-quality equipment is likely to be better designed as well as better made. This means smooth corners, no dirt-catching crevices, dent-free surfaces that clean easily. It all makes for better sanitation and better appearance.

Like everything else in life, quality cannot always be judged by price. Nor do we mean that you should go Cadillac all the way; you will certainly not buy lead crystal glassware unless your entire operation sustains this level of luxury. For equipment quality, look at weights or gauges of metals (the lower the gauge, the thicker the metal); look at energy requirements, horsepowers of generators, insulation of ice bins and refrigerated storage, manufacturer's warranties and services. Look at the design features of each item in relation to its function; look at sizes and shapes and capacities in relation to needs.

Keep it simple. The number of bar gadgets available today, both large and small, is mind-boggling. They range from trick bottle openers to computerized drink-pouring systems costing many thousands of dollars. Each one has its bona fide uses and some are highly desirable for some operations. But the wise buyer will measure his or her purchases by these criteria:

- Does it save time or money or do a better job?
- Is it worth the time and money it saves?
- Is it maintenance-free? If not, how upsetting will it be to your operation if it malfunctions? If it needs repairs, is local service available?

It is easy to go overboard on hand tools; there's a gadget on the market for every little thing you do. It is better not to clutter up your bar with tools you seldom use.

On the other hand, it is wise to have a spare of every tool that is really essential, from blenders down to ice scoops. This way no time will be lost if something breaks or malfunctions. Keep the spares in a handy place, easily accessible for emergencies.

Summing up . . .

The bar's equipment must be suited to the drink menu of an enterprise, just as kitchen equipment must service a food menu. All equipment must meet health department requirements and must be kept constantly in top condition, with special attention to temperatures and pressures and the right conditions for machinery to work properly. It is easy to forget that these small details affect such things as the taste and the head of the draft beer, loss of the bubbles in carbonated mixes, and the rate of production of an icemaker. These matters are just as important as the initial selection of equipment.

The right equipment, arranged for maximum efficiency and used and maintained with respect for its function, can be one of the best investments a bar owner can make.

chapter 4 . . .

THE STAFF

The people who deal with the customers are a crucial element in the profitability of any beverage enterprise. It is they who represent the enterprise to the public, who create the human side of its ambience, who sell its wares. They also often represent the largest percentage of its cost and usually the greatest percentage of its headaches. High turnover, poor performance, and unreliability are frequent complaints among employers.

There is no doubt that people problems tend to plague an industry using mostly hourly employees, where hours may be irregular and pay variable and there are few ladders for advancement and many temptations along the way. But certain approaches to these problems do work, while certain other approaches do not work at all. It is a subject worth more attention than it often gets. This chapter suggests points where such attention pays off.

One of the major ingredients in the atmosphere of any bar or restaurant is the staff. Staff members can make customers feel welcome and important, or small and unwanted. They can create a feeling of fun and festivity or they can throw a wet blanket on any party. They can please customers with their efficiency or turn them off with inattention, carelessness, bad manners, or dishonesty.

Employees affect your profits in many other ways. They are important links in any cost-control system. They are your best merchandising agents. To your customers, they represent you and your philosophy.

How does one go about finding the right people and putting them all together to function in a smooth operation? Who are the right people anyway, and where does one find them?

Staff needs vary. Let's look at the whole spectrum of positions first, and then consider how to figure your staffing needs, where to find your people, and how to pick them, schedule them, train them, supervise them, and pay them.

Staff positions . . .

The staff needs of bars are unique to each establishment, and there is probably none that has all the positions described below. At one extreme is the small owner-operated bar in which the owner is manager, bartender, and everything else. At the other end of the spectrum is the beverage service of a large hotel or restaurant chain. The accompanying organization charts illustrate typical positions and their relationships in two types of beverage operation (Figure 4.1).

The duties and responsibilities of the job described may also vary greatly. In a small operation a single individual may handle the functions of three or four jobs, while only very large operations need a full-time steward or beverage manager. Even the job of bartender or cocktail waitress varies from one bar to another.

Bartender. The central figure in any beverage operation is the ***bartender***. The primary function of the bartender is to mix and serve drinks for patrons at the bar and/or to pour drinks for table customers served by waiters or waitresses. The bartender is usually responsible for recording each drink sale, for washing glassware and utensils, for maintaining a clean and orderly bar, for stocking the bar before opening, and for closing the bar. In many operations the bartender also acts as cashier. The bartender is typically a host and a promoter whose public-relations talents can build goodwill and good business.

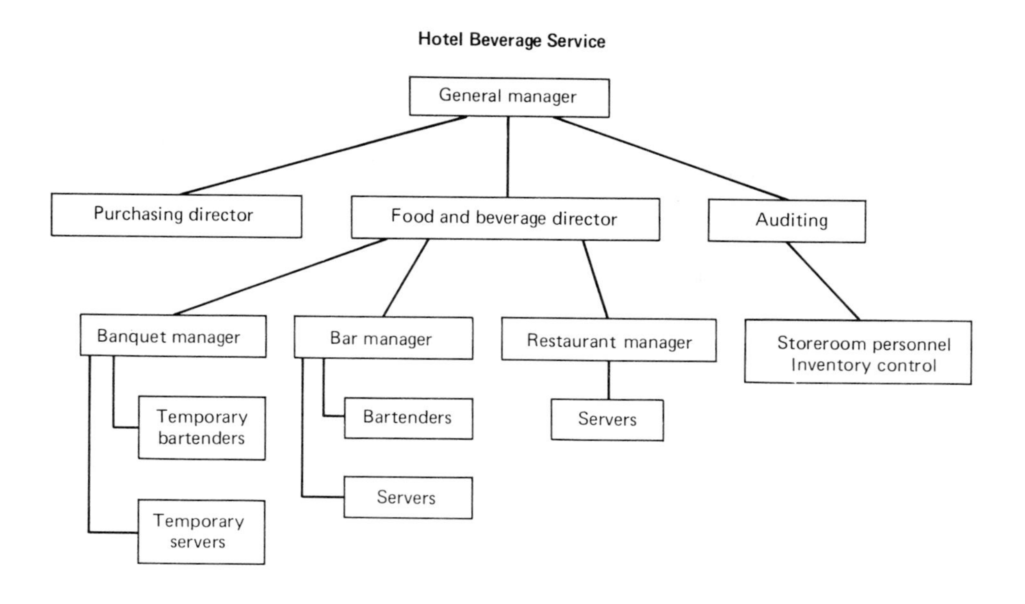

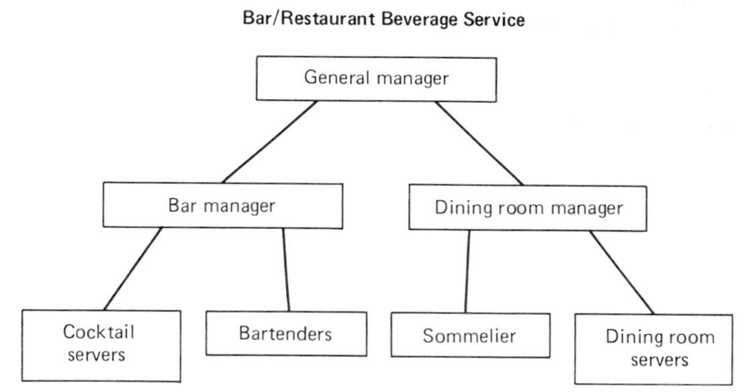

figure 4.1 Organization charts for two different types of beverage service.

Bartending requires certain skills and aptitudes. The bartender must know the recipes for whatever drinks the house serves (from dozens to hundreds depending on the house) and the techniques for mixing them, and be able to work quickly and accurately. To this end he or she must be dexterous—ambidextrous if possible—and have a good short-term memory. In a high-volume bar (often referred to as a speed bar), the ability to work with great speed under pressure is essential. A pleasing appearance and a pleasant personality are essential in any bar, though less so in a service bar out of customer sight. Honesty is the most important quality of all and—if you can believe some restaurateurs—the hardest to find.

In a small operation one bartender may perform all the tasks listed above. In larger operations there may be several bartenders and one or more helpers, often called ***bar backs***. This situation calls for a ***head bartender*** who has overall responsibility for the bar function and acts as supervisor for the other bar personnel. The head bartender may also participate in hiring and training new bar employees and in teaching cocktail servers how to describe drinks to customers, how to transmit orders, how to garnish drinks, and how to use the special vocabulary of the bar.

Bar back. A bar back typically relieves the bartender of everything except pouring the drinks and handling the customers and the cash register. Among the bar back's duties are setting up the bar, preparing garnishes and special mixes and syrups, filling ice bins, washing glassware and utensils, maintaining supplies of towels, napkins, picks, straws, stir sticks, matches, keeping bar surfaces and ashtrays clean, washing fixtures, and mopping floors. The bar back is also a runner or "go-fer," going for liquors, beers, wines, and other supplies requisitioned by the bartender. Often a bar back is an apprentice bartender and may serve beer or mix simple drinks under the bartender's supervision.

Waiter/waitress/server. Beverage service at tables, whether in a cocktail lounge or in a dining room, is handled by waiters or waitresses (hereinafter grouped together as ***servers***). They record the customers' drink orders, transmit them to the bartender, pick up the drinks, serve the customers, present the tab, and collect payment. They also keep the serving areas clean and return the empty glasses to the bar. In heavy-volume bar lounges servers may also help prepare drinks by putting ice in the glasses and adding the garnishes. Sometimes they ring up their own tabs; in other places the bartender does this. Like the bartender, the server is also a host and a promoter.

Table service requires a person with a pleasant personality, a neat and attractive appearance, poise, and a head for detail. Serving personnel must be able to deal with both customers and bartenders pleasantly and efficiently. They must know a broad variety of drinks and their variations so that they can pin down the customers' exact preferences: Does he want his Martini with an olive or a lemon twist? Will her sour be straight up or on the rocks? The best cocktail servers have almost as extensive a knowledge of drinks as bartenders do. They may even mix drinks in some circumstances, such as during a bartender's relief break.

In many restaurants waiters and waitresses serve both food and drinks, including wine by the bottle with the meal. Servers must be able to open a wine bottle, carry out the rituals of service, and answer questions about wines, specialty drinks, and recipes.

Like bartenders, servers are involved in the system of controls by which management keeps track of beverages. Their drink tabs, or order checks, literally keep tabs on what has been ordered, while a cash register uses the checks in recording the payments. Quick, accurate, and honest carrying out of check routines is an essential part of the server's job.

Wine steward, sommelier. Luxury restaurants featuring elegance of service may have a special person to handle the ordering and serving of wine. This is the ***wine steward*** (also called ***winemaster*** or ***wine waiter***, all terms applying to both sexes), who presents the wine list with appropriate recommendations and commentary and also serves all wines. This person must be thoroughly versed in the wines on the list and must be able to answer customer questions.

If the wine steward is a true connoisseur he or she merits the title ***sommelier***. A sommelier traditionally wears a chain around the neck with a tasting cup and sometimes a key to the wine cellar (Figure 4.2). A leather cellar apron is sometimes added to the costume, depending on the character of the restaurant.

Beverage steward. High-volume establishments, large hotels, private clubs, and restaurants with extensive wine cellars may have a ***beverage steward*** in charge of all wine and liquor purchasing, storage, receiving, requisitioning, and inventory control. This person must know a great deal about wines and spirits, the wine and spirits market, and the entire beverage operation. This is a position of considerable responsibility. In some instances the beverage steward may be in charge of the entire beverage side of an establishment, hiring, training, scheduling, and supervising all beverage-related personnel.

Beverage director, beverage manager. A ***beverage director*** or ***beverage manager*** is usually part of a management team in a corporate operation such as a hotel, club, high-volume restaurant, chain of restaurants, or hotel or motel chain. (Whether the title is director or manager depends on the corporate hierarchy.) The beverage director/manager is the team member in charge of the beverage operation—hiring, training, and supervising all beverage-related personnel; purchasing all beverages and beverage equipment; establishing and maintaining inventory and control systems; setting standards and making policy on matters relating to beverage operation.

A beverage director usually reports to top management. In some large or-

figure 4.2 Wine stewards from Arthur's, Seascape Inn, Old Warsaw, and Mario's in Dallas. Note the silver wine-tasting cups and the cellar key. (Photo courtesy Universal Restaurants)

ganizations responsibility for food and beverage service is combined in one position—that of ***food and beverage director*** (***manager***).

Such positions require several years of industry experience, preferably firsthand experience in each area of responsibility. A beverage director must also have management training and/or experience. In addition he or she must have the appearance, personality, poise, and wardrobe that are generally associated with management positions, in order to have credibility in dealings with other management personnel both within and outside the organization.

Manager. An individual bar or restaurant usually has a ***manager*** who is in charge of all aspects of its operation. Sometimes the owner is the manager. If not, the manager is a surrogate, or stand-in, for the owner, running the show on the owner's behalf, whether the owner is an individual or a corporation. The owner sets the goals, establishes policies, and delegates to the manager the authority and responsibility for carrying them out. Within this framework the manager must make any and all decisions necessary to running a profitable operation.

A manager's overall responsibilities may include hiring and firing; training, scheduling, and supervising personnel; forecasting and budgeting; purchasing beverages and related supplies or requisitioning them from a corporate commissary; maintaining records; carrying out control systems (the manager typically has the only key to the storeroom); handling cash and payroll; maintaining quality; promoting the enterprise and the merchandise—all this in a way that will meet the owner's profit goals. (These activities form the subject matter of later chapters.)

On the day-to-day level the manager keeps everything running smoothly, settling staff problems, dealing with difficult customers, and coping with emergencies, sometimes pitching in to do someone else's job. A typical day for a manager is crowded with major and minor decisions, because every day is different and things are always happening that have never happened before.

A manager should be well grounded in every aspect of the enterprise and experienced enough at all jobs to substitute in any one of them and to relate to the people who do these jobs daily. He or she must be the kind of person who can make decisions easily and deal effectively with all kinds of people, both patrons and staff. Essential personal qualities are a good mind, a cool head, a positive attitude, sensitivity to people, leadership ability, self-confidence, and honesty.

Figuring staff needs . . .

The type and size of the enterprise and the projected volume of business will immediately put certain limits on the kinds of jobs you need to fill. Nearly every enterprise has bartenders. A bar lounge requires cocktail servers. Restaurant table service has somewhat different needs in service personnel. Few restaurants need a sommelier unless they have a long, expensive wine list or they want that touch of class.

Every operation needs a manager. In a large hotel it may be a beverage steward. In a chain of hotels or restaurants it may be a beverage director at the top, with unit managers in each individual facility. In a restaurant it is often one person who manages both the food and the beverage side of the operation. If you and your spouse are planning to open a small neighborhood bar you may not even need a manager, and you yourselves may fill some of the other jobs as well.

The numbers of bartenders, serving personnel, and people in charge depend on the projected volume of business and the days and hours of operation.

Many employers guesstimate staff needs, run an all-purpose ad in the daily paper, and, relying on intuition, hire the first warm bodies having a promising degree of personality and experience. Sometimes it works. But often it results in friction, high turnover, and a catch-as-catch-can operation that has a hard time building and holding customers.

Careful staff planning can reduce turnover, friction, and rough edges. The goal of planning is to put the right worker into the right job at the right time. The more detailed your planning before you hire your first employee, the less it will cost you in the end for a better staff, and the better things will run.

The first step in planning is to define each job in your enterprise. The next step is to determine the qualifications a person needs in order to do that job. The third is to figure the numbers of persons you need and the times you need them. Only then are you ready to begin hiring.

Developing job descriptions. The job descriptions in the previous section can function as a jumping-off point, but you'll want to adapt them to the needs of your enterprise, and you'll need more detail. For example, will your bartenders be working a service bar or public bar? Free pour, measured pour, or metered gun? How many drinks per hour minimum? What kind of person will maintain your image? And so on. The answers belong in the job description.

Large organizations often do a systematic analysis of each job, listing each small task performed, its purpose, how it is done, what equipment is used in doing it, and the skills needed. The same technique can profitably be used in any size operation. Examining each job in such detail often reveals things you hadn't thought of before—too much work for one person, not enough work to fill a shift, ways of combining or dividing jobs, gaps in coverage, overlapping responsibilities, skills needed, and whether you should hire skilled people or train them on the job.

Analyzing jobs in this way will give you considerable insight into the personal qualities each job requires. The next step, then, is to list the qualifications for each job—skills and aptitudes, physical characteristics and health requirements, mental abilities, attitudes, minimum age requirement (state laws prohibit persons under age from handling liquor).

The final step is to combine the data from your job analysis with your list of personal requirements to write a concise job description, as in Figure 4.3. In addition to duties, tasks, and qualifications, it should include

JOB DESCRIPTION

POSITION: Bartender/Cashier

SCOPE: Full charge, evening shift, servicing 12 servers

WORK STATION: Service bar

SUPERVISOR: Restaurant manager

SUPERVISES: 1 bar back (weekends only)

DUTIES AND RESPONSIBILITIES:

- Mixes and garnishes drinks for table service, using measured pour and standard house recipes (from memory)
- Sets up and closes bar according to posted routines
- Prepares garnishes
- Washes glasses
- Maintains par stock
- Maintains bar supplies of draft beer, soda tanks, condiments, napkins, straws, picks, etc.
- Maintains temperatures and pressures of coolers, beer and soda systems
- Sets up cash register
- Records all drink sales (precheck and total)
- Closes register and prepares checkout slip
- Takes weekly bar inventory with assistant manager

EQUIPMENT USED: mixing glass, jiggers, blender, shake mixer, frozen drink machine, postmix soda system, draft beer, electronic cash register

PERSONAL QUALITIES: Over 21, honesty, speed, dexterity, accuracy, calm under pressure, good short-term memory, must know drinks.

figure 4.3 Job description developed after analyzing individual tasks.

(1) the scope of the job, (2) work station or area, (3) title of supervisor, and (4) positions supervised, if any.

The job description forms a sound basis for an employment interview. It also informs the applicant of all aspects of the job so that he or she understands it fully before agreeing to take it. Used consistently, it will ensure that all your employees have the same concept of the job—*your* concept.

Planning a staff schedule. It is impossible to staff a beverage operation without planning a detailed schedule. This means matching the days and hours of business and the peaks and valleys of customer demand with work shifts that make sense to employees and to your budget. A chart for each day of the week you are open, showing each hour of the day you must staff, is an indispensable planning tool. On it you can plot the highs and lows of customer demand along with personnel needed to handle the volume at each hour of the day (Figure 4.4).

Many would-be employees are looking for a full-time job with full shifts. Regular part-time jobs (fewer full shifts or several shorter shifts) are attractive to students, moonlighters, and others who do not depend entirely on your wages for their living, and they can help you deal with peak demand periods.

Eight-hour shifts should have scheduled breaks for meals and short rest periods. Breaks should be scheduled during periods of low volume if possi-

Bar Schedule

	10 a.m.	11 a.m.	12 noon	1 p.m.	2 p.m.	3 p.m.	4 p.m.	5 p.m.	6 p.m.	7 p.m.	8 p.m.	9 p.m.	10 p.m.	11 p.m.	12 m.	1 a.m.
Bar manager						XXX	XXX	XXX	XXX	XXX	XXX	XXX	XXX	XXX	XXX	
Asst. bar manager	XXX	XXX	XXX	XXX	XXX	XXX	XXX	XXX	XXX							
Bartender	XXX	XXX	XXX	XXX	XXX	XXX	XXX	XXX								
Bartender							XXX	XXX	XXX	XXX	XXX	XXX	XXX			
Bartender									XXX	XXX	XXX	XXX	XXX	XXX	XXX	
Bar back	XXX	XXX	XXX	XXX	XXX	XXX	XXX	XXX								
Bar back									XXX	XXX	XXX	XXX	XXX	XXX	XXX	
Cocktail server		XXX	XXX	XXX	XXX	XXX	XXX	XXX								
Cocktail server			XXX	XXX	XXX											
Cocktail server							XXX	XXX	XXX	XXX	XXX	XXX	XXX			
Cocktail server									XXX	XXX	XXX	XXX	XXX	XXX	XXX	
Sales per hour	0	0	$100	$200	$300	$100	$100	$200	$300	$300	$300	$300	$400	$300	$200	$200

figure 4.4 Staff schedule for a busy bar on a typical weekday.

ble; if not, your schedule must include someone who can take over the job during the break. An experienced server, for example, can tend bar for 15 minutes, or breaks can be staggered where there are several bartenders.

Personnel on full shifts may have periods of time when there is little to do. You may be able to schedule tasks from other jobs during such periods, such as requisitioning or purchasing, training new personnel, developing promotional materials, answering the phone. (You may not have this flexibility in a union house.) Another way to handle peak volume without idle time at both ends is to stagger shifts.

A manager's hours of work must be scheduled systematically like everyone else's. The day is past when managers will work 80 hours a week, unless they are owners too. Someone must be in charge in the manager's absence—an assistant manager or head bartender, for example—and the detail of such duties and extent of responsibilities must be carefully worked out on a job description.

In figuring the number of service personnel you need you may find the following ball-park figures useful:

- A good cocktail server can handle 40 customers or 10 tables of 4.
- A good restaurant server can handle beverage orders for 4 or 5 lunch or dinner tables along with food service.
- A bartender can pour 60 to 150 drinks an hour depending on dexterity, experience, types of drinks, method of pour, and efficiency of bar design.

Job descriptions and a tentative schedule form a sound, well-organized basis for recruiting and selecting the right people for the right jobs. For a small operation it may seem like a lot of paperwork, expensive in time and effort. But whether or not it is formalized in writing, the same careful planning and analysis should take place. In hiring for a large enterprise the written process is essential.

Finding the right people . . .

Where in the world do you find people who want to spend hour after hour on their feet, working under pressure to satisfy demanding and unpredictable customers? And if you find them will they have the pleasant personalities, service skills, and dependability so carefully spelled out in your job descriptions?

To hear some restaurant people talk, you would think it is impossible to find good people to work in bars and restaurants. They would have you believe that the only people available are lazy or incompetent, thievish or peevish, and will leave for a better job as soon as you get them trained.

It is perfectly true that there are these personality types and that many of them gravitate to the restaurant industry because it is easy to get hired—usually by employers who assume everybody is like that. But there are also hard-working, honest, personable individuals who find bar and restaurant work attractive. They are not likely to go to work for employers who have a negative or contemptuous view of the people they interview and the jobs they are offering. And it doesn't take long for word to get around that such-and-such is a good place to work and the one across the street is the pits. To find good employees, you have to look to your own attitudes and your own image as an employer.

For some people the positive aspects of the jobs you offer will outweigh the hard work, pressures, and difficult customers. Others truly enjoy the exhilaration of working under pressure, the challenge of pleasing customers, and the special qualities of life after dark in a setting dedicated to the pursuit of happiness. The hours of work are very attractive for many people, including students, people seeking a second job, husbands or wives who take turns working in order to care for families, people who want only part-time work. On-job training may be attractive for persons looking for their first job. Money is certainly a drawing card, especially if tips are involved.

Recruiting applicants. How do you reach these people? And can you reach them selectively so that you won't be overwhelmed with unsuitable applicants? There are several likely sources. Pick the places where your kind of person might be shopping for jobs. For selectivity, state the essential qualifications for each job, such as legal age, skill level, and appearance. Emphasize the positive things you have to offer.

Here are some promising hunting grounds:

- Many experienced managers say that friends of present employees are among the best sources. If you and your present employees are happy with each other, they are likely to bring in others like themselves, and they are not going to risk your displeasure by suggesting someone unsuitable. They may want to bring relatives too. But hiring two members of the same family has definite drawbacks. If there is friction or disagreement it is multiplied by two. If you lose one you lose the other as well, and you have two positions to refill. In many places it is a stated policy not to hire members of the same family.
- State employment agencies often have a reservoir of potential applicants, since anyone applying for unemployment insurance must register there. If their employment counselors are familiar with your needs they can often do a good job of screening applicants for suitability, and there is no fee.
- Private employment agencies may supply prescreened managerial candidates. These agencies charge a fee, and for high-level personnel it is

usually the employer who pays it. The right person is probably worth it. You won't find service-level personnel listed with such agencies because at these levels the fee is charged to the applicant.

- Schools offering food and beverage service courses may be a source of students who are interested in the field and have some training and experience.
- Bartender schools are often looking for placements for their graduates.
- Unions can usually furnish applicants for bartender and server positions if you meet union requirements.
- Placement offices in nearby colleges may be a source of suitable part-time help.
- Selective advertising can often draw the kinds of applicants you are looking for without flooding your office with unsuitable types. Run your ad in the kind of paper your prospects read—maybe even the local campus weekly. In wording an ad it is important to avoid anything that might suggest discrimination on the basis of race, sex, age, national origin, or religion. State the qualifications for the job in a way that intrigues the reader, and play up the advantages. If you use the daily paper, a display ad with a headline will catch the eye (Figure 4.5).
- Daytime radio can reach people thinking of entering the labor force who haven't yet reached the point of combing through the ads.

Interviewing and hiring. If the recruiting has been done well, a good interviewer with a good job description in hand has a good chance of selecting the right person to fit the job. At this point an additional tool is needed—an application form.

A written application is a necessary prelude to an interview. Not only does it provide a concise summary of the necessary data, but it gives the interviewer a point of departure for further questioning. In addition, the way the applicant fills it out can provide clues to personality traits and habits that are relevant to the job. Did he or she follow instructions? Write clearly? Leave questions unanswered? Take forever to finish? How will these characteristics relate to the job you are filling?

The data required fall into three categories: personal data, employment history, and references.

Again it is important to avoid asking for information that could be used to discriminate on the basis of race, sex, age, religion, or national origin. Such questions are violations of various federal laws. It is also illegal to discriminate against the handicapped, although inability to perform some aspect of a job is a legitimate reason for not hiring. Questions about police records should be handled carefully. Federal law allows you to ask whether a person has been convicted of a felony, but you cannot refuse employment on

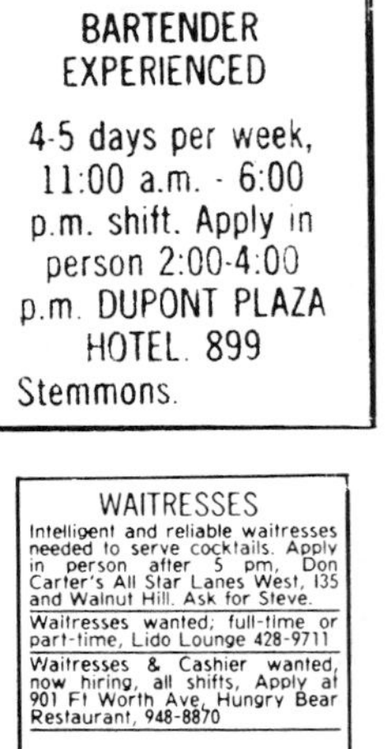

figure 4.5. Recruiting ads for bartenders and servers. Which would you respond to?

the basis of conviction unless the crime is pertinent to the duties of the job—as embezzlement is to bartending/cashiering.

A personal interview is the next step. Managers of small operations often find themselves too busy to take the time, relying instead on intuition or taking a chance on the basis of a written application. Yet it is in the oral interview that the applicant's personality and potential can best be assessed. The manner of response to questioning will tell you whether this person is friendly, poised, intelligent, alert, open. Nonverbal clues are also important—the way the person sits, stands, moves, gestures, and speaks can indicate capacity for speed and dexterity as well as confidence and poise. And of course a personal interview is essential in judging appearance and ability to relate to others.

The interview should have two phases. In one phase you as an interviewer should explain clearly the job being offered, using the job description as a basis for discussion. As a result of this interchange both of you should be able to judge whether the prospect has the necessary skills or the capacity to develop them quickly enough to fill your needs.

The other phase of the interview should amplify the data on the application. Are there gaps in past employment? If so, why, and are the reasons good ones? What did this person like or dislike about previous jobs, and are there parallels in your job? If he or she did not give a former employer as a reference, why not? Would there be any transportation problems, any conflicts of schedule (such as school classes or spouse's schedule), any difficulties with basic skills required for the job (such as handwriting or arithmetic)? Is there any history of frequent or chronic illness?

When you have narrowed the field to one or two, you should check the applicant's references. Include former employers, whether they are on the applicant's reference list or not, especially if the job requires handling money. Though most people are honest, it is the ones who are not that you want to avoid hiring.

You may not always find the perfect person for every job, and you are the only one who can judge when a compromise is in order and when to go on looking. In any case, if you have done a thorough job of interviewing and checking, you know just what kind of person you are hiring and you have minimized your risks.

Managing personnel for success . . .

Hiring is only the first step in the management of people. The manager is also responsible for explaining jobs and assigning responsibilities, for making sure that people carry out the jobs assigned, and for taking disciplinary action when they do not. It seems like a simple and clear-cut assignment.

Yet personnel problems seem to be a constant in the restaurant industry, including the beverage side. People come and go, people don't do as they are told, people bicker and complain, they are late, they are slow, they are rude to the customers—you name it. The manager is frustrated and blames the employees. The employees resent the manager. They become enemies. People are fired, people leave, and the process begins over again.

This is an obvious simplification, but it is all too common. The whole subject of personnel management is of crucial importance to running a profitable operation. For one thing, a high rate of turnover means high labor costs—costs of hiring and training new personnel plus the generally lower productivity of new employees. Even more important, high turnover and employee dissatisfaction are bound to affect the quality of customer service. When the quality of service drops, customers go somewhere else.

Let's look at some of the things large enterprises are doing today to improve the management of people. The same principles will apply in smaller operations.

Orientation and training. The first part of the manager's personnel responsibility—explaining jobs and assigning responsibilities—begins right after employees are hired. They must learn exactly what is expected of them, whom they must report to and work with and supervise, how to use the equipment and follow the house routines, what the beverage menu includes, what size drink is poured, what glasses are used for what drinks, what wines are served, all the beverage prices, and what the rules are about dress and smoking and drinking and reporting in and out.

When you list out all these essentials, you can see it is quite a lot for a new person to absorb. Yet many a manager puts a new employee right to work, counting only on a bit of coaching as mistakes are made and questions asked. Often the coaching is left to another employee, who may resent the extra burden and who may not be the world's best teacher.

There are at least four good reasons for taking the time and effort to give employees a basic, all-inclusive introduction to their jobs, their place of work, and their colleagues *before* they start work:

- They will be more productive, faster, with less confusion.
- They will feel more confident in their jobs, and this will be reflected in their attitudes toward their work and in the way they relate to your customers.
- They will more easily establish good relations with their colleagues, and the relationships will work more smoothly over the long run.
- They will be more likely to stay if you see to it that they have a good experience on Day One.

Some enterprises develop a formal orientation manual to give to new employees. Certainly it is a sound idea to have all the rules and information on paper, so that there is no possibility of misunderstanding or leaving something out. A completed form similar to Figure 4.6 is the absolute minimum of written information for a new employee.

However, handing a new employee a sheet of paper or a manual is not enough. There is no guarantee that he or she will understand it, remember it, or even read it. Everything should be discussed step by step on a one-to-one basis. To make sure the employee understands, ask to have the information repeated back to you. Then go over it again if necessary.

TO OUR NEW EMPLOYEE:

Welcome!

We are glad you are joining us and we hope you will enjoy working here.

Your hours are: ____________________ Punch in and out on clock beside kitchen door. Your days of work are: ____________________

Your pay rate is: ____________ ($1\frac{1}{2}$ regular rate after 40 hours)

Payday is: ______________________________

Your supervisor is: ____________________

You supervise: ____

You work with: ____

Dress: Conservative blue blouse, navy blue skirt or slacks, closed-toe shoes

Meals: You may buy your meals here for $2 each. You must eat on your own time.

Breaks: ______________________________

Rules: Wash your hands as soon as you come to work. Make this a habit!

No smoking or drinking on the job.

Hair, nails, and clothing must be neat and clean.

Do not serve a minor any alcoholic beverage.

Do not serve an intoxicated person any alcoholic beverage.

Your duties are:

A menu, drink list, and wine list are attached. Please study these. You will receive further personal orientation. Meantime, do not hesitate to ask questions.

figure 4.6 Sample information handout for a new employee.

Even experienced personnel will need some skills training in the way you do things—how to pour and garnish drinks your way, how to set up your bar and close it, how to requisition the beverages, how to use your check system, how to operate your cash register. Less experienced personnel will need more training. It is important to gear the training to what the employee can absorb. Experts recommend teaching one task at a time, one step at a time, spread over several days.

A formula for job instruction training developed during World War II to train people quickly in industry proved so successful it has been used widely ever since. It applies a four-step method of instruction to each task:

step 1: *Explain the task.* Tell the employee about the task itself, its importance, and its place in the job and the operation as a whole.

step 2: *Demonstrate the task.* Show, explain, step by step. Emphasize key points, techniques, standards. Repeat the demonstration until the employee is ready to try it.

step 3: *Have the employee perform the task.* Encourage and correct. Repeat until both you and the employee are comfortable with the performance.

step 4: *Follow up.* Once the new employee is on the job, supervise as needed. Check frequently; coach and correct. Taper off as performance meets standards consistently. Check periodically to see that the employee is following standard procedures.

If you do not have the time or the personnel to do a proper job of training, then you might be better off to hire skilled workers and pay the extra wages. One way or another you will have to pay for trained personnel. Many operators feel it is better to do one's own training; it is easier than retraining someone accustomed to a different way of doing things.

Some principles of handling people. Getting employees to perform their duties to the required standard is a product of many complex human factors. Large enterprises today are giving their managers and supervisors special training directed to understanding these factors and putting them to work. Among the most important factors are:

- The manager's attitude toward the employees and their work.
- Good communications.
- Methods of maintaining standards.
- Teamwork.

The manager's attitude toward employees and their work is probably the one most influential factor in employee performance. Nothing is more devastating to an enterprise than a manager's negative attitude toward the workers. Nothing is more constructive than a positive attitude.

If you as a manager look down on the people you supervise and the work they do, you create resentment and anger. If on the other hand you respect each employee as a person and recognize each job as important (and every job *is* important to the enterprise), then you will establish valid person-to-person relationships. You are encouraging the employees to feel good about themselves and about you.

If you expect people not to do their jobs well, they won't. If you expect the best of them, they will generally give their best. If they think you consider them capable and hard-working, they will try hard to maintain that image. This has been proven time and again.

Corollaries of these positive attitudes are fairness and concern for individuals. Employees respect a boss who treats them fairly, without favoritism, and who understands their point of view even when not agreeing with it. Concern for individuals does not mean relaxing standards or becoming involved in employees' personal lives; it simply means appreciating them as human beings and being aware of human needs and feelings.

A second factor in maintaining good employee performance is good communications. There are many pitfalls between the sending and receiving of verbal instructions. The speaker may send one message and the listener may hear another. The speaker may not say what he or she meant, or may use unfamiliar words or words that have more than one meaning. The employee may not grasp the meaning because the expression on the boss's face is frightening and that is the only message received.

Usually a perceptive supervisor can tell when something is going in one ear and out the other. It pays to be sure your messages are sent clearly and received and understood clearly.

Another very important aspect of communication is to make sure that employees are told in advance of all changes affecting them and their jobs. Explain, if you can, the reason for the change. They will appreciate your concern and will be more ready to adapt.

Another vital aspect of good communications is to be receptive to what employees have to say. Often they know more than you do about their jobs and can come up with good suggestions. Sometimes they have a legitimate complaint that you were not aware of. Be appreciative of their ideas and criticisms, even when you can't do anything about them. A word of warning: never promise anything unless you can and will deliver.

Be vigilant in maintaining the standards you have set, yet considerate in the way you point out lapses and shortcomings. Never chew out an employee in front of customers or co-workers. If workers leave things undone, don't do their work for them and then fume about it; insist that they do it. If a problem keeps recurring, ask the employee for a solution, or work one out together; then follow through.

Give positive feedback; too often the boss's remarks are all critical. Express your appreciation for jobs well done. And make sure your own performance meets the same high standards you set for others. Your employees are watching you.

The truly effective manager of people builds a spirit of teamwork among the personnel. It begins with the hiring process—the selection of people who will get along together. It continues with the careful training of each new employee, to integrate that person into the operation as a whole. It is nurtured by a positive attitude toward the workers and their work, a spirit of respect and concern for individuals. People who work together in a positive atmosphere usually develop a feeling of belonging together, even people who have nothing else in common. Receptiveness to ideas and a free interchange between boss and subordinate lead to a spirit of sharing and working together. Some managers find time for staff problem-solving meetings; if these are free and open they can encourage team solutions to mutual problems.

The manager who can build a team can hardly lose. And every manager needs a team to win.

Compensation and benefits . . .

An employer may pay the members of a staff in several different ways. Some of them will receive additional income in tips. The compensation picture is often made more complex by overtime pay, bonuses, commissions, incentive pay, and perquisites, while wage and hour laws, both federal and state, have many provisions that come into play when payroll is figured. Every employer must know the requirements well. Let us look first at the usual ways in which various positions are compensated.

Methods of compensation and rates of pay. Bar personnel are usually paid by one of two methods: straight-time hourly wages or a fixed salary. Tips add substantially to the earnings of bartenders and cocktail servers, while a supervisor or beverage director may receive a monthly bonus.

Table 4.1 shows in general terms the typical methods of compensation for typical beverage-related positions. As usual, no one formula applies to all enterprises.

Wages and salaries differ widely from one enterprise to another and from one locality to another. Rate of pay is related in a general way to sales volume, number of employees, and number of seats; in other words, the bigger and busier a place is, the better the base pay (but not in direct proportion). To a lesser degree pay is related to type of ownership, with sole proprietors generally paying less than partnerships and corporations. Other differences

table 4.1 **HOW BAR PERSONNEL ARE PAID**

position	*base pay*	*supplemental pay*
Bar back	Hourly wage	
Cocktail server	Hourly wage	Tips (customer-paid)
Bartender, public	Hourly wage or weekly salary	Tips (customer-paid)
Bartender, service	Hourly wage or weekly salary	
Unit manager	Salary, weekly/monthly	Periodic bonus (employer-paid)
Beverage director (corporate)	Annual salary	Periodic bonus (employer-paid)

in pay may come from local and regional differences in prevailing wages and cost of living, local labor supply and demand, and union activity.

Federal laws require every employer to give equal pay for equal work without regard to age, sex, race, national origin, creed, or color.

Federal minimum-wage requirements. In determining compensation, federal and state minimum-wage laws provide a floor. As of December 31, 1981, the federal Fair Labor Standards Act applies to any food/beverage enterprise grossing $362,500 a year. The Act set the minimum wage at $3.35 per hour as of January 1, 1981.

A special provision of the Act concerns wages of ***tipped employees***, who are defined as persons receiving at least $30 per month in tips. The employer may pay such persons a base rate of 60% of minimum wage ($2.01 in 1981), taking a 40% ***tip credit***. If, however, an individual's tips plus wages amount to less than the minimum wage in any workweek, the employer must make up the difference. It is the employer's responsibility to advise employees of the tip-credit provisions of the Act.

Suppose, for example, a waitress working a 40-hour week makes $80.40 in wages at $2.01 per hour and $80 for the week in tips, an average of $2 per hour. Her tip average is well above the employer's 40% credit of $1.34. The next week, however, a spell of bad weather cuts her tips in half, netting her only $40 for the week, an average of $1 per hour. The employer must then pay her an additional $0.34 per hour to bring her total pay for the week up to the minimum wage.

The Act requires that minimum wage be computed on the basis of a workweek, whether the employee is paid weekly, biweekly, monthly, or at some other interval. A ***workweek*** is defined as a fixed and regularly recurring period of 168 hours—seven consecutive 24-hour periods beginning any day of the week at any hour of the day.

The Act specifies that all tips belong to employees; the employer cannot claim any tip money. However, the employer may require ***tip-pooling*** for bartenders and other service personnel having direct customer contact (but not persons in jobs for which tipping is not customary).

The tip-pooling requirement cannot apply to an employee until his or her tips have satisfied the employer's 40% tip credit. In other words, the tipped employee does not pay into a tip pool until after he or she has made enough in tips to bring the total hourly pay up to the minimum wage for the week.

Where a bar has a policy of adding a compulsory service charge of, let's say, 15% of the bill, this charge is not considered a tip but is defined by the Act as part of the employer's gross receipts. In this case the employer must pay at least the full minimum wage to service personnel and must pay applicable sales taxes on the 15%.

Most bars, as part of their ambience, specify uniforms for both bartenders and bar servers, and some employers charge uniform costs to employees. The Fair Labor Standards Act requires that such charges must not reduce an employee's wages below the minimum wage. The employer must also reimburse employees the costs of laundering uniforms (as necessary business expenses for the employer) if such cost reduces employee wages below the minimum wage level.

Federal overtime-pay requirements.[1] The Fair Labor Standards Act requires that employees must receive overtime pay for hours worked over 40 in one workweek. The Act further specifies that the rate of overtime pay must be at least 1½ times the employee's regular rate.

Regular rate for figuring overtime is always an hourly rate. No matter whether the employee's rate of pay is tied to the hour, the shift, the week, the month, or a percentage of sales, pay for a given week is translated into an hourly rate by dividing the week's pay by the number of hours worked that week. A week's pay includes all remuneration—wages or salary, commissions, attendance bonuses, production bonuses, shift differentials, and tips credited to wages.

Let us see how this regular-rate computation works out in differing methods of payment:

- ***The hourly employee.*** The regular rate for figuring overtime is the regular hourly rate. To figure the overtime rate, the hourly rate is multiplied by 1.5.

  Example: An employee's hourly wage is $4.00. She works 47 hours in one workweek.

Regular rate: $4.00	40 hours @ $4.00 =	$160.00
Overtime rate: $4.00 × 1.5 = $6.00	7 hours @ $6.00 =	42.00
	Week's pay:	$202.00

[1] This section is based on U.S. Department of Labor WH Publication 1325, *Overtime Compensation Under the Fair Labor Standards Act*, Washington, D.C., revised January 1979.

Another way to figure this:

Regular rate: $4.00	47 hours @ $4.00 = $188.00
Overtime premium: $4.00 × 0.5 = $2.00	7 hours @ $2.00 = 14.00
	Week's pay: $202.00

- ***The salaried employee paid monthly.*** The regular rate is computed by multiplying the monthly rate by 12 and dividing by 52 to find the weekly salary. This is divided by the number of hours worked to find the hourly rate.

  Example: An employee's monthly salary is $1000 at 40 hours a week. In one workweek he works 48 hours.

  $1000.00 × 12 = $12,000.00 ÷ 52 = $230.77 ÷ 40 = $5.77
  $5.77 × 1.5 = $8.66 overtime rate
  $8.66 × 8 = $69.28 overtime pay for week

- ***The salaried weekly employee working less than 40 hours a week.*** The regular rate is computed by dividing salary by hours regularly worked. The employee is paid this hourly rate up to 40 hours and 1.5 times this rate thereafter.

  Example: This employee's salary is $160 for a 35-hour week. In one workweek she works 45 hours.

$160.00 ÷ 35 = $4.57 regular rate	40 hours @ $4.57 = $182.80
$4.57 × 1.5 = $6.86 overtime rate	5 hours @ $6.86 = 34.30
	Week's pay: $217.10

- ***The employee paid a fixed salary for varying hours per week.*** For this employee the regular rate for computing overtime will vary with total hours worked and must be computed for each workweek. Since the straight salary applies by definition to all hours worked, only the overtime premium, or 0.5 the regular rate, is added.

  Example: An employee is paid a straight salary of $200 a week. She works 42 hours the first week, 48 the next, 45 the third, and 38 the fourth.

Week 1:	$200.00 ÷ 42 = $4.76 regular rate	straight salary: $200.00
	$4.76 × 0.5 = $2.38 overtime premium	2 hours @ $2.38 = 4.76
		Week's pay: $204.76
Week 2:	$200.00 ÷ 48 = $4.17 regular rate	straight salary: $200.00
	$4.17 × 0.5 = $2.09 overtime premium	8 hours @ $2.09 = 16.72
		Week's pay: $216.72

Week 3: $200.00 ÷ 45 = $4.44 regular rate — straight salary: $200.00
$4.44 × 0.5 = $2.22 overtime premium — 5 hours @ $2.22 = 11.10
Week's pay: $211.10

Week 4: Her pay for the week is $200 because that is her fixed salary.

- ***The minimum-wage employee with tip credit and extra pay.*** For computing the regular rate the hourly wage is figured at minimum wage, since the employee will receive it either in tips or from the employer. This wage is multiplied by the hours regularly worked. To this figure extra pay during the week is added. (This will include such things as bonuses and commissions, but not the balance of tips earned.) This total is then divided by the number of regular hours to arrive at the regular rate for computing overtime for that week.

  Example: A tipped employee whose base pay is minimum wage less 40% tip credit works 45 hours in one workweek and makes $95 in tips. In that same week he earns a commission of $18 on wine sales and also a bonus of $25 for the highest overall sales for the week.

  $3.35 (minimum wage) × 40 = $134.00 + $18.00 + $25.00 = $177.00
  $177.00 ÷ 40 = $4.43 regular rate — week's earnings: $177.00
  $4.43 × 1.5 = $6.65 overtime rate — 5 hours @ $6.65 = 33.25
  $210.25
  less 40% tip credit = 53.60
  Week's pay: $156.65

Supervisory employees are exempt from the overtime-pay requirements of the Act, and may have to work extra hours without extra pay when their duties require it. To be exempt the employee must meet all the following criteria:

- Manages a department as his or her primary duty.
- Directs the work of at least two other employees.
- Has authority to hire and fire.
- Devotes no more than 40% of work hours to activities not directly concerned with managerial duties.
- Is paid a salary of at least $155 per week (1980 figure).[2]

If an employee does not meet all five of these criteria, the employer must pay for overtime work at 1½ times the employee's regular rate.

Hourly employees must be paid from the time they report to work until the time they go off duty, whether they are working or not. Even if they go

[2] U.S. Department of Labor WH Publications 1281 and 1363.

on working voluntarily beyond their normal hours, they must be paid for the extra time.

Payroll taxes, benefits, and perquisites. In addition to wages and salaries, two other forms of compensation add to labor costs: payroll taxes and fringe benefits. Payroll taxes go to Social Security and unemployment funds —both forms of deferred compensation to employees, while to the employer they are real and present labor costs. In 1982 the employer's contribution to Social Security was 6.70% of payroll. Federal and state unemployment taxes together varied from 0.8 to 4.7%, depending on the state rate and the rate of unemployment claims against the employer. State rates vary, so check them out for your state.

As of 1982 the employer was required to withhold 6.70% of each paycheck up to $32,400 of income as the employee's contribution to Social Security and to withhold federal income tax as well. These deductions from paychecks cause a lot of employee discontent because of the wide disparity between base pay and the much-reduced "take-home" pay. Every time an income-tax rate or a Social Security rate increases, the take-home pay drops accordingly. Yet at the same time the payroll expense to the employer reaches considerably beyond the base pay. It is generally estimated that government social insurance programs plus company benefits add 12 to 17% to the wage/salary cost of an employee.

Look at the following example: A bartender who has a base pay of $200 per week represents a total expense of $224 or more per week. Yet the bartender understandably feels that "real" pay is the $160 or less take-home pay left after deductions.

What are these benefits that don't show up in the paycheck? The term ***fringe benefits*** refers to all tangible or significant compensation other than money wages. The most common benefits are free meals and paid vacations. According to a National Restaurant Association survey 85% of enterprises surveyed granted employees, including bar personnel, paid vacations. The number of weeks of vacation time increased as number of years of service increased. At least one week seemed to be the norm for persons with one to 20 years of service, while the majority of enterprises granted three weeks' vacation to employees with 20 or more years. Many employers also offer group medical insurance and group life insurance.

Some enterprises reimburse their management employees for job-related educational expenses and pay professional dues. Benefits that are related to specific jobs or job levels are known as ***perquisites***. A car or car expenses for use on the job is a common perquisite. Sometimes interest-free loans are provided to top-level personnel.

Benefits and perquisites are part of an employer's labor cost, while for the employee they are tax-free additions to income.

Summing up . . .

Heading off personnel problems before they begin would be the ideal way to go if one could achieve it. Even though the best-laid plans do not always work out, it does pay to analyze your personnel needs carefully, to pick the right person for each job, and to give due attention to orientation, training, and follow-up.

Beyond that, the employer who is fair, who takes care to communicate clearly, who respects employees as individuals while demanding their best work at all times, who succeeds in building a team of loyal workers—this is the one whose personnel build up the business. Pay, perquisites, and incentives are important too—no doubt about it—but not so important as the spirit conveyed from the top down.

chapter 5 . . .

THE BEVERAGES: SPIRITS

Most people know the contents of liquor bottles by linking names to tastes, but haven't any idea what they really are or what makes them different from one another. These are fundamentals every beverage manager should know. This chapter explains the difference between fermented beverages and distilled spirits and then goes on to examine the different kinds of spirits—how they are made and how they differ. It is all important. It helps you to understand what you are buying and selling, why some spirits cost more than others, which ones are most intoxicating, where they come from, and all sorts of incidental information that makes good conversation.

Which of the following well-known beverages are spirits: rum, brandy, champagne, scotch, sherry, bourbon, vermouth, gin, ale, vodka?

Many people cannot distinguish spirits from other alcoholic beverages. Many, in fact, do not know what a spirit is. In the list above, rum, brandy, scotch, bourbon, gin, and vodka are spirits. So are the other whiskies, tequila, and all liqueurs and cordials. Champagne, sherry, and vermouth are not spirits but wines. Ale is not a spirit but a beer.

What is the difference, and what difference does it make?

Beers, wines, and spirits taste different, are served differently, have different alcoholic contents, and tend to have different uses. In order to provide for your customers properly you need to know about these differences and how to handle each kind of beverage from purchase to pouring.

Types of alcoholic beverages . . .

All beers, wines, and spirits are alcoholic beverages. An ***alcoholic beverage*** is any potable (meaning drinkable) liquid containing ethyl alcohol. It may have as little as ½% by volume or as much as 95%. The half percent was set by the federal government at the time of the Prohibition Amendment as it was groping to define an intoxicating beverage. At ½% you'd have to drink 4 to 5 gallons of a beverage to become intoxicated, but the figure remains in the government's definition.

Fermented beverages. All alcoholic beverages begin with the fermentation of a liquid food product containing sugar. ***Fermentation*** is the action of yeast upon sugar in solution, which breaks down the sugar into carbon dioxide and alcohol. The carbon dioxide, a gas, escapes into the air. The alcohol, a liquid, remains behind in the original liquid, which thus becomes a ***fermented beverage*** (Figure 5.1).

Beers and wines are fermented beverages. Beer and ale are made from fermented grains. Wines are made from fermented grapes and other fruits. The alcoholic beverages our early ancestors made from honey, dates, rice, milk, sugarcane, molasses, palms, peppers, berries, seeds, and pomegranates were fermented beverages. Any liquid with sugar in it could be fermented if a yeast was handy to start the action. When the sugar was converted to alcohol and carbon dioxide, the result was a beverage with an alcohol content of about 4 to 14% depending on the amount of sugar in the original liquid.

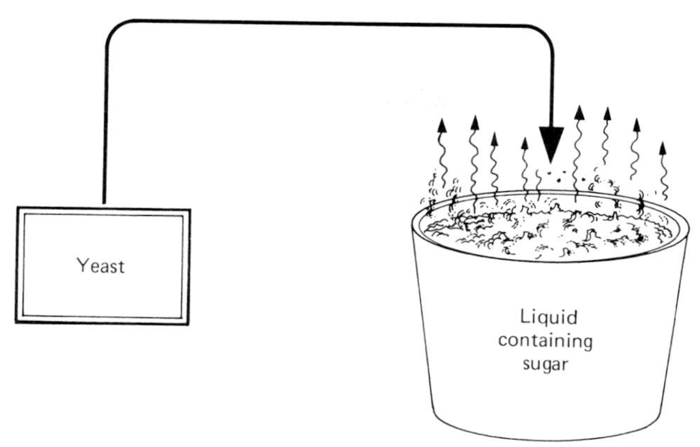

figure 5.1 Fermentation. Yeast breaks down the sugar into carbon dioxide and alcohol. The carbon dioxide evaporates; the alcohol remains in solution.

Distilled spirits. If you can separate the alcohol from a fermented liquid you have what you might think of as the essence or the spirit of the liquid. This is exactly what spirits are and how they are made.

The process of separation is called ***distillation***. The liquid is heated in an enclosed container, called a ***still***, to a temperature of at least 173° Fahrenheit (78.5° Celsius). At this temperature the alcohol changes from a liquid to a gas and rises. Most of the water of the liquid remains behind; water does not vaporize until it reaches 212°F (100°C). The gas is channeled off and cooled to condense it into a liquid again. The result is a ***distilled spirit***, or simply a ***spirit***. All the spirits we use today are made by this basic process, which is illustrated in Figure 5.2.

Distillation came down to us through the alchemists of medieval Europe. These secret dreamers of great dreams, looking for an elixir or potion that would cure all diseases and give everlasting life, produced what they called *aqua vitae*, water of life. Though it was not quite the miracle drug they were looking for, it did gain respect as a medicine. When it was flavored to make it more palatable, people discovered they liked it. Soon the technique of distilling was applied to all kinds of fermented products to produce beverages much stronger than the simple fermented drinks.

Beverages classified. Which of those bottles at the bar are fermented and which are distilled? You can put the familiar names into place by examining Table 5.1.

The federal government has established Standards of Identity for the various classes of alcoholic beverages—that is, the types of spirits (gin, vodka,

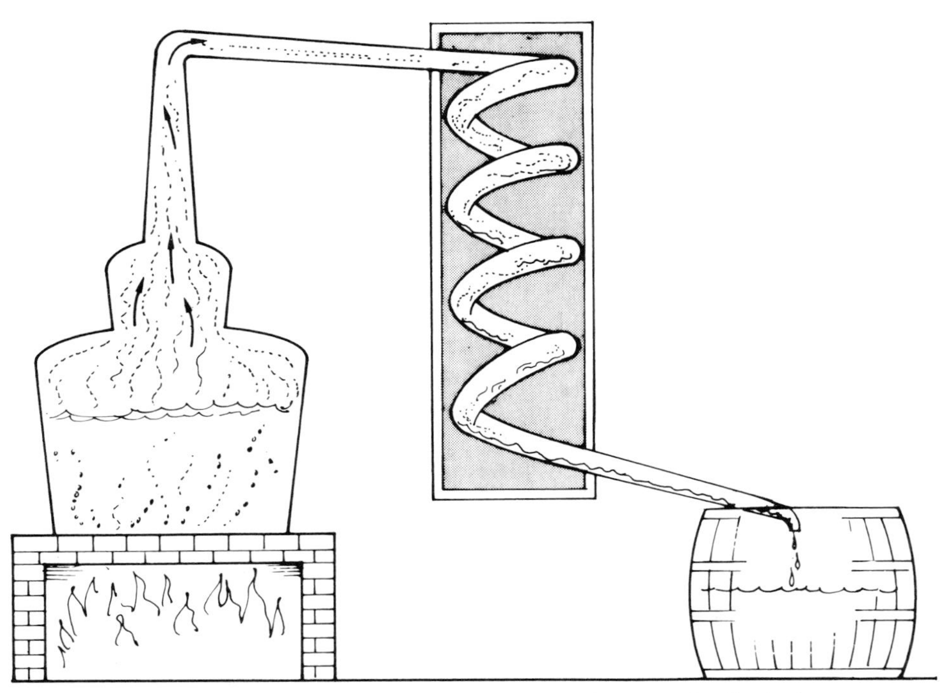

figure 5.2 Distillation. Fermented liquid is heated in a still. Alcohol vapors rise and are carried off through a coil that passes through cold water, condensing them into a liquid spirit.

brandy, rum, tequila, the various whiskies), types of wine, and types of malt beverages. For example, behind the word bourbon on a label is a Standard of Identity stating certain requirements as to what it is made of and how, the type of container it is aged in, and the alcoholic content. These standards, rigidly enforced, produce a beverage having the distinctive characteristics everyone recognizes as bourbon. If the name is on the bottle, you'll know what's inside, because there's a federal inspector making periodic compliance investigations at each distillery in the country. Imported products must meet similar standards in order to enter the country.

These Standards were developed after Repeal as part of the strict control system imposed on the new beverage industry to avoid the chaos of the Prohibition era. Their purpose is twofold: to provide the base for assessing and collecting federal taxes, and to protect the consumer. Beyond this the Standards can be helpful to you in learning to read bottle labels so you know what you are buying, and in understanding the differences between similar products. The three beverage chapters in this book draw heavily on Standards of Identity.

table 5.1
ALCOHOLIC BEVERAGES

fermented		*distilled*							
Beers and ales	***Wines***	***Whiskies***	***Gins***	***Vodka***	***Rum***	***Tequila***	***Brandies***	***Liqueurs***	***Others***
Lager beers	Table	Scotch	London dry				Cognac	Amaretto	Aquavit
Light beers	Reds	Irish	Hollands				Armagnac	B & B	Bitters
Ales	Whites	Bourbon					Calvados	Bénédictine	Neutral spirits
Porter	Rosés	Rye					Kirsch	Chartreuse	Others
Stout	Aperitif	Blends					Brandy (U.S.)	Chéri-Suisse	
Bock beer	Vermouth	Canadian					Applejack	Cointreau	
Steam beer	Dubonnet	Light					Slivovitz	Crèmes	
	Lillet	Others					Pear William	Curaçao	
	Byrrh						Metaxa	Drambuie	
	Dessert						Pisco	Galliano	
	Sherry[a]						Others	Grand Marnier	
	Port							Irish Mist	
	Madeira							Kahlúa	
	Marsala							Ouzo	
	Málaga							Peter Heering	
	Muscat							Pernod	
	Sparkling							Sabra	
	Champagne							Sloe gin	
	Sekt							Tia Maria	
	Sparkling burgundy							Triple sec	
	Spumante							Fruit liqueurs	
	Saké							Fruit-flavored brandies	
								Others	

[a] Often served as an aperitif.

Alcoholic content. There are some differences between fermented and distilled beverages in addition to the way they are made. One is their alcoholic content. Beers and ales contain 2½ to 8% alcohol by weight (3.1 to 10% by volume). Table wines may be 7 to 14% by volume, aperitif and dessert wines 14 to 24%. Spirits run usually from 35 to 50% by volume, with a few liqueurs as low as 18 to 20% and one or two rums as high as 75½%. There are even neutral spirits available at 95%, but they are never used as bar liquors.

For spirits the alcohol content is expressed in terms of ***proof***, which is twice the percentage figure. Thus a 100-proof whisky is 50% alcohol by volume. The rest of the spirit is water except for a tiny percentage of impurities. Some of this water was carried along by the alcohol vapors during distillation, and some distilled water was added later. The term proof comes from the early days of spirits when the distiller tested, or proofed, his product by mixing it with an equal amount of gunpowder and setting it afire. If it didn't burn it was too weak. If it burned too fiercely it was too strong. If it burned with a steady blue flame it was just right for drinking—100 proof. This turned out to be about 50% alcohol. Proof is sometimes indicated by the same symbol used for degrees of temperature—80°, for example.

It is the alcohol in any beverage that is the intoxicating element as it runs through the veins relaxing the nerves, warming the heart, and making the conversation brilliant. Looking at the percentage figures one might conclude that a 90-proof gin is 10 times as intoxicating as beer. Ounce for ounce this is true. But comparing percentages gives a statistical picture with very little meaning, since the typical serving size varies widely from one beverage to another.

Let's translate the statistics into the drinks you might be serving at the bar. A 10-ounce serving of 4% beer would contain 0.4 ounce of alcohol. A 4-ounce serving of 13% wine would have 0.52 ounce alcohol, or 30% more than the serving of beer. A Gin and Tonic made with 1½ ounces of 90-proof gin would have 0.675 ounce alcohol, or 30% more than the glass of wine and 69% more than the beer. A Martini made with the same amount of the same gin and a 19% vermouth would contain 0.745 ounce alcohol, about ¾ ounce—10% more than the Gin and Tonic, 43% more than the wine, and 86% more than the beer. Table 5.2 condenses this information.

Comparing these beverage types, you can see that the alcoholic content has a lot to do with the size of a serving. Imagine the consequences of pouring gin in 4-ounce servings as though it were wine! So we serve a beer in a large glass and a Martini in a small one (and it still takes roughly two beers to equal one Martini in alcohol). It is more than tradition that dictates a certain glass for a certain type of beverage.

Mixed drinks. Speaking of Martinis, we need to define a beverage item of central importance to most types of bar, the mixed drink. A ***mixed drink*** is

table 5.2 **ALCOHOLIC CONTENT OF DRINKS COMPARED**

drink	*alcoholic content (oz)*	*more than beer (%)*	*more than wine (%)*	*more than Gin and Tonic (%)*
10 oz of 4% beer	0.4			
4 oz of 13% wine	0.52	30		
Gin and Tonic with 1½ oz of 90° gin	0.675	69	30	
Martini with 1½ oz 90° gin, ⅜ oz 19% vermouth	0.745	86	43	10

a single serving of two or more beverage types mixed together, or of one beverage type mixed with a nonalcoholic mixer. The cocktail and the highball are the two most common types of mixed drinks, but there are many others—coffee drinks, collinses, cream drinks, for example.

Most mixed drinks are made with spirits, though a few wines are cocktail mainstays, notably the vermouths. In fact, the most common use of spirits is in mixed drinks. The rest of this chapter takes an in-depth look at spirits, those inviting and expensive bottles of liquid gold and silver that are so important to your profit that you have to keep track of every ounce.

About spirits in general . . .

All spirits are alike in several ways. They are all distilled from a fermented liquid. They all have a high percentage of alcohol in comparison to other alcoholic beverages—most of them are nearly half alcohol and half water (80 to 100 proof except for some liqueurs). They are usually served before or after dinner. Yet there are several distinct and familiar categories of spirits.

The primary differences between them are differences of flavor and body. Each type has a characteristic taste—whiskies have a whisky taste, gins a gin taste, rums a rum taste, and so on. Within categories of spirits there are further taste differences: bourbon whisky, for example, tastes very different from scotch, and Irish tastes different from both. There are also taste differences between brands. And there are differences in body—we have full-bodied spirits and light spirits. How do these differences come about?

There are three main factors that determine flavor and body: (1) the ingredients in the original fermented liquid, (2) the proof at which it is distilled, and (3) what is done with the spirit after distillation. To understand how these work we need to look at the distillation process more closely.

Congeners. The core of the process, as we've seen, is evaporating the alcohol by heating it until it separates itself from the fermented liquid by vaporizing. If it were the only thing that vaporized at distilling temperatures, we would have 100% pure ethyl alcohol—a 200-proof spirit. It would be colorless and have a raw sharp taste with no hint of its origin. That taste would be the same no matter what it was distilled from—wine, grain, molasses, whatever.

But other substances may join the alcohol as it vaporizes. One is water, as we have noted. In addition there are minute amounts of other volatile substances, and it is these that provide flavor, body, and aroma in the beverage. Called ***congeners***, they come from ingredients in the original fermented liquid. Chemically they have such identities as acids, other alcohols, esters, aldehydes, trace minerals. In the final product they translate into the smoky malt taste in scotch, the full-bodied pungency of bourbon, the hint of molasses in rum, the rich aroma of fine brandy, and so on.

Distillation proof. By varying the distillation temperature, the length of time, the type of still, and other factors, the amounts of water and congeners can be controlled. The less water and the fewer the congeners, the higher the proof of the distilled spirit and the purer the alcohol. And, since the congeners are the flavorers, the higher the distillation proof, the less pronounced the flavor and the lighter the body. Conversely, the more congeners and the lower the distillation proof, the more distinctive the flavor and the more there is of it. To experience the difference, taste a vodka and a bourbon: vodkas are distilled at 190 proof or above; bourbons are usually distilled at 110 to 130 proof.

Spirits distilled at 190 proof or above are known as ***neutral spirits***, since they are so close to pure alcohol that they have no distinctive color, aroma, character, or taste. They are used to make vodka and gin and to blend with spirits distilled at lower proofs.

All neutral spirits, as well as many lower-proof spirits, are distilled in ***column stills***. This type of still is used to make most spirits in this country. On the other hand cognac, malt scotch, Irish whiskey, tequila, and some rums, gins, and liqueurs are made in ***pot stills*** that have not changed much in design since the early days of distilling. The still in Figure 5.2 is a pot still; so is George Washington's still in Figure 1.1 (page 4). Pot stills are limited in the degree of proof they can achieve; consequently the pot-still product always has a lot of flavor, body, and aroma.

The column still (also called ***continuous*** or ***patent still***), on the other hand, can be controlled to produce spirits at a wide range of proofs up to about 196°. It consists of a tall column, or series of columns, in which the fermented liquid is heated by steam inside the still instead of heat from below. The alcohol vapors can be drawn off at various heights and redistilled in a continuous process, making it possible to separate nearly all the water and congeners from the alcohol if neutral spirits are the goal, or to produce spirits at almost any lower proof.

Aging, flavoring, blending, and bottling. A newly distilled spirit is raw, sharp, and biting. How is it turned into the mellow and flavorful product we sell at the bar?

Our less sophisticated ancestors drank it as it came from the still. The story goes that someone noticed that a batch of spirits shipped a long distance in wooden barrels tasted better on arrival than it did when it left the still. However the discovery was made, most of today's spirits distilled at less than 190 proof are ***aged*** in wooden barrels for periods ranging from 1 year for some light rums to 20 or so years for choice brandies. Two things happen in the barrels: (1) the spirit undergoes changes as the congeners interact with air filtering through the porous wooden casks; and (2) new congeners are absorbed from the wood itself, adding other flavoring agents. In due course all the flavors are "married"—blended—and mellowed to the desired final taste. Aging in wood adds color as well as flavor to the spirit.

Not all spirits are aged; sometimes the sharp bite of a raw but flavorful spirit is part of its appeal—think of gin, for example, or kirsch.

There are other means of producing or modifying flavors after distillation. One is by introducing new flavors, as is done with gin and with liqueurs. Another is by blending two or more distillates, as is done with many whiskies. A spirit taste may also be modified by filtering through charcoal, as is done in making vodka, or by other special ways of removing congeners.

At bottling time, spirits are diluted to drinking levels of taste, usually 80 to 100 proof, by adding distilled water. This lessens the flavor but does not change it. Once in the bottle, a spirit does not undergo any further change. No matter how old it gets sitting on a shelf, it does not age, since it is not exposed to air or wood.

All spirits produced in the United States are stored and bottled in bonded warehouses under strict government regulation. At bottling time the bottler checks for full bottles, correct proof, accurate labeling, and purity. The federal revenue stamp across the top means compliance with all government standards and payment of the federal tax.

All this does not mean that the government guarantees the quality of the product. It does not. Many people, however, think of the label ***Bottled in Bond*** as a guarantee of quality conferred by the government. What it really means is that a given spirit meets certain special conditions: it is straight (i.e., unblended), distilled at 160 proof or less at one plant by one distiller, aged at least four years, and bottled at 100 proof in a bonded warehouse under federal supervision. Since all spirits are now bottled in bonded warehouses, the phrase has lost much of its meaning.

You can see that there are literally hundreds of ways in which a beverage that is roughly half alcohol and half water can be made in thousands of different versions. Every ingredient, from the grape to the grain to the water to the yeast, can make a difference in taste. Distillation methods are critical. Different aging times and conditions produce different tastes. The type of wood in the barrel, and whether the barrel is new or used, charred or uncharred, has a definite effect on flavor. Blending can produce an almost infinite number of products, and flavoring can too. All these different factors explain why each brand of each spirit is unique.

Fortunately there are only a few basic spirit types, and they are easily recognizable by general taste, aroma, and character. We will now look at each one in some detail.

Whiskies . . .

The earliest spirit makers started with whatever fermentable product there was plenty of. In the southern countries of Europe it was wine, already fermented and available. In northern climates such as Scotland and Ireland, grapes did not grow well but grain did, and beer and ale were plentiful. So the first whisky makers started with a fermented mash of grain, similar to the early stages of making beer, and they distilled that. It was a raw, biting drink they produced, called *uisgebeatha* in Scotland and *uisgebaugh* in Ireland, Celtic translations of *aqua vitae*, water of life. Later the last syllables were dropped and the name became *uisge* and eventually whisky.

To get a grain product to ferment, there is an extra step that starts off the whisky making: the starch of the grain must be converted to sugar. This is done by adding a malt. ***Malt*** is sprouted grain, usually barley. It contains an enzyme called ***diastase***, which changes the starch to sugars. Malt, grain, and hot water are mixed together until conversion takes place. The liquid is then fermented by adding yeast. After fermentation it is distilled. Figure 5.3 shows the sequence of steps.

After distilling, the raw whisky is stored in barrels, usually of oak, for at least two years. The length of the aging depends on the character of the raw product; some take longer to mellow than others. For this reason a 12-year-old whisky is not necessarily better quality than, say, a 5-year-old or one whose age isn't given on the label. It all depends.

For straight whiskies, the making of the product ends at this point. The most common straight whiskies include bourbon, rye, and corn, each containing 51% or more of a single grain type. Straight whiskies account for about half of U.S.-made whisky sales.

The majority of whiskies marketed in the United States (this includes imports) undergo still another process—blending. Whiskies of different grains or different batches or different stills or different ages are blended together, and sometimes with neutral spirits, to produce the standard of flavor and quality that represents a particular brand of whisky. Usually the formula is a house secret, and the final blend is perfected by skilled tasters.

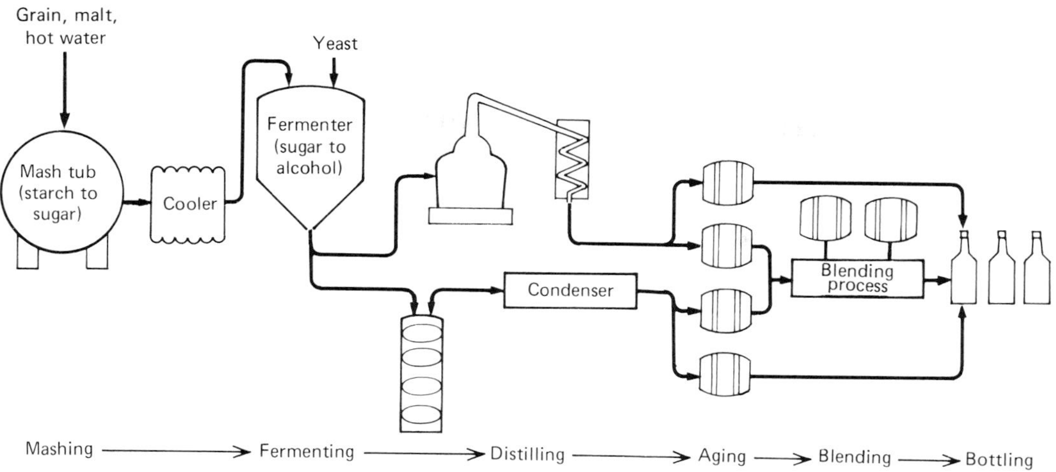

figure 5.3 Making whisky.

Scotch. Scotch whiskies are made-in-Scotland blends of malt whiskies and high-proof grain whiskies. The malt whiskies are made in pot stills, mainly from sprouted barley that has been dried over peat fires, giving it a smoky flavor that carries over into the final product. The grain whiskies are made chiefly from corn and are distilled at around 180 proof in column stills, somewhat below neutral spirits but very light in flavor. The two types of whiskies are aged separately for several years and then blended, with sometimes as many as 30 or 40 different malt and grain whiskies in a given brand. Straight malt scotches such as Glenfiddich and Glenlivet are also beginning to appear in the U.S. market.

The popularity of scotch stems from Prohibition days, when it was smuggled into the States from Canada, the Caribbean, and ships at sea. After Repeal, Scotland's distillers tailored their production to American preferences. Today the most popular brands are light-bodied, with subtly distinctive tastes. A light-bodied scotch is not necessarily light-colored, since all scotches have caramel (burnt sugar) added to assure color uniformity. Nor does light body mean low alcoholic content. All scotches are bottled at a minimum of 80 proof; most are 86 proof.

Irish whiskey. Irish whiskey (the Irish spell it with an *e*) is probably most familiar as the essential ingredient in Irish Coffee. It has an ancestry going back to the twelfth century. It may have been the first whisky in the world, as the Jameson label says, but the Scots make the same claim. Today's production techniques for Scotch and Irish whiskies are similar, but with some differences that definitely affect the flavor.

The main flavor difference in Irish comes from the fact that the freshly malted barley is not exposed to peat smoke when it is dried, so there is none of the smoky taste of scotch. Another difference is that it is made from several grains in addition to malted barley. A third is a triple distillation process that takes the Irish product through three separate stills (most pot-still whiskies go through two). The result, after aging seven or eight years, is a particularly smooth, mellow whisky of medium body. This is the traditional Irish—unblended, and seldom called for in America. The more familiar Irish is blended with high-proof grain whisky, as in making scotch, to create a lighter drink for today's market.

Bourbon. Bourbon is the best-known straight (unblended) American whisky or whiskey (several brands of bourbon carefully put the *e* in the name). It is distilled at 160 proof or less from a fermented mash of at least 51% corn, and aged at least two years in charred new oak containers. These requirements are spelled out in the federal government's Standards of Identity. Most bourbons are aged for about six years, or for however long it takes to reach their desired mellowness. At the usual distillation proofs of 110 to 130, the whisky produced has a strong flavor component and a full body. The charred barrels add a special flavor.

Bourbon was named for Bourbon County, Kentucky, where it was first distilled by a preacher named Elijah Craig in 1789. In those days having whisky on hand was an ordinary part of life, and many a farmer had his own still or raised grain for his neighbor's still. Today most bourbons are made in Kentucky, though at this writing there is no working distillery in Bourbon County!

Most bourbons use a ***sour-mash*** yeasting process. Along with the fresh yeast, a portion of the leftovers from a previous distilling is added to the mash. This encourages yeast growth, inhibits bacterial contamination, and provides a certain continuity of flavor, though definitely not a sour taste.

Familiar examples of bourbons are Jim Beam, Old Grand Dad, Early Times, I.W. Harper, Old Forester, Old Crow, Wild Turkey.

A sour-mash whisky similar to bourbon is Jack Daniels, a Tennessee whisky. Many people drink it as their favorite bourbon, and it meets the grain requirements of bourbon, but Tennessee whisky has a special twist to its production: the distillate is filtered through maple charcoal before it is barreled for aging. This bit of regional tradition eliminates some harsher elements in the whisky and adds its own touch of flavor and romance. George Dickel is the only other Tennessee whisky; there are only two legal distilleries in the state.

Rye. A rye whisky is one that is distilled at 160 proof or less from a fermented mash of at least 51% rye and aged in charred new oak containers at least two years. Notice that this description is just like that of bourbon except for the kind of grain.

In the early days of American whisky making, rye was the grain of choice. As America expanded westward, corn and other grains took the place of rye. Today few people drink straight rye whisky—perhaps in a Rye Manhattan. Along parts of the East Coast, blended whiskies are referred to as "rye." These are not ryes, however. Straight rye is a full-bodied spirit with the strong flavor of its parent grain, whereas most blended whiskies are lighter and less defined.

Other straight whiskies. There are other straight whiskies made from 51% or more of wheat, malted barley, malted rye. There is corn whisky made from 80% or more of corn and aged in uncharred containers. There is also something called simply straight whisky, which is whisky made from a mixture of grains with no one grain predominating. And there are blended straight whiskies that are mixtures of straight whiskies of the same type—blended straight bourbon, for example.

Blended American whiskies. Blended American whiskies are combinations of straight whiskies or of whiskies with neutral spirits. A blend must contain at least 20% straight whisky. Beyond that the blend is wide open to whatever combinations of whisky and neutral spirits will achieve the balance, taste, aroma, and body that characterize the particular brand. Small quantities of certain blending materials—sherry, prune juice, peach juice—are also allowed, up to 2½%. Often the neutral spirits are aged in used oak barrels to remove harshness. The process is like the blending of scotch from various straight malt whiskies and grain spirits, and the blending has a similar effect: the product is lighter-flavored and lighter-bodied than the original unblended whiskies, though *not* lower in proof.

Blended whiskies made in this country have the words "American Whisky" on the label. Close to half the American whiskies we drink are blends—Seagram's 7 Crown, Four Roses, Schenley Reserve, for example. There is no aging requirement for blended whisky, which means that the cheapest brands can be pretty raw, and definitely not suitable as bar whiskies.

Light whisky. Made-in-USA light whisky is a new whisky class. It came into being in order to let American distillers compete with the lighter imports that are distilled at higher proofs than those allowed in the United States. Thus the federal government created a new category, effective in 1972, for whiskies distilled at above 160 proof but below 190 proof. These whiskies may be stored in used or uncharred new oak. Aging in seasoned (used) wood permits good development of the lighter flavor intensities of the high-proof spirits.

There are several light whiskies on the market, though none so far has caught on with the drinking public, who seem to be turning to entirely different beverages.

Canadian whisky. Canadian whiskies are blended whiskies, light in body, delicate yet mellow in flavor. Canadian law requires only that the whiskies be made from cereal grains and aged at least three years, leaving the rest up to the distiller. The grains usually used are corn, rye, barley malt, and wheat, and each brand's formula is a trade secret. The whiskies are distilled at 140 to 180 proof and most are aged six years or more. Their lightness keeps them popular in today's "light"-minded market. Familiar examples are Canadian Club and Seagram's V.O.

Uses and service. Whiskies are served straight, on the rocks, with water or club soda or another mixer such as 7-Up, or in cocktails. They are ordered by type (scotch, bourbon, etc.) or by brand name, except sometimes in cocktails. Whisky drinks are served before, after, or between meals but are usually not offered with the meal.

A whisky ordered straight (straight up, neat) is served in a shot glass or other small glass with a glass of ice water beside it.

A whisky on the rocks is served in a 5- to 7-ounce rocks glass filled with ice cubes. The ice is put in the glass first and the whisky is poured over it.

A whisky and water, soda, or other mixer is served in a highball glass. The glass is first filled with ice; then the whisky is poured and the glass is filled with the water, soda, or mixer and swirled with a barspoon.

Vodka, gin, and aquavit . . .

Neutral spirits form the basis of vodka, gin, and aquavit, three of the spirits often referred to as "white goods." Each of these very different beverages is produced by modifying (***rectifying***) the base spirit.

Vodka. Today's largest-selling spirit, with 18% of the market, vodka was almost unknown in this country 35 years ago. Yet it has been around since the fourteenth century, and at one time there were 4000 brands in Russia alone.

The earliest vodkas were Russian or Polish *aqua vitae* made from grain or sometimes potatoes, distilled at fairly high proof but not aged, and so strongly flavored that they were often spiced to mask the raw grain taste. Then in the early 1800s it was discovered that charcoal would absorb the congeners, and modern-day vodka—a tasteless, colorless spirit—was born. Drunk neat, chilled, usually with spicy foods—caviar, smoked salmon, anchovies, at least by the upper classes—it became the rage in Russia. Peasants drank it too. Everyone did.

In 1914 the Smirnoff family, the first distillers to use charcoal, were producing a million bottles a day. Three years later the entire family of over 100 members was wiped out by the Russian Revolution except for one, who escaped with the family formula. An associate brought the family name and formula to the United States after Repeal, but in spite of valiant efforts he was unable to sell any vodka to speak of, even to American Russians.

In 1946 the owner of the Cock 'N Bull restaurant in Hollywood put together a drink made from vodka, which wasn't selling, and ginger beer, which wasn't selling either, added half a lime, served it in a copper mug, which also wasn't selling, and christened it the Moscow Mule. With skillful promotion the drink and the vodka caught on, helped by a brilliant advertising slogan, "It will leave you breathless." Then came the Screwdriver, the Bloody Mary, and a host of others. Vodka turned out to be the perfect partner for all sorts of juices and mixers, having no taste of its own. It was also, with its breathlessness, the perfect drink for the three-Martini business lunch.

Vodka is defined in the U.S. Standards of Identity as "neutral spirits so distilled, or so treated after distillation with charcoal or other materials, as to be without distinctive character, aroma, taste, or color." It is the ultimate light spirit. It can be distilled from any fermented materials, since neutral spirits from any source taste pretty much alike—tasteless. U.S. vodkas and the best imports are made from grain.

Many vodka producers in this country buy neutral spirits from distillers who specialize in them. There are two types of further processing: one is filtering through charcoal, and the other is to use such techniques as cleansing vapors that absorb the remaining congeners or centrifugal purifiers that spin them out. The goal is to come as close as possible to Nothing and to bottle it at 80 to 100 proof. But not all vodkas are alike, though many people think they are. Quality, or lack of it, is definitely perceptible.

Gin. Gin is a spirit having the predominant flavor of the juniper berry. It was invented in the 1600s by a Dutch professor of medicine known as Doctor Sylvius, who made an *aqua vitae* from grain flavored with juniper berries, combining their medicinal benefits. The result was so potable that it swept the country under the name of Geneva or Genever (from the French *genièvre*, meaning juniper). It crossed the English Channel via British soldiers, who called it "Dutch courage" and shortened Genever to gin.

Cheap gin was soon being made in London from almost anything—"make it in the morning and drink it at night"—and sold in hole-in-wall dram shops all over London. Even the best of it must have tasted terrible, but the poor of England drank it on a national scale to the point of disaster.

The story repeated itself in the United States during Prohibition, when "bathtub gin" was made at home. It, too, was a poor, sometimes lethal product, and the custom arose of mixing it with something else to kill the taste, thus popularizing the cocktail.

It took gin some years to outlive its history as a cheap road to drunkenness and degradation, but today it is a highly respected favorite. The British officer's Gin and Tonic (to prevent malaria) and the post-Repeal adoption of the Martini as *the* fashionable cocktail had a lot to do with changing gin's image. Today, as one of the so-called light spirits, gin is definitely "In."

Today's gins are of two types, descended from the Dutch and English gins of the seventeenth century. They are quite different, and only the English type, whether imported from England or made in this country, is used in mixing drinks.

Dutch gin—Hollands, Genever, or Schiedam—is made starting with a mash of several grains and distilling and redistilling at low proofs, with the juniper berries included in the final distillation. The product is a full-bodied gin with a definite flavor of malt along with the juniper. It is a flavor that would overpower almost anything it might be mixed with. This is not a bar gin. It is drunk straight and cold.

English-style gin always has the word Dry on the label and in fact is often called London dry even when it is made in the United States. The term ***dry*** means lacking in sweetness. Gins may be labeled very dry or extra dry, but these are not any drier than the others.

In England gin is made from nearly neutral grain spirits distilled at 180 to 188 proof. These are then redistilled in pot stills with juniper berries either in the spirits (Figure 5.4*a*) or suspended above them on mesh trays so that the rising vapors pass through and around the berries (Figure 5.4*b*). Often other herbs and spices (known as ***botanicals***) are included with the

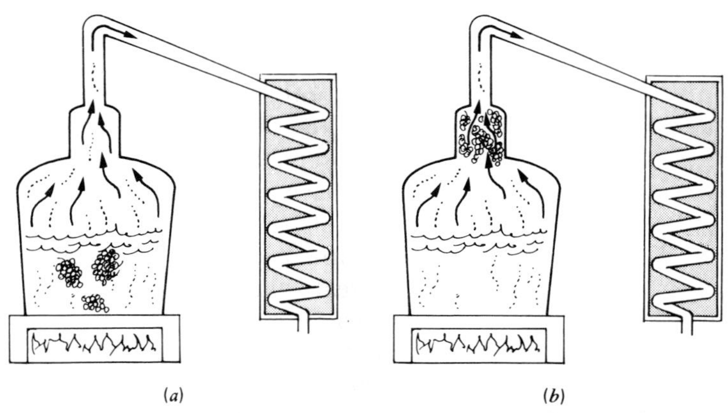

figure 5.4 Making gin. (*a*) Distilling or redistilling with juniper berries and other botanicals. (*b*) Redistilling to pass vapors through juniper berries and other botanicals.

juniper. They account for the subtle flavor variations from one brand of gin to another. Many English gins are made by British distillers in the United States.

American gins are made in two ways: distilling and compounding. Distilled gins are made in much the same way as the English gins, by redistilling neutral spirits with juniper berries and other botanicals. Compound gins are made by simply mixing high-proof spirits with extracts from the juniper berry and other flavorings. Distilled gins are allowed to use the word Distilled on the label; compound gins are not identified on the label.

Gin does not need aging. It is stored in stainless-steel or glass-lined tanks until bottled. One type known as golden gin is aged briefly for color.

English and American gins are used in mixed drinks and are almost never drunk straight—except in the very dry Martini.

Aquavit. Aquavit is the Scandinavian version of the old *aqua vitae*, water of life. Like gin, it is distilled at 190 proof and then redistilled with flavorings, but in place of gin's juniper aquavit has caraway. It is bottled at 86 to 90 proof. It is served ice cold, straight, usually with a beer chaser and with food. It is downed in a single gulp, usually in drinking a toast.

Rum . . .

Rum, another of today's light spirits, has a long and not always savory history. It is made from sugarcane, which Columbus brought to the West Indies, or from molasses, a by-product of making sugar. A seventeenth-century writer reported that "the chief fuddling" (intoxicant) made in the island of Barbados "is rum-bullion, alias kill-devil, and this is made from sugar-canes distilled, a hot, hellish, and terrible liquor." Rum was the drink of the English pirates who scourged the shores of the Spanish Main and the Caribbean islands, and of the British Navy, who issued it to the sailors to prevent scurvy. It was also *the* drink in seventeenth-century Europe and the eighteenth-century American colonies.

Rum making became a flourishing industry in New England at this time, and here appears one of the darkest phases of rum's history. For years rum manufacturers and New England ship captains carried on a highly profitable triangular trade with Africa and the West Indies. They exchanged New England rum for slaves in Africa, and slaves for molasses in the West Indies, and turned the molasses into rum for the next go-round. British taxes and restrictions on the rum trade were as much a cause of the American Revolution as the taxes on tea were.

In the nineteenth century Temperance leaders made Demon Rum the symbol of the evils of alcohol. The image stuck right through Prohibition when smugglers were called rumrunners and the offshore ships were known as Rum Rows, though they sold as much scotch as rum.

Today rum holds a respectable place as one of the light spirits. Perhaps a hint of its dark past lingers in the Zombie, a superpotent drink topped with 151° rum.

Most of the rum consumed in this country is made in Puerto Rico. There are several types, differing somewhat in flavor according to the amount of aging. They are made from molasses and distilled at 160 proof or higher, resulting in a dry, lightly flavored, light-bodied spirit.

Puerto Rican rums labeled White or Silver are aged at least a year. Amber and Gold rums are aged at least three years and are colored with caramel. They have somewhat more flavor and are a bit mellower than the Whites and Silvers. Red Label and Heavy Dark rum labeled For Planter's Punch are usually aged six years or more and have a dry, mellow, full-bodied flavor and bouquet.

Rum made in Jamaica is characteristically full-bodied and pungent, with a dark mahogany color it owes mainly to caramel. It begins with molasses that is fermented by yeasts from the air, called natural, wild, or spontaneous fermentation. It is distilled at 140 to 160 proof, producing a spirit with full flavor and body. This is aged five to seven years. It is bottled usually at 80 and 87 proof and occasionally at 151 proof.

Demeraran rum is made in Guyana along the Demerara River. Darker in color but lighter in flavor than Jamaica rum, it is bottled at 80, 86, and 151 proof. Until the Zombie was invented, the chief market for 151° Demerara was among lumbermen and fishermen in far northern climates, who drink it half and half with hot water as a grog to warm the bloodstream. High-proof Demerara is often used to flame drinks.

You may also encounter Barbados rum or medium-bodied rums from Haiti and Martinique which are made from sugarcane juice instead of molasses. Then there is arak, distilled and aged across the world in Java and then aged another six years in The Netherlands until it resembles brandy. Puerto Rico and Jamaica also make "liqueur" rums, aged up to 15 years like fine brandies.

Rums are used mostly in mixed drinks, the drink recipe specifying the type of rum to be used. Today rum is a popular base for frozen drinks and house specials.

Tequila . . .

A distinctive product of Mexico, tequila comes from a small, officially defined district surrounding the town of Tequila. It is a special type of mezcal, the chief Mexican spirit, and is made only from one species of agave plant, the "blue" variety.

The heart of this agave, weighing 50 to 200 pounds, is filled with a sweet sap that is fermented and then distilled and redistilled at 104 to 106 proof, producing a strongly flavored spirit. Some tequila is aged. White tequila is usually shipped unaged. Silver may be aged up to three years. Gold is aged in oak two to four years, and special tequilas are aged longer.

Tequila is on the rise, notably in the Southwestern states bordering Mexico, where the Margarita cocktail is a strong competitor of the Martini, especially in summer weather. In some places the Tequila Sunrise is well known as a favorite of the Rolling Stones rock group. In Mexico tequila is drunk straight, along with salt and lime. A lick of salt, a shot of tequila, a bite of lime—rather like some of the vodka rituals practiced in Czarist Russia. In the United States the trend is more likely to be the invention of more and more tequila mixed drinks. Tequila seems to qualify as one of the light beverages, in spite of its heavy flavor.

Tequilas that have met the Mexican standards of quality carry the initials DGN on the bottle. It's wise to look for them if you don't recognize the brand, since the whole export industry is so new and has grown so fast. Among the best-known brands are Cuervo, Sauza, and Arandas.

Brandies . . .

Brandy began as an *eau de vie*, the French version of that *aqua vitae* that in other countries became whisky, vodka, and gin. It was thought of as the spirit or soul of French wine. One story credits its origin to a Dutch ship captain who wanted to save on the cost of shipping wine from France to Holland. He figured he could take the water out of the wine in France and put it back again in Holland, saving valuable cargo space. His Dutch customers liked the spirit better than the wine. They called it *brandewijn*, meaning burned wine, which was shortened in time to brandy.

Federal Standards of Identity define brandy as the distilled product of any fruit, but what we call just brandy must be made from grapes. Other fruit brandies must carry the name of the fruit. All brandies must be bottled at 80 proof or higher. At the bar we use two types—a good domestic brandy in the well for mixed drinks, and premium brands, usually imported, for after-dinner service.

American brandies. Most brandy used in the United States is made in California. It is made in column stills at up to 170 proof and is aged in white-oak barrels at least two years and usually longer. Most of these brandies are blends and may have up to 2½% of sweetening and flavoring added; therefore they vary in taste. There are also straight brandies having no additives except caramel coloring, and there are some premium brands that qualify as after-dinner brandies.

American brandies from other fruits—apple, apricot, blackberry, pineapple—include the name of the fruit on the label. They may or may not be aged. Apple brandy, also called applejack, was one of the earliest and most favorite spirits of our New England ancestors. Today's applejack is aged in wood at least two years. It is bottled at 100 proof as a straight brandy or blended with neutral spirits and bottled at 80 proof.

Cognac and other French brandies. Of all the brandies in the world, cognac is the most famous and prestigious. It is made only in the Cognac region of France, where chalky soil, humid climate, and special distillation techniques produce this distinctive brandy. It is distilled by the farmers of the region under strict government control. Only certain kinds of white grapes may be used, and specific distillation procedures must be followed, including traditional copper pot stills known as alembics and precise control of temperatures and quantities. The farmers sell their freshly distilled spirits to shipper-producers, who age and blend them to meet the standards of their particular brands. Cognac is aged in special oak casks at least 1½ years. Most are aged two to four years and some are aged longer. Caramel is added for uniform color.

During aging, the alcohol evaporates through the porous casks at an average of 2% per year. In the warehouses the escaping vapors—known as the angels' share—are noticeable. There is a story of an old woman, a strict teetotaler, who lived above a cognac warehouse and died at a ripe old age from alcoholism as a result of breathing those vapors over the years.

A cognac label may carry cryptic letters, special words, and varying numbers of stars. The stars may mean somewhat different things for different brands. By French law a three-star cognac must be at least 1½ years old; most are around four. Since cognacs are blends of brandies of various ages, no age is allowed on the label. The cryptic letters are symbols of relative age and quality:

V = very

S = superior

O = old

P = pale

E = extra or especial

F = fine

X = extra

A cognac specified VS (very superior) is similar to a three-star cognac. A VSOP (very superior old pale) has been aged in wood at least 4½ years, and probably seven to ten. The words Extra, Vieille Réserve, and Napoléon may not appear on the label unless the cognac is at least 5½ years old. Contrary to legend, there is no cognac around dating from Napoleon's day, though some shippers have stocks up to 50 years old and more to use in blending their finest cognacs. After a certain age cognac does not improve but actually loses quality.

A cognac labeled Grande Champagne or Fine Champagne has nothing to do with the bubbly beverage. The French word *champagne* means field, and the French bubbly and the Champagne cognac both take the name from the common word. Grande Champagne is the heart of the Cognac district, whose grapes are considered best of all. Grande Champagne on the label means that the cognac was made from these grapes. Fine Champagne means that 50% or more of its grapes came from Grande Champagne and the balance came from the next-door and next-best area known as Petite Champagne.

Armagnac is another French brandy familiar to Americans. Like cognac, it comes from its own restricted region and is made from white grapes. It has the same kinds of label indications as cognac, with the same meanings. It is permitted to have its age on the label.

A fine apple brandy known as calvados is also made in France. It may be aged up to 40 years.

There are other good French brandies, less fine, less well known, and mostly less expensive. They are seldom called for, and unless you know your way around brandies it isn't wise to buy them.

Imported fruit brandies. The best-known nongrape brandy is kirsch or kirschwasser, made from the wild black cherry that grows in the Rhine valley. A colorless liquid, often called a white brandy, it is made in pot stills from a mash that includes the cherry pits and skins. A low distillation proof of 100 or less allows the bitter almond flavor of the pits to be carried into the final spirit. It is bottled immediately to retain the maximum flavor and aroma of the fruit. Although production is relatively simple, the cost is high because of the large amounts of wild fruit needed.

A liqueur carrying the same name is also made; it is sweetened and carries the word liqueur on the label.

Another fruit brandy familiar to many is slivovitz, a plum brandy made in central Europe. It is distilled in the same way as kirsch but is aged in wood to a golden color.

A colorless, unaged brandy is made from pears in Switzerland and France. It is known here as Pear William and there as Poire Williams.

Fruit-flavored brandies are not true brandies but are sweetened liqueurs of lower proof with a brandy base. In Europe the word "flavored" is omitted from these spirits. Don't be fooled: an imported "strawberry brandy" is not a brandy but a liqueur of considerable sweetness. The word "flavored" is required for the comparable American product.

figure 5.5 Serving brandy in a snifter.

Other brandies. Metaxa is *the* Greek brandy. Slightly sweetened, it is technically a liqueur but is generally thought of as brandy.

Pisco is a strong-flavored brandy from Peru. It was popular in California in Gold Rush days and is still drunk in San Francisco in the Pisco Sour cocktail.

There are also several Mexican brandies on the market; Pedro Domecq is perhaps the best-known brand.

Uses and service. Brandy served straight is a traditional after-dinner drink, presented according to custom in a large rounded brandy glass, or snifter (Figure 5.5). The glass is cupped in the palm of the hand, to warm the brandy slightly, and rolled about to release the brandy's rich aroma—an important part of the sensual pleasure of the drink. It can also be served straight up in a pony glass. Brandy is also served with soda or water as a highball, in coffee, and in many mixed drinks.

White brandies ordered straight should be served icy cold in a pony glass. Their most common use, however, is in mixed drinks.

Liqueurs and cordials . . .

Liqueur and ***cordial*** are two words for the same thing: a distilled spirit steeped or redistilled with fruits, flowers, plants, or their juices or extracts, or other natural flavoring materials, and sweetened with 2½% or more of sugar. To simplify things from here on we'll use the word liqueur. Liqueurs are natural after-dinner drinks, sweet and flavorful.

The makers of liqueurs are today's alchemists, with their secret formulas of herbs and spices and flowers and fruits and exotic flavorings. No longer looking for the elixir of life, they deal in flavor, color, romance, and recipes. New liqueurs are constantly being developed, and both old and new are promoted with recipes for new drinks in the hope that something like the Moscow Mule miracle will happen. And it does happen now and then—the Harvey Wallbanger put Galliano in every bar, and other success stories are popping up so fast there will probably be several new ones by the time this book reaches print.

Of the ingredients in a liqueur the distilled spirit is the starting point. It may be any spirit—brandy, whisky, rum, neutral spirits, others. The distinctive flavorings may be any natural substance such as fruits, seeds, spices, herbs, flowers, bark. Many of them are complex formulas containing as many as 50 ingredients. One of them (Cointreau) uses oranges from five different countries.

The flavorings may be combined with the spirit in different ways. One way is ***steeping*** (soaking) the flavorers in the spirit—called ***maceration***. Another is by pumping the spirit over and over flavorers suspended above it, as in a coffeepot—called ***percolation***. Still another is by redistilling the spirit with the flavorers.

The sugar may be any of several forms, including honey, maple syrup, and corn syrup. The sugar content is the main thing that distinguishes liqueurs from all other types of spirit. It varies from 2½% to as much as 35% by weight from one liqueur to another. A liqueur with 10% or less sugar may be labeled Dry.

Color is often added to colorless spirits, as in the cases of green crème de menthe and blue curaçao. Colors must be natural vegetable coloring agents or approved food dyes.

Table 5.3 gives you an alphabetical listing of some of the most popular liqueurs, along with their spirit base, flavor, color, and bottling proof. Notice how much the proof varies from one to another. Generic types are made by more than one producer and vary in flavor and proof.

Liqueurs are served straight up in a pony glass, on the rocks, and especially in mixed drinks.

Bitters . . .

Bitters are spirits flavored with herbs, roots, bark, fruits, and so on, like the liqueurs, but they are unsweetened, and bitter is the right word for them. There are two kinds, concentrated flavorers and beverage bitters. Among the flavorers are Angostura, various orange bitters, and the lesser-known Peychaud's, a New Orleans product. They are used in minute amounts to flavor mixed drinks.

The best-known of the beverage bitters is Campari, a low-proof bitter red Italian spirit. It is usually drunk with soda or tonic or in a cocktail. It is a fashionable drink all over Europe and is becoming well known in this country, especially in California. Another of these bitters is Amer Picon, a qui-

table 5.3
LIQUEURS

liqueur	spirit base	flavor	brand or generic	color	proof
Amaretto	Neutral spirits	Almond-apricot	Generic	Amber	48–56°
Anisette	Neutral spirits	Anise, licorice	Generic	Red, clear[a]	40–60°
Apricot liqueur or cordial	Neutral spirits	Apricot	Generic	Orange-amber	60–70°
Bailey's Irish Cream	Irish whiskey	Irish-chocolate	Brand	Pale café au lait	34°
Bénédictine	Neutral spirits	Herb-spice	Brand	Dark gold	86°
B & B	Neutral spirits, cognac	Herb-spice	Brand	Dark gold	86°
Blackberry liqueur or cordial	Neutral spirits or brandy	Blackberry	Generic	Red-purple	60°
Chartreuse	Brandy and neutral spirits	Spicy herb	Brand	Yellow Green	80°, 86° 110°
Chéri-Suisse	Neutral spirits	Chocolate-cherry	Brand	Red-pink	52°, 60°
Cherry liqueur or cordial	Neutral spirits or brandy	Cherry	Generic	Red	30–60°
Cointreau	Neutral spirits	Orange	Brand	Clear	80°
Cordial Médoc	Neutral spirits, cognac, armagnac	Brandy and fruit	Brand	Dark amber	80°
Crème de bananes	Neutral spirits	Ripe banana	Generic	Yellow	50–60°
Crème de cacao	Neutral spirits	Chocolate-vanilla	Generic	Clear, brown	50–60°
Crème de cassis	Neutral spirits	Black currant	Generic	Red-black	30–50°
Crème de menthe	Neutral spirits	Mint	Generic	Clear, green	60°
Crème de noyaux	Neutral spirits	Almond	Generic	Clear, red, cream	50–60°
Crème d'Yvette	Neutral spirits	Violet, jellybean	Generic	Blue-violet	36–40°
Curaçao	Neutral spirits, (rum or brandy)	Orange peel	Generic	Clear, orange, blue	54–80°
Drambuie	Scotch	Scotch-honey-herb	Brand	Gold	80°
Forbidden Fruit	Brandy	Grapefruit	Brand	Red-brown	60–64°
Galliano	Neutral spirits	Anise-vanilla, licorice	Brand	Bright yellow	80°
Grand Marnier	Cognac	Orange peel, cognac	Brand	Light amber	80°
Irish Mist	Irish whiskey	Irish-honey-herb	Brand	Amber	80°
Jeremiah Weed	Bourbon	Bourbon	Brand	Gold	100°
Kahlúa	Neutral spirits	Coffee	Brand	Brown	53°
Kirsch liqueur	Kirsch	Sweetened kirsch	Generic	Clear	90–100°

table 5.3
LIQUEURS *(continued)*

liqueur	*spirit base*	*flavor*	*brand or generic*	*color*	*proof*
Kümmel	Neutral spirits	Caraway	Generic	Clear	70–100°
Lochan Ora	Scotch	Scotch-honey-herb	Brand	Gold	70°
Mandarine	Brandy	Tangerine	Generic	Bright orange	80°
Maraschino	Neutral spirits	Cherry-almond	Generic	Clear	60–80°
Midori	Neutral spirits	Honeydew	Brand	Ice green	46°
Ouzo	Brandy	Anise, licorice	Generic	Clear	90–98°
Peach liqueur or cordial	Neutral spirits	Peach	Generic	Amber	60–80°
Peppermint schnapps	Neutral spirits	Mint	Generic	Clear	40–60°
Pernod	Neutral spirits	Licorice, anise	Brand	Yellow-green	90°
Peter Heering	Neutral spirits, brandy	Cherry	Brand	Dark red	49°
Raspberry liqueur or cordial	Neutral spirits	Raspberry jam	Generic	Red-purple	50–60°
Rock and rye	Rye and neutral spirits	Rye-fruit	Generic	Gold-brown	60–70°
Sabra	Neutral spirits	Chocolate-orange	Brand	Deep brown	60°
Sambuca	Neutral spirits	Licorice	Generic	Clear	40–84°
Sloe gin	Neutral spirits	Wild plum	Generic	Red	42–60°
Southern Comfort	Bourbon	Bourbon-peach	Brand	Gold	80°, 100°
Strawberry liqueur	Neutral spirits	Strawberry	Generic	Red	44–60°
Strega	Neutral spirits	Herb-spice	Brand	Gold	80–85°
Tia Maria	Rum	Coffee	Brand	Brown	63°
Triple sec	Neutral spirits	Orange peel	Generic	Clear	60–80°
Tuaca	Brandy	Eggnog-cocoa	Brand	Yellow-brown	84°
Vandermint	Neutral spirits	Chocolate-mint	Brand	Dark brown	52°
Vieille Cure	Neutral spirits	Herb-vanilla	Brand	Green, yellow	60°
Wild Turkey liqueur	Bourbon	Bourbon	Brand	Amber	80°
Yukon Jack	Canadian whisky	Light whisky	Brand	Light golden	80°, 100°

[a]Often referred to as white.

nine-laced French bitters with a brandy base, said to have been what the French Foreign Legion in Algeria added to the water in their canteens. It is served with ice and water or used in cocktails. Still another is Fernet Branca, known chiefly as a hangover treatment—perhaps as a counter-irritant.

Summing up . . .

Distilled spirits form the largest part of the bartender's stock in trade. Except for liqueurs they are all around 40 to 50% alcohol (80 to 100 proof). All are made by extracting alcohol from a fermented liquid. What makes them taste different from one another is whatever else is extracted from the fermented liquid along with the alcohol and what is added or subtracted after distillation. Contrary to popular misconception, "light" or "white" applied to spirits does not mean less alcohol but less taste or less color; since color is usually added in the form of caramel, it means nothing about the beverage itself.

In practice, "knowing" distilled spirits is largely a brand-name game. But people selling liquor should know a great deal more than that about what they are selling.

chapter 6 . . .

THE BEVERAGES: WINES

The popularity of wine today is a good reason to study the subject, especially since many people on both sides of the bar know little about it. But you need not master all its complexities in order to sell wine successfully. The important things are to start slowly and to know how to sell and serve.

This chapter gives you this basic knowledge, concentrating on American wines, which account for about 80% of sales. It describes the different types of wine and explains how they are made, then gives you basic information you need in order to buy, store, and sell wine intelligently. It moves then to wine service, wine lists, and wine tasting. Finally, for those who want to delve further, there is a section on imported wines.

Wine is the wave of the present. In the past few years its sales have burgeoned, and in 1980 wine outsold spirits. A glass of white wine is threatening to replace the Martini as America's Number One drink, and more and more people have discovered the pleasures of drinking wine with their meals. Wine is appearing on menus in hospitals and nursing homes because it is good for the patients. A hospital in Portland, Oregon, even has its own sommelier. A researcher-nutritionist calls wine in moderation "one of the great panaceas of mankind."[1]

Yet 10 years ago few Americans drank wine. We are still in a stage of discovery and enthusiasm and—let's face it—a certain amount of confusion. What wine goes with what food? How can you tell from the label what's inside the bottle? Which is a better buy, imported wine or U.S. wine, expensive wine or cheap wine, old wine or new, red wine or white? Why would anyone pay thousands of dollars for a bottle of wine made 150 years ago?

At a 1980 auction a restaurateur from Memphis paid $31,000 for a bottle of Château Lafite, vintage 1822. After you have read this chapter you still may not know why he did it, but at least you will begin to understand what it all means.

Wine is as old as history. It figures in the Bible, old Babylonian law codes, Egyptian wall paintings, the literature of Greece and Rome. The Romans spread winemaking to lands they conquered; some of today's most famous French vineyards were established by well-to-do Romans. In medieval Europe, monks developed grape growing and winemaking to a fine art.

American colonists brought with them a taste for wine, but American grapes did not make good wine. The settlers imported a few wines, made homemade wines from berries, and drank hard cider, beers, and spirits. So it went with wine in this country. People drank mostly rum, then whisky and beer, until Prohibition put a lid on everything.

Wine did not really catch the public fancy until a few years ago, when American wines began to make news for their excellence and people began looking for lighter and lighter drinks. Today no bar can afford to be without wine, while all kinds of restaurants are finding it very profitable to offer wine with meals.

Types of wine . . .

The wines we serve in restaurants and bars are the fermented juices of the grape. The ripe grapes are crushed for their juice. The yeast on the grape skins converts the sugar in the juice into alcohol and carbon dioxide. The

[1] Janet B. McDonald, "Not by Alcohol Alone" in *Nutrition Today*, January/February 1979.

CO_2 escapes into the air; the remaining solution—juice plus alcohol—is wine.

Table wines. Three types of wine account for nearly all the wine sales in restaurants and bars—***red wine***, ***white wine***, and ***rosé***. Collectively they are classified as ***table wines***; they are also sometimes referred to as ***dinner wines***. They typically have an alcoholic content of 10 to 13% by volume; the percentage must be stated on the label. They are sold by the bottle, carafe, or glass.

Red wines are distinguished first by their color. Different wines are different shades of red, ranging from orange-red to ruby red to deep purple, depending on the type of grapes they came from. Second, red wines as a group have characteristics of taste that are somewhat different from those of whites and rosés. They tend to be heartier, tarter, more pronounced in flavor, fuller in body, and drier. Red wines are nearly always dry.

In wine the word ***dry*** means lacking in sweetness. There are degrees of sweetness, ranging along a scale from bone-dry (totally lacking) to very sweet. Nearly all red wines are found near the dry end of the scale.

This dryness is one of the qualities that makes a red wine a suitable accompaniment to a hearty meat course, such as steak or lasagna or game. Its pronounced flavor and full body also go well with such foods. However, today there are no rules about which wine to drink with which food: the customer's taste is the only criterion. Many customers do not care for red wine at all, particularly if they are beginners in the wine game. Wine connoisseurs, on the other hand, are likely to be especially interested in fine red wines.

Red wines are served at a cool room temperature of 65 to 70°F—the temperature of water from the cold tap.

White wines range in color from pale straw to bright yellow to gold. They are generally more delicate and less pronounced in flavor than the reds. This is why they have traditionally been chosen to accompany fish and other foods of delicate flavor and texture.

White wines range from very dry to very sweet. It is the drier whites that complement fish dishes; they also go well with other entrees and with appetizers and, in fact, with any foods. The sweet white wines are generally served only with desserts, or as dessert.

Today white wine is popular not only as a mealtime accompaniment but as a drink in its own right in place of a mixed drink. Usually a medium-dry white is a good choice for a house wine to be served by the glass. White wines are always served chilled, whether by the glass or by the bottle.

The rosés are various shades of pink, from very pale to nearly red. In appearance you might think of them as pale red wines, but in character and taste they are more like whites. Wine buffs tend to look down on rosé almost as a form of cheating, since it is neither red nor white. But for a young

American tasting wine for the first time, a glass of rosé can be a good beginning. It is light and fresh, with a fruity flavor, and many rosés have a touch of sweetness that provides a link between soft drinks and the world of wine.

Rosés are served chilled, like white wines. They make good wines-by-the-glass in informal restaurants, especially for lunch and to accompany between-meals refreshment for fashion shows, club meetings, and the like.

Sparkling wines. Another category of wine commonly offered in restaurants and bars is ***sparkling wine***. This is likely to be champagne, though there are other similar wines. The outstanding characteristic of a sparkling wine is that it bubbles. (A wine that does not bubble is called a ***still wine***.)

A Benedictine monk named Dom Perignon first discovered champagne some 300 years ago and is said to have cried out, "I am drinking stars!" He had happened on a bottle of white wine in which a second fermentation had taken place, trapping carbon dioxide bubbles in the wine. He and others set out to duplicate this remarkable beverage, and so began the French Champagne industry.

Sparkling wines may be white, red, or rosé. They generally contain 10 to 13% alcohol by volume. The best-known red is sparkling burgundy.

The sparkling white wine we know as champagne is called by other names in other countries, such as sekt in Germany and spumante in Italy. By agreement between France and other countries of Europe, the name Champagne has been reserved for wines produced in the Champagne district of France. In the United States we use the name champagne for any sparkling white wine made by the French Champagne method, but the label of the American product must clearly indicate its geographic origin.

Champagne is the classical wine of celebration. It and the other sparkling wines should always be served well chilled. Champagne blends well with almost any food and is equally good by itself. Usually it is sold by the bottle, but it is often served by the glass with Sunday brunch as a promotional feature. It is also an ingredient of certain cocktails and other mixed drinks.

Fortified wines. Still another category of wines consists of those that have alcohol added in the form of brandy or other grape spirits. Among these are such old friends as sherry and the vermouths. Adding the extra alcohol is known as ***fortifying*** the wine. Fortified wines are usually sold by the drink, not by the bottle. Some of them are used in mixed drinks.

The government does not allow the word fortified to be used on the label. The percentage of alcohol is the clue; a wine above 15% has been fortified; the allowable limit is 24%. Generally the alcohol content is around 18% or 19%.

The federal Standards of Identity divide this group of wines into two classifications—***aperitif wines*** and ***dessert wines***. According to this classification, aperitif wines are not only fortified with spirits but are flavored with aromatic herbs and spices, or ***aromatized***. The best-known of these aromatized wines are the vermouths.

You may think of vermouths as bar liquors rather than wines. French vermouth (dry) is put to good use in Martinis, while Italian (sweet) goes into Manhattans. Vermouths and other aromatized wines such as Dubonnet and Lillet from France are also served as wines by the glass—straight up (well chilled), or on the rocks, or mixed with soda and a lemon twist, or half-and-half (half dry vermouth, half sweet).

In everyday usage the term aperitif is also applied to several other wines. The word is a Latin derivative that means to open—to open the meal with a wine that stimulates the appetite. In this broader sense the term includes white wine and dry sherry when served as openers to a meal.

Most dessert wines, the other group of fortified wines, are sweet, rich, and heavy, appropriate to the end of a meal. They include port, sherry (all kinds), madeira, marsala, angelica, muscatel. Some of these are more familiar in the kitchen than in the glass—in veal marsala, for example, or madeira sauce. Sweet dessert wines are usually served at the end of the meal in the manner of a liqueur. Such wines may be served either chilled or at a cool room temperature.

The term dessert wine is also applied in daily usage to other wines that are not fortified but are very sweet. Among them are several European white wines famous for their rich, sweet taste.

How wines are made . . .

All the wines we have been discussing begin in the same way, with the grapes, the soil, the weather, and the winemaker turning the same basic product into an infinite variety of forms. Figure 6.1 shows the flow of events.

Grapes grow best in rather poor soil on well-drained land (often hillsides) in a temperate climate with enough sun and warmth to develop their sugar, and a little rain at the right times. As they mature, the grapes increase in sugar content and decrease in acidity. In due course—about 100 days of sunshine from blossom to harvest—the grapes ripen to the precise balance of sugar and acid that makes the best possible wine. The moment of harvest is determined by frequent testing for sugar content and an eye to the weather (a heavy rain on ripe grapes can be a disaster).

The winemaking process. As soon as possible after picking—within 12 hours—the whole ripe grapes, minus their stems, are put into tanks for

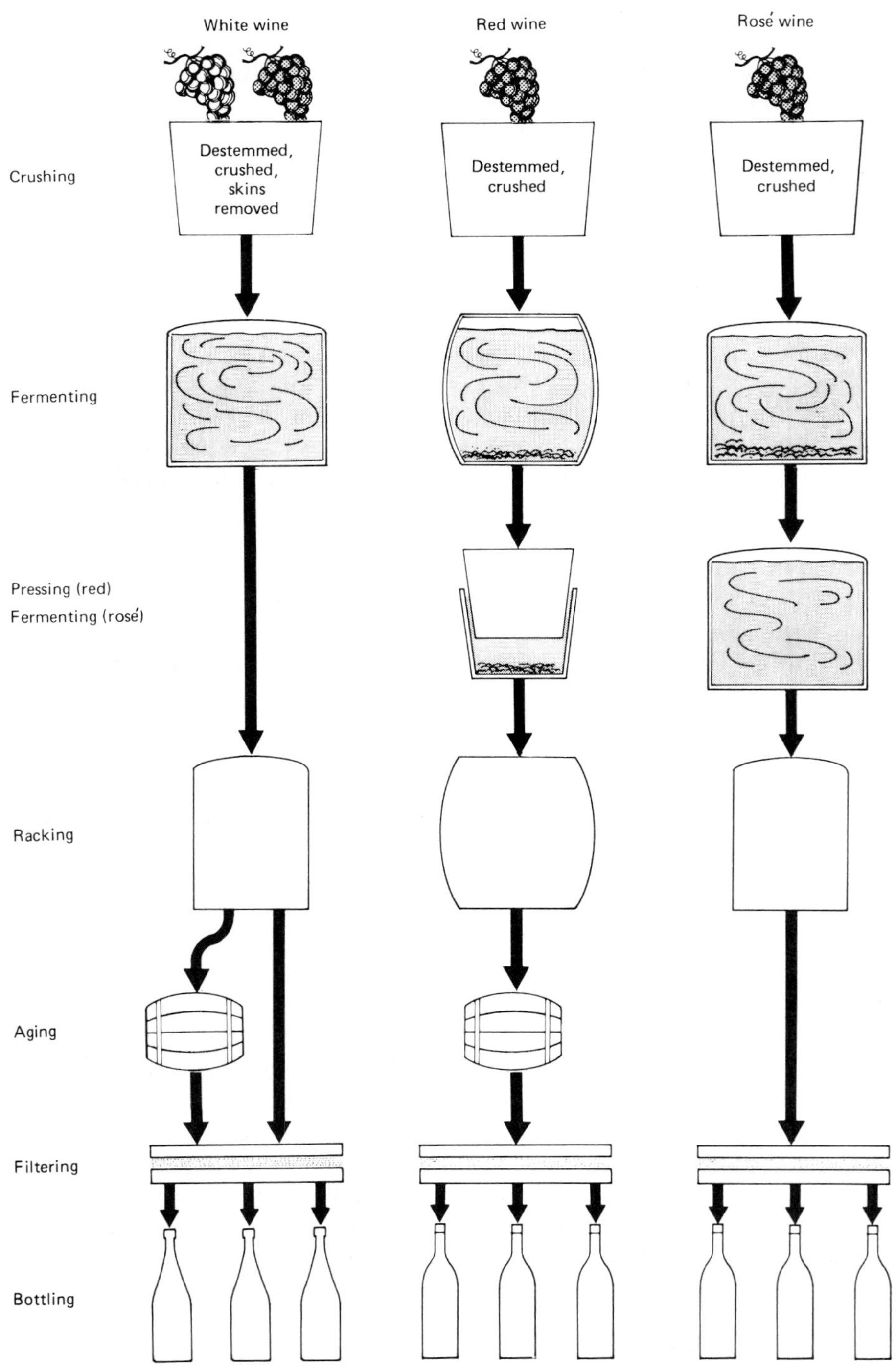

figure 6.1 Making wines.

crushing. As the grapes are crushed, the yeasts from their skins mingle with the juices. Today most winemakers introduce laboratory-produced yeast cultures rather than using the wild yeasts. The latter are killed either by the heat of fermentation or by the laboratory yeast.

The juice, now called ***must***, is channeled into fermentation tanks. If it is for a white wine, the juice alone goes into the tank. For a red wine the skins go along too, since it is the red or black or purple skins that yield the color as well as much of the wine's character. A rosé wine is made from dark-skinned grapes by leaving the skins in the fermenting must briefly—12 to 24 hours according to the color desired. The must is then pumped into another tank, leaving the skins behind.

In the fermentation tank, the yeasts feed upon the grape sugar and break it down: one molecule of sugar yields two molecules of alcohol and two molecules of carbon dioxide. Fermentation continues for one to two weeks or even longer, until the sugar is consumed or until the alcohol content becomes high enough to kill the yeasts—around 14%. The process stops automatically unless the winemaker intervenes. If the sugar is used up, the wine is dry. If sugar remains, there is sweetness in the wine.

When fermentation stops, the new wine is placed in large casks or vats. The skins in red wines are pressed out in the transfer. Each wine is stored until residues settle out and the wine stabilizes. Periodically it is drawn off the residues (the ***lees***) and placed in a fresh cask to settle further. This is known as ***racking***. When the wine "falls bright"—becomes clear—it is moved to other vats or casks for maturing.

White wines and rosés are often matured in vats lined with plastic or glass or in stainless-steel containers. They are usually ready for bottling in a few months. Most red wines and certain whites are aged in wood casks (usually oak) for six months to two years or more. Here they undergo changes that mellow and smooth them and develop their special character. Some elements of flavor are absorbed from the wood. Some changes come from interaction with the air that seeps through the porous wood. Other mysterious chemistry occurs within the wine itself, contributing complexities of flavor especially to wines that age slowly. Some of what happens is a mystery.

The moment when the wine is ready for bottling is determined by taste. The wine must have lost enough acidity, developed enough balance, to be moved from cask to bottle. Before bottling, the fine particles still remaining in the wine are filtered out in one way or another, or spun out by centrifuge. The wine is then bottled and corked. The cork, being porous, permits tiny amounts of air to reach the wine, permitting further changes in it. A ***capsule***, or cap of foil or plastic, is added over the cork to protect it. Many wines not intended for aging and most fortified wines have screw tops instead of corks.

Some wines are ready to drink as soon as they are bottled; others must undergo further aging in the bottle before they are drinkable. This bottle aging is called ***finish*** and is best done at the winery. The bottles are stored on their sides in a cool dark place until the wine is ready to drink. In this bottle position the wine keeps the cork moist so that it does not dry out, shrink, and become loose, which can ruin the wine. Sometimes the wine is sold before it is ready, and the buyer ***lays down*** the wine—stores it in the same way until it is ready to drink. A vintage date identifying the year the grapes were harvested is essential to wines that mature slowly.

Most wines are blends—blends of different grape varieties, or grapes from different vineyards, or wines of different vintages or degrees of maturity. The blending may be done at different points of the winemaking (this is why you don't see it in Figure 6.1). Sometimes the grapes themselves are blended before crushing. Sometimes the new wines are blended during racking. Sometimes wines are blended after they mature.

Blending serves several purposes. It may be done to produce the finest wine possible, or to maintain product consistency from one year to the next, or to tailor a wine to a special market, or to make more wine for less money, or to make the best of a bad year.

How sparkling wines are made. Sparkling wines and fortified wines are made by adding something to the process we have described. To make a sparkling wine, yeast and sugar are added to still wine to induce a second fermentation. This is done in closed containers, so that the carbon dioxide produced cannot escape and thus becomes part of the wine. In the French Champagne method the closed containers are the heavy glass bottles that eventually reach the customer. It is a long and complicated process involving handling each bottle many times, removing sediment bottle by bottle, and other refinements. All this makes the champagne very expensive and very very good.

Two other methods have been developed to shortcut the process. In the Charmat or bulk process, refermentation is carried out in large closed containers and the wine is then bottled under pressure. Such wines can be very modestly priced, but they do not begin to approach the quality of those made by the Champagne method—for one thing they lose their bubbles quickly. Their labels must clearly state the process.

The other alternate method referments the still wine in bottles but then transfers it under pressure to other bottles, filtering it in the transfer. This produces a moderately priced product of a quality that may satisfy popular tastes but not wine buffs. Sparkling wines made in this way are labeled "fermented in the bottle." Wines made by the French process may say "fermented in this bottle" or "Champagne process."

Figure 6.2 shows a French Champagne label (*a*) and an American champagne made by the transfer process (*b*). The latter clearly states its geographic origin, as required by the federal government to distinguish it clearly from French Champagne.

(a)

(b)

figure 6.2 Two champagnes. (*a*) Champagne from the Champagne district of France. (*b*) An American champagne made by the transfer process. Notice "*the* bottle" and the geographic reference.

How fortified wines are made. To make a fortified wine, extra alcohol is added at some point in the winemaking process. It can be added to stop fermentation before the grape sugar is used up, thus producing a sweet wine. In a dry wine it might be added during aging. To make aromatized wines, aromatic flavorers are added by soaking herbs, barks, flowers, seeds, spices in a fortified wine or by adding them in the form of extracts.

What's in a name? . . .

The bewildering variety of table wines on the market is enough to inhibit any novice, whether it be a restaurateur introducing wines for the first time or a customer struggling with a long wine list. The picture may become simpler when we examine why wines are named as they are.

Every label must carry a name to identify the product inside the bottle. In this country wines are named in three ways: (1) by the predominant variety of grape used (varietal), (2) by broad general type (generic), or (3) by brand name. Imported wines may also be named by these methods, but a fourth method is more common—place of origin.

Varietal wines. A ***varietal*** wine is one in which a single grape variety predominates. The name of the grape is the name of the wine, and that grape will give the wine its predominant flavor and aroma. Well-known examples are Cabernet Sauvignon, Chardonnay, Chenin Blanc, Zinfandel. In the United States at least 51% of a varietal wine must come from the grape named, and after January 1, 1983, this figure goes up to 75%. In France the figure is 100%; in Germany it is 85%.

A list of grapes commonly used in varietal wines is included in Table 6.1. Figure 6.3 (*a* and *b*) shows labels of two varietal wines.

Varietals are very popular in this country. They are well worth exploring for your wine list. The names, once learned, are quickly recognized and the better-known varietals almost sell themselves. Varietals range in price from moderate to high, depending to some extent on the wine quality. Taste them before buying, because they can vary greatly from one producer to another and one vintage to another. (This, by the way, is true of all wines.) The name and fame of the grape alone do not guarantee the quality of the wine.

Generic wines. A ***generic*** wine is an American wine of a broad general style or type. The best-known generics are burgundy, claret, and chianti among the reds, and chablis, rhine wine, and sauterne among the whites. Their names are borrowed from European wines that come from well-known wine districts, but their resemblance to these European wines is slight to nonexistent. Federal law requires all American generics (it calls them ***semi-generics***) to include a place of origin on the label (such as California, Washington State, Napa Valley, American). This distinguishes them clearly from

table 6.1 **AMERICAN VARIETAL WINES**

white	*red*
Blanc-Fumé	Baco Noir
Chardonnay	Barbera
Chenin Blanc	Cabernet Sauvignon
Delaware	Carignane
French Colombard	Concord
Gewürztraminer	Gamay
Johannisberg Riesling	Gamay Beaujolais
Pinot Blanc	Grenache
Riesling	Grignolino
Sauvignon Blanc	Merlot
Sémillon	Petite Syrah
Sylvaner	Pinot Noir
White Riesling	Zinfandel

the European wines whose names they have borrowed. Figure 6.3 (*c* and *d*) shows labels of two generic wines.

The best of the generics make good house and cocktail wines. These are pleasant and uncomplicated wines that can go with almost any food. There is little about a generic name to indicate the character or quality of the wine. In fact, there are instances of a winery taking its chablis and its sauterne from the same vat. If you are exploring generics to serve at your bar or restaurant, the only way to determine character and quality is by taste.

Generics often come in large-size bottles—1½ to 3 liters or even 4 liters. These are known as jug wines. They are a good buy if you sell a lot of wine by the glass, but they do not keep well once opened (no wine does). You can't count on using a red wine more than 24 hours, nor a white wine more than 2 days.

Some wineries have begun to use the names Red Table Wine and White Table Wine instead of the old generic names. These are inexpensive blends like the generics and can be used like them as house wines if they pass your taste test. If all wineries would give up the generic names it would clear up a great deal of confusion.

Brand-name wines. A ***brand-name*** wine may be anything from an inexpensive blend to a very fine wine with a prestigious pedigree. A brand name, also called a ***proprietary*** name or in France a ***monopole***, is one belonging exclusively to a vineyard or a shipper who produces and/or bottles the wine and takes responsibility for its quality.

A brand name distinguishes a wine from others of the same class or type. It is a means of building an identity in the mind of the customer, who is

Region

Name of wine (varietal = 51% or more from this grape)

Winery

Grower/producer/bottler

Alcoholic content

San Luis Obispo

CABERNET SAUVIGNON

Estrella River Winery

Grown, Produced and Bottled by Estrella River Winery
Paso Robles, California. Alc. 12 % by Vol.

(*a*)

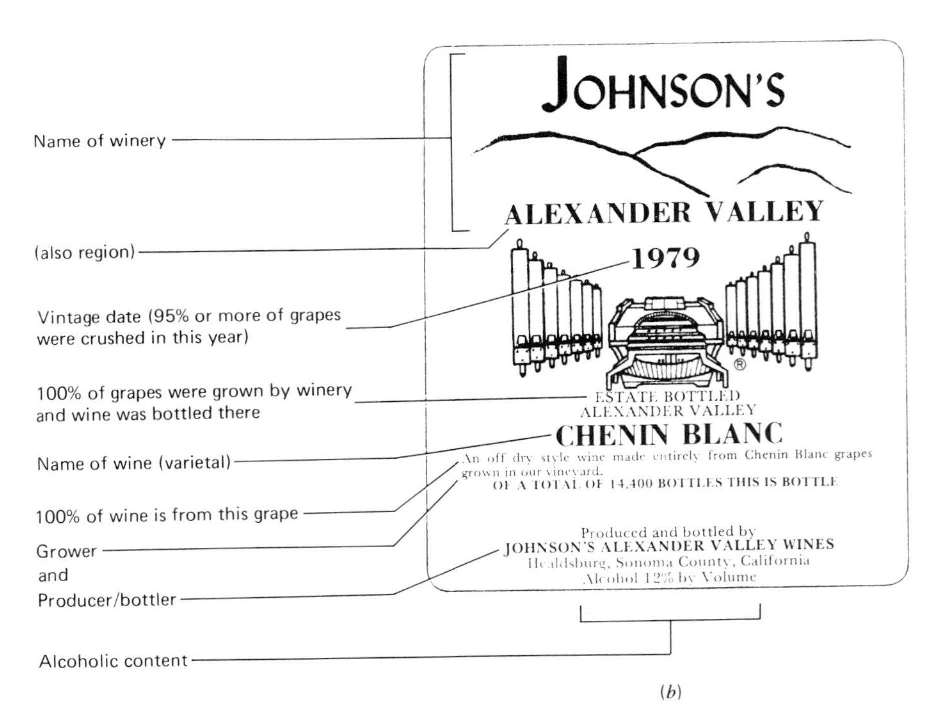

(*b*)

figure 6.3 Varietal wines [(*a*) and (*b*)] and generic wines [(*c*) and (*d*)].

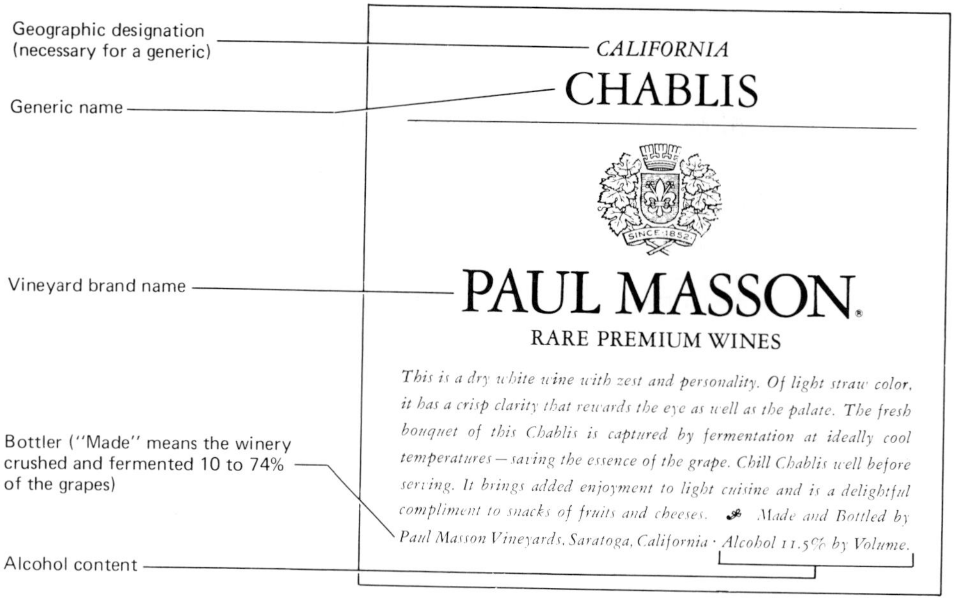

Geographic designation (necessary for a generic)

Generic name

Vineyard brand name

Bottler ("Made" means the winery crushed and fermented 10 to 74% of the grapes)

Alcohol content

CALIFORNIA

CHABLIS

SINCE 1852

PAUL MASSON®

RARE PREMIUM WINES

This is a dry white wine with zest and personality. Of light straw color, it has a crisp clarity that rewards the eye as well as the palate. The fresh bouquet of this Chablis is captured by fermentation at ideally cool temperatures—saving the essence of the grape. Chill Chablis well before serving. It brings added enjoyment to light cuisine and is a delightful compliment to snacks of fruits and cheeses. Made and Bottled by Paul Masson Vineyards, Saratoga, California · Alcohol 11.5% by Volume.

(*c*)

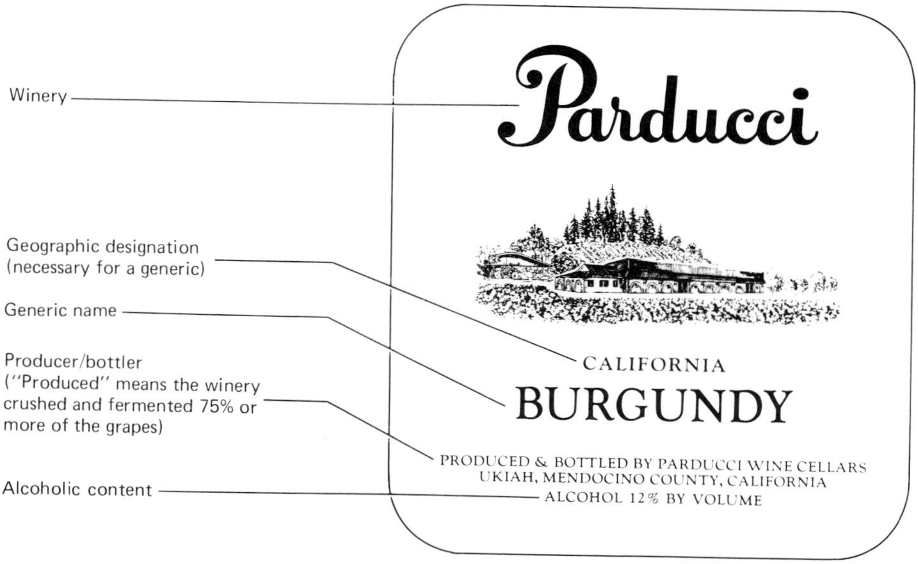

Winery

Geographic designation (necessary for a generic)

Generic name

Producer/bottler ("Produced" means the winery crushed and fermented 75% or more of the grapes)

Alcoholic content

Parducci

CALIFORNIA

BURGUNDY

PRODUCED & BOTTLED BY PARDUCCI WINE CELLARS
UKIAH, MENDOCINO COUNTY, CALIFORNIA
ALCOHOL 12% BY VOLUME

(*d*)

(*a*)

(*b*)

figure 6.4 Brand-name wines. Each label also states the class or type, a government requirement. (*a*) This wine emphasizes a brand name although it is a varietal wine. (*b*) This brand name is identified with rosé, though a red wine and a white also carry the name.

used to choosing liquors and beers by brand. Brand naming coupled with brand advertising may be a strong trend of the future. It is a way to reach the customer confused by the profusion of wines to choose from, who would rather pick one and stay with it, like a favorite beer.

A brand name alone does not tell you anything about the wine. The reputation of the producer and the taste of the wine are better keys to choice. Figure 6.4 shows examples of brand-name wines.

Wines identified by place of origin. Many imported wines use their place of origin as the name on their label. The place of origin is usually a rigidly delimited and controlled area that produces superior wines of a certain character deriving from its special soil, climate, grapes, and production methods. Wines from such an area must meet stringent government regulations and standards in order to use the name. The defined area may be large (a district, a region) or small (a commune, a parish, a village, a vineyard). Generally the smaller the subdivision the more rigorous the standards and the more famous the wine.

Along with the area name on the label is a phrase meaning "controlled name of origin"—*Appellation Contrôlée* in France, *Denominazione de Origine Controllata* (DOC for short) in Italy. Other countries have similar requirements for using the name of a delimited area. We explore the various

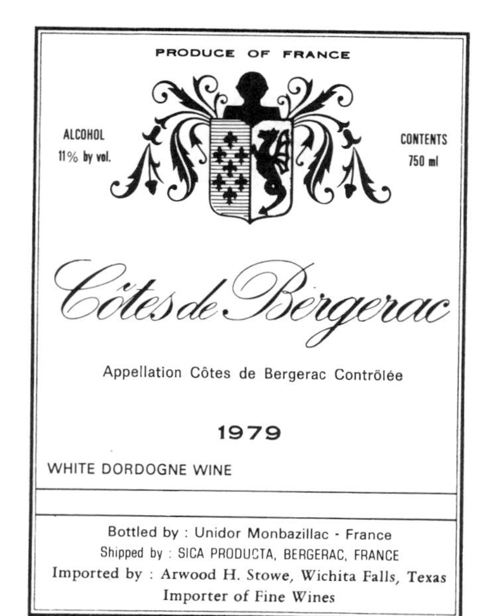

(a)

(b)

figure 6.5 Wines known by place of origin. Type of wine is also stated on the label. (*a*) Côtes de Bergerac is a delimited wine district in France. (*b*) Piesporter means wine from the town of Piesport in Germany. Goldtröpfchen is a very famous vineyard. Both are controlled areas.

systems more fully later in this chapter. Meanwhile we give you typical labels in Figure 6.5.

Generally a wine from a controlled area has a certain claim to quality, and the best wine-growing areas have the best claim. But the name is not a guarantee, and all wines from the same area are not the same wine. Picking the right wines from the right places is a job for an expert.

In the United States some wines use a place of origin along with another name for the wine, such as Napa Valley Chablis or Sonoma County Pinot Noir, as a way of linking their product with a famous wine region. The product does indeed come from the region named, but in this country wine districts do not yet have the same strict regulation of product and process that European delimited areas do, and hence there is not necessarily the same correlation with character. A number of wine regions are currently developing such regulations and standards.

Maturity, life span, and vintage . . .

You need to know about certain characteristics of wine as a product in order to choose wines sensibly, care for them properly, and use them profitably.

A table wine is not a standard product like a spirit. You can buy a certain brand of scotch and it will be the same as the last bottle you bought. Unopened, it will keep indefinitely; opened, it stays the same day after day.

The same thing cannot be said for table wines or sparkling wines. The same wine may be different from one year to the next depending on the weather while the grapes were growing and many other factors. Wines also change within the bottle. They do not last indefinitely, and each has its own life span. When you open a bottle of table wine you may as well finish it off because it won't be good tomorrow.

Wines vary in their rate of maturity. Most white wines and rosés mature early and are put on the market ready to drink within a few months after harvest. Most have a life span of two to five years depending on the wine. You can tell how old they are if there is a vintage date on the bottle. This is the date of harvest, the year the grapes were pressed and the wine began.

All this means that you can safely buy most young white wines and rosés and sell them right away. The older the vintage date, the more skeptical you should be. You should not overbuy on whites because of their short life span.

Fine white wines are a different matter. They are not ready to drink as early but they last longer. The better the wine to begin with, the longer the life span.

Red wines mature more slowly than whites, and the rate of maturity can vary greatly from one wine to another. Be wary of buying too young a red; on the other hand you may be able to keep it longer.

Many American wines do not have vintage dates, though the trend is toward adding them to the bottle. In addition to telling how old the wine is, for the better wines it can be a clue to quality. There are great and poor and average years for wines, depending on the weather. A great vintage year produces a wine of finer taste and longer life span than an ordinary wine.

But a great year in one region may be a poor year in another: 1975 produced excellent wines in the Médoc district of France and poor wines in the Beaujolais district not too far away. So a vintage chart is useless unless it applies specifically to the region of the wine you are thinking of buying.

Some of the finest wines may take 8 or 10 years or even longer to reach their peak and may hold their own for decades; these are the rare ones. (Let us hope the $31,000 bottle sold at auction was one of these.) But even such wines will not last forever.

Maturity rate, life span, improving, and declining all happen because wine is a living substance. Its chemistry changes particularly when it is in contact with air. This is why it deteriorates once opened, why you must take care to store it on its side in a cool storeroom at ***cellar temperature*** —50 to 65°F—and why you should deal with a wholesaler who takes equally good care, both in the warehouse and on the delivery truck.

Serving wines . . .

The manner in which a wine is served is important in presenting the wine at its physical best. It is also an important psychological ingredient of its flavor. Whether it is a jug wine at the bar or the finest wine in the finest restaurant, it deserves the proper service.

Wine at the bar. Both red and white wines are served at the bar by the glass. These are usually inexpensive but good-quality California jug wines. The wine is usually poured behind the bar rather than on the glass rail, since the jugs are harder to handle than liquor bottles. Bulk wine is also available in small barrels, dispensed from a tap.

Wine by the glass is usually served in an 8- or 9-ounce stemmed glass at 5 to 6 ounces a serving. This makes a fuller glass than is usually poured at table, but it is right for the wine and the customer. Jug wines are not among those greats that require swirling around in the glass, and the customer may feel cheated if the glass looks less than full. However, service should not be casual. The glass should be sparkling. It should be held by the stem and set down before the customer in a manner that dignifies it as the graceful drink it is. Most important, the wine should be the right temperature—45 to 55° for white wines and rosés, 65 to 70° for reds.

If wine is ordered on the rocks, it is still best served in a wine glass. You can fill the glass one-third full with ice and pour wine over the ice to within about half an inch of the top.

Preparation for table service. Proper table setting for wine service is important, both for efficiency of service and for etiquette. The wine glass belongs to the right of the water glass (Figure 6.6). If more than one wine is to be served—as at a formal multicourse banquet—the wine glasses are arranged in order of service with the first wine at the right, and each glass is removed at the end of the course the wine accompanied. If wine glasses on the table are not appropriate to your image and your clientele, your servers should place them in the correct position when the wine is brought to the table.

A wine glass should always be handled by its stem. In formal service the glasses are brought to the table on a tray held at waist level. A more informal way of carrying them is to hold them upside down by the base, with the stems between the fingers—you can carry as many as four this way. In either case the bowl is not touched at all.

The best wine glass is a plain 8- or 9-ounce stemmed glass with a rounded bowl tilted slightly inward at the top, to conserve the wine's aroma (Figure 6.7). If you are setting a banquet table with several wine glasses, the glasses for the different courses should be different from one another.

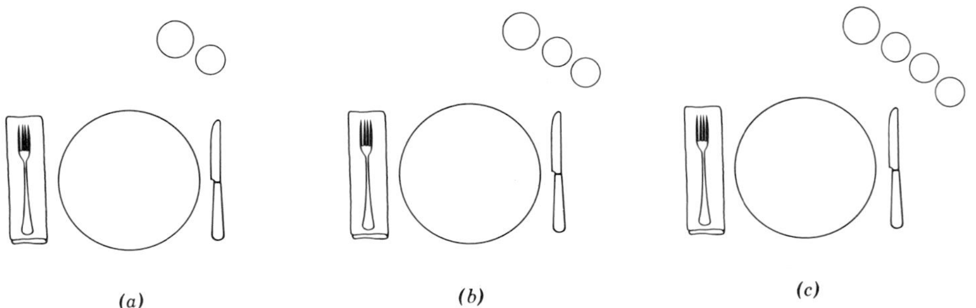

figure 6.6 Table set with wine glasses. (*a*) One wine, a single, all-purpose glass. (*b*) Two wines—the white-wine glass at the right, to be served first. (*c*) Three wines—an aperitif (white) at bottom right, a white above it for the fish course, a red at left for the entrée. If more than three wines are to be served, each additional glass is set in place at time of service. Empty glasses are removed.

Dessert wines are served in small after-dinner glasses, larger than a liqueur glass but much smaller than an ordinary wine glass. The serving size is 2 to 3 ounces. It should not quite fill the glass.

Wines by the glass for table service are served the same way as wine at the bar—that is, a 6-ounce serving that more or less fills the glass. Wines by the glass or carafe should be prechilled to the proper temperatures. White wines by the bottle should be kept in a cool place and chilled as ordered. This takes 10 to 20 minutes in a wine cooler (Figure 6.8). The most efficient procedure is to put some crushed ice in the bottom of the cooler, put the bottle in, surround it with crushed ice, and add a little water and some table salt. If you do not have crushed ice, use a layer of cubes in the bottom, put the bottle in, fill the bucket two-thirds full with cubes, and add cold water. Bring the wine to the table in the cooler.

If you offer a white wine by the bottle that sells at a steady, predictable rate, you can keep one or two days' supply in the refrigerator. No wine should be kept chilled for more than a week. And no wine should be put into a freezer; overquick chilling may cause it to precipitate solids that are in solution.

Red wines should never be served over 70°F. A fine expensive red wine should be served at 65 to 68°F; younger wines can be served a few degrees colder. Sometimes a bit of chilling takes the harsh edge off a thin or rough wine. Sparkling wines should be served well chilled. Sweet white wines should also be very cold, except for certain German wines, which should not go below 55°F.

A bottle of red wine should be opened at the table and allowed to "breathe" a few minutes before pouring. Red wines change slightly when exposed to air, and these delicate chemical changes rid the wine of any aroma of storage or mustiness it may have developed.

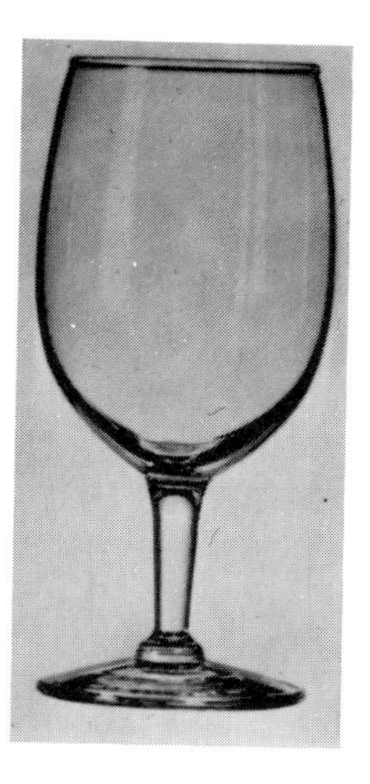

figure 6.7 A standard, all-purpose wine glass. (Photo courtesy Libbey Glass, Toledo)

figure 6.8 Wine cooler at left is an inexpensive model suitable for family restaurants. Cooler at right is silver. The same cooler is shown on a stand in Figure 6.10. Wine basket is used to carry old wines for decanting. (Photo by Pat Kovach Roberts)

Serving wine at the table. The ceremony of wine service is one part efficiency and one part showmanship. There is a ritual that seasoned wine drinkers expect, each step of which has a practical reason. These steps are illustrated in Figure 6.9. (Our wine expert is left-handed so that you can read the pictures as though watching yourself in a mirror.)

With the glasses in place on the table, the server, standing at the right of the host, removes the bottle from the cooler and wipes it with a clean nap-

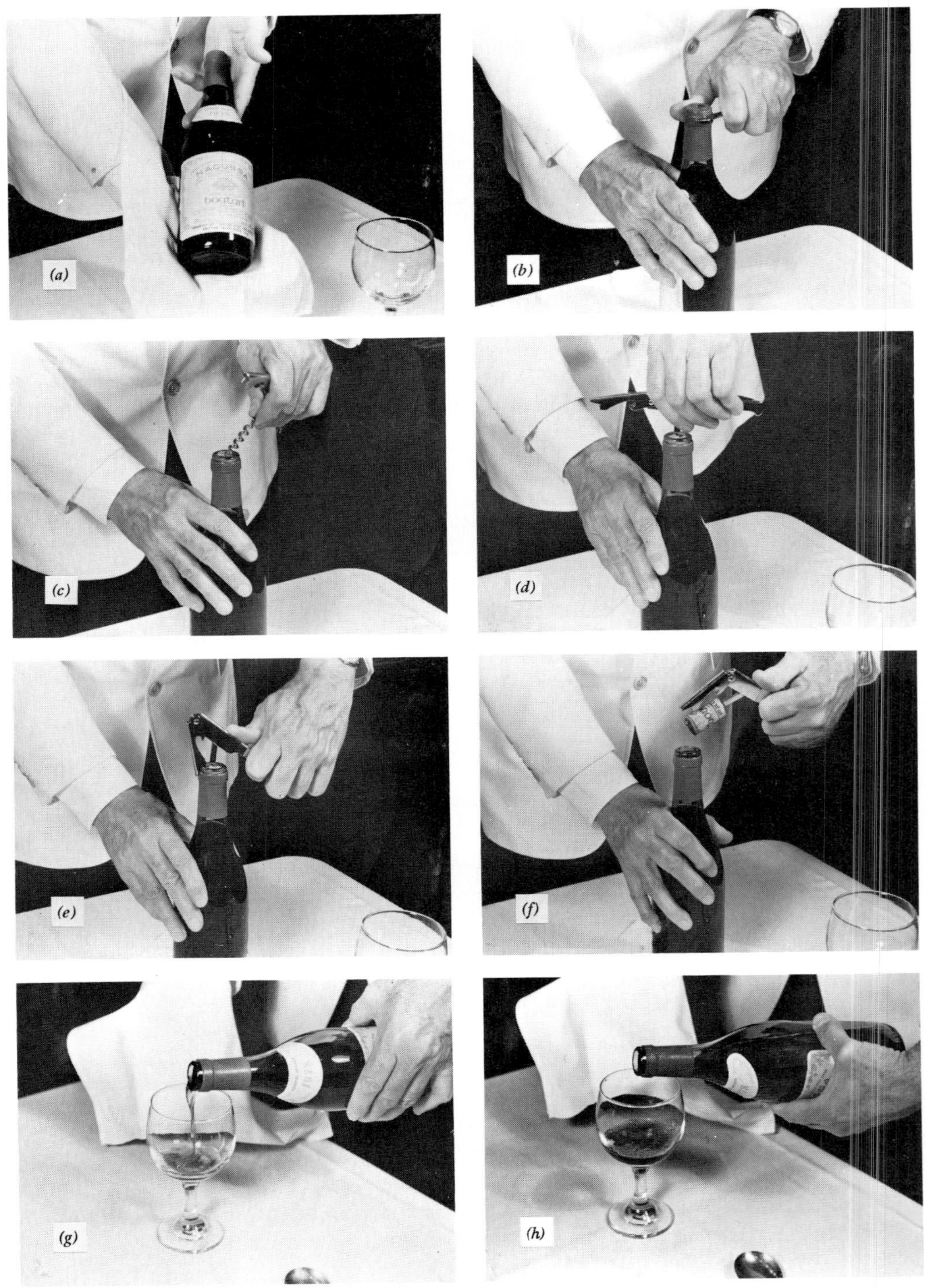

figure 6.9 Wine service. (Photos by Pat Kovach Roberts)

kin. Then, holding the body of the bottle from underneath, with the label toward the host, the server presents it for the host's approval (*a*). The server then opens the bottle.

The most practical bottle opener is the flat jackknife type known as a waiter's corkscrew, which fits easily into the pocket. With the knife the server cuts through the capsule along the ridge at the top of the neck (*b*) and removes the top of the capsule. If there is mold (there may be, but it has not hurt the wine), the server wipes it off the cork and the lip of the bottle with the napkin.

The next step is to insert the corkscrew with the point slightly off center (*c*), so that the worm as a whole is directly over the middle of the cork. Keeping the worm completely vertical, the server turns it clockwise until all of it has disappeared into the cork up to the shaft and the tip has appeared at the bottom (*d*). The server then moves the lever into position on the rim of the bottle (*e*) and, holding the bottle firmly, slowly raises the knife end (the knife has been closed, of course). This will bring the cork out of the bottle (*f*).

The server wipes the lip of the bottle inside and out with a clean corner of the napkin, removes the worm from the cork, and presents the cork to the host. The server then pours an ounce or so of wine into the host's glass for tasting and approval (*g*). If a bit of cork has fallen into the glass, it will appear at this point and can be easily discarded. A slight twist of the wrist as the pour is ended and the neck is raised will spread the last few drops on the lip of the bottle so that they don't drip (*h*).

When the host has approved the wine, the server pours wine for the other guests, going counterclockwise around the table and serving from the right (if possible), completing the host's serving last. Each glass should be poured not more than half full. The best way to hold the bottle is by the body from the top, not by the neck. When everyone is served, the wine should be returned to the cooler or, if no cooler is used, placed on the table near the host.

As the meal progresses the server should replenish the glasses as they are emptied. When the wine is gone, it is the custom to put the bottle upside down in the cooler—a "dead soldier"—to signal the host that it is empty.

Serving champagne. Champagne and other sparkling wines, because they have a special mushroom-shaped cork and the wine is under great pressure, are opened and served in a special manner. Champagne is always served in a wine cooler, well chilled (Figure 6.10). The warmer the wine, the more it will fizz and the more effervescence will be lost. It is important to handle sparkling-wine bottles gently so as not to agitate the wine, which will make it fizz.

Figure 6.11 shows you how to open the bottle. Over the champagne cork is a wire hood with a twist fastener, covered by foil. The first step is to re-

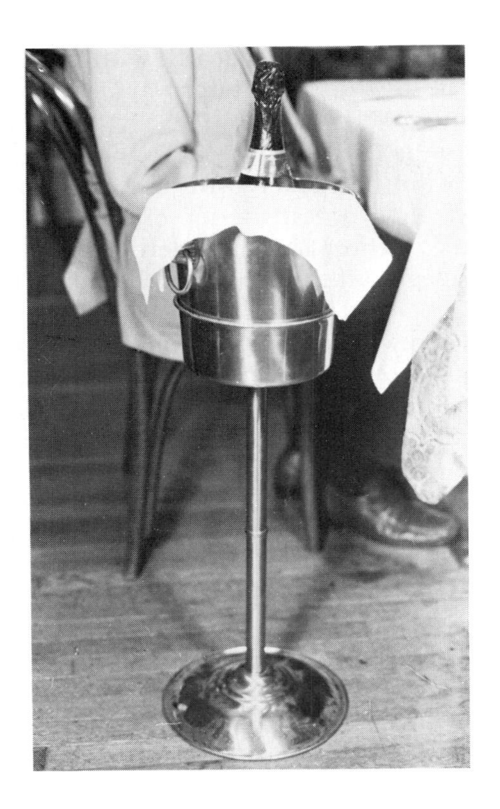

figure 6.10 Champagne awaiting service. (Photo by Pat Kovach Roberts)

figure 6.11 Champagne service. (*a*) Removes foil. (*b*) Untwists fastener, removes wire hood. (*c*) Twists cork, holding bottle, or vice versa. (*d*) Holds open bottle at angle. (*e*) Pours first frothy installment. (*f*) Completes pour to two-thirds full. (Photos by Pat Kovach Roberts)

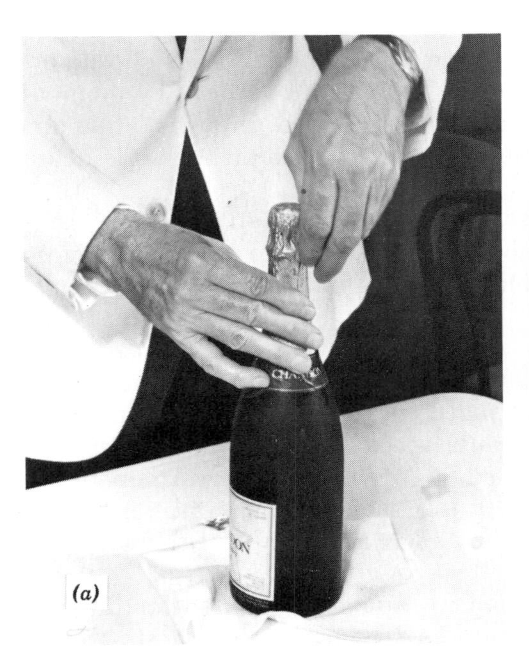

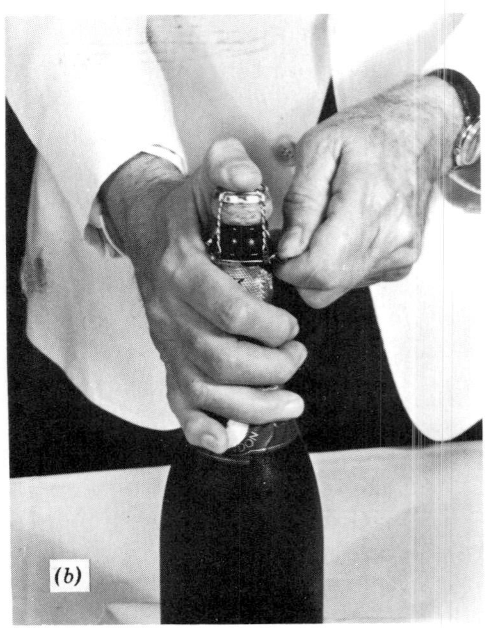

(c) (d) (e) (f)

move the foil (*a*). Then, with a thumb over the cork, the server untwists the fastener and removes the wire hood (*b*).

Now the server holds the bottle at a 45° angle, pointing it away from the guests. The other hand holds the cork. The choice here is to hold the bottle firmly and twist the cork (*c*) or to hold the cork firmly and twist the bottle

figure 6.12 Champagne glasses. (*a*) Saucer (popular though not recommended). (*b*) Tulip. (*c*) Flute.

in one direction. Either way works better than twisting them both, which agitates the wine. The pressure inside the bottle, plus the twisting, will ease the cork out.

The server should keep the bottle at an angle for at least five seconds before pouring. This equalizes the pressure, letting gas escape without taking the champagne along with it. No gush, no fizz, just delicate chilly little wisps (*d*).

But just in case . . . have an empty champagne glass in front of you. If you happen to have a wild bottle, pour some immediately and it will stop gushing.

The wine is poured into the guest's glass in two motions. The first one brings froth (*e*), sometimes reaching the rim of the glass. When this subsides, the second pour should slowly fill the glass about two-thirds full(*f*). When everyone is served, the bottle goes back into the cooler to conserve the effervescence.

A fine champagne should be served to a discerning clientele in a tulip or flute glass, to better conserve the effervescence and aroma. For many years the shallow saucer-shaped cocktail glass was traditional for champagne, and if it symbolizes champagne and celebration to your clientele you may want to use it, even though it dissipates the bubbles faster. Chances are good the glass will be emptied before the bubbles are gone. Figure 6.12 shows champagne glasses of different types.

Decanting old wines. Red wines that are five years old or more and rare old white wines may develop sediment in the bottle and need to be ***decanted***—poured into another container in such a way that the sediment remains in the bottle and the wine served is clear.

figure 6.13 Decanting wine. (Photo by Pat Kovach Roberts)

The bottle to be decanted must be handled very gently, keeping it on its side just as it was stored, so as not to disturb the sediment and mingle it with the wine. Sometimes the bottle is carried to the table in a special wine basket (shown with the coolers in Figure 6.8). If a basket is used, the wine is presented, opened, and decanted while still in its basket, keeping the bottle in an almost horizontal position.

To decant a wine, a wide-mouthed decanter or carafe is placed on the table, with a lighted candle positioned a little behind the shoulder of the bottle as illustrated in Figure 6.13. In this way, when the wine is poured from bottle to decanter it can be seen clearly as it passes through the neck. The server then pours the wine slowly and steadily in a single motion without stopping, until the candlelight shows sediment approaching the neck of the bottle. The remainder of the wine—a small amount—is not served, and the clear wine is served from the decanter.

Since only a small proportion of wines served will need decanting, only the wine steward or one or two experienced servers need this special training. It goes without saying, however, that your entire serving staff should be carefully trained and practiced in all the other routines of wine service.

Developing a wine list . . .

There are two aspects to developing a wine list—choosing the wines and presenting the choices to the customer.

As in everything else, the planning you do will be customer-oriented. The wines you offer and the way you present them on your list is nine-tenths merchandising and the starting point of all your wine sales efforts.

Choosing the wines. Many restaurants serve only house wines by the glass or carafe and make a very good profit in this uncomplicated and inexpensive way. Usually there are just two or three wines—a red, a white, and perhaps a rosé, California generics of good quality and general appeal. This simple wine selection, usually combined with beer and soft drinks, is very appropriate for certain types of restaurants and certain clienteles.

If you want to offer a choice of wines by the bottle, your first consideration is your clientele. This will decide the overall character of your list—whether you keep it short, simple, and inexpensive, or extensive, expensive, and loaded with prestigious choices, or somewhere in between.

The second consideration is your food menu. If you serve fine French cuisine you will choose well-known French wines. If you have a steak house you will concentrate on red wines, with a few whites for those who don't like reds and a rosé for people who want to compromise. If you specialize in seafood you will specialize in white wines, with a few reds and again a rosé.

The third guideline is price compatibility. Your wine choices should suit not only your food menu but also your menu prices. The average customer does not want to pay more for the wine than for the dinner check. In fact, sales are best when the wine costs only about one-fourth as much as the total check. You should also have a spread of prices, and a choice of wines in each price range.

Another point to consider is availability. Are you going to invest in a large stock that sits on cash in the cellar until it is sold, or can you get your selections regularly from your suppliers in small quantities? You don't want to print a list that will soon be full of wines you can no longer obtain.

If you are just beginning to serve wine, it is better to start with a short list of readily available wines, and to buy frequently a little at a time until you find out what sells. A long wine list does not necessarily generate more sales than a short one. It is easy to add to your list; it is not easy at all to get rid of a wine customers don't care for. A limited, moderately priced wine list usually assures favorable sales and inventory turnover.

Taste the wines you are considering for your list. Most suppliers will arrange comparative tastings at least of moderately priced wines. But beware of imposing your own taste on your customers. You might hold a wine tasting for a group of your regular customers to see which wines are well liked. Have your serving personnel taste the wines too. It will increase their interest in selling wines and help them describe wines to their customers.

If you want an extensive wine list, it is wise to find an expert to help you develop your list and advise you on purchasing. A wine cellar is a big and sometimes risky investment, and identifying and selecting good wines is an immensely complicated business.

Presenting the selections. The wine list is your silent salesman for wine, just as your menu card is for your food. There are several formats. The chalkboard (Figure 6.14*a*) is the simplest; it is typically used in wine bars, especially for wine by the glass. It has a nice air of continental informality and a hint of the individual proprietor giving daily personal attention to the wine menu. For the proprietor the chalkboard has the added advantage of not making long-range commitments in print.

The table tent has similar advantages. It is appropriate in informal restaurants serving only a few wines. It can also be used to promote specials or new offerings.

A more formal list—and a more permanent one—is a printed list given to one person at the table, usually the host, since wine by the bottle is typically ordered for the whole party. Such a list comes in two common formats—a printed card (or carte) like a menu (Figure 6.14*b*), or folded pages held inside a nice-looking cover with a cord. The latter format has several advantages: You can change the inside leaves easily as your offerings change, reusing the more expensive cover. This type of list is often easier for the diner to handle than a large single card. And depending on your state laws, your wine dealer may provide you with the permanent cover, and you pay only for the printing of the inner pages.

Expensive restaurants, especially those featuring fine wines, often use a multipage wine list inside an embossed cover, typically designed by a graphics specialist to coordinate with the establishment's overall ambience. Sometimes each page has only one listing with the wine label inserted in plastic (easily changed). The most extensive list we know of belongs to Bern's Steak House in Tampa, Florida—500 pages, 5000 wines.

figure 6.14a Blackboard list of wines by the glass at The Wine Press, Dallas. (Photo by Pat Kovach Roberts)

WINES

WHITE WINES

		Full	Half
CALIFORNIA			
2.	**Napa Valley White - Daniele** .... *Crisp, young, fine white wine blend*	7.50	
4.	**Gewurztraminer - Rutherford Hill** .... *Spicy, Alsatian-type wine with a perfume quality*	8.75	
5.	**White Riesling - Trefethen** .... *Crisp, dry flowery with lingering finish*	10.50	
6.	**Johannisberg Riesling - Joseph Phelps** .... *Rich fairly dry wine with a hint of apples*	10.75	
7.	**Chardonnay - Rutherford Hill** .... *Dry, youthful, rich white to go with rich foods*	14.50	7.50
9.	**Chardonnay - Chateau Chevalier** .... *Dry, austere, fine, limited wine from Napa Valley*	17.50	
BURGUNDY			
11.	**Blanc de Blanc - L'Epayrie** .... *Light, dry blend from regions outside Burgundy*	6.50	
12.	**Macon - Villages - Armand Roux** .... *Dry, all purpose white for seafood and poultry*	12.50	6.75
13.	**St. Veran - Chateau Beauregard** .... *A very dry white similar to Pouilly-Fuisse*	12.50	
BORDEAUX			
21.	**Chateau Rosechatel - Schroder & Schyler** .... *Quite dry, goes well with both fish and chicken*	7.25	
LOIRE			
24.	**Muscadet - Pierres Blanches** .... *Light, young, both for an aperitif and with food*	7.50	4.00
25.	**Vouvray - Chateau Moncontour (Demi-sec)** .... *Fuller, slightly sweeter wine, good with scallops*	9.25	
26.	**Pouilly - Fume - Gaudry** .... *Very dry, full bodied wine with lots of character*	15.25	
ITALY			
31.	**Frascati - Superiore - Pallavicini** .... *Light, dry, the favorite wine of the Romans*	7.25	
32.	**Verdicchio Classisco - Garofoli** .... *Fastest growing favorite of the Italian whites*	8.50	
33.	**Soave - Bolla** .... *Light, dry, delicate white from Verona*	7.50	4.00
34.	**Bianco Corvo - Salaparuta** .... *Most famous of the dry white wines of Sicily*	8.25	
GERMANY			
35.	**Bernkasteler Riesling - Madrigal** .... *Fruity, goes well both with and without food*	8.25	

ROSE WINE

		Full	Half
CALIFORNIA			
41.	**Zinfandel Rose - Pedroncelli**	6.75	3.50

figure 6.14b A printed wine list. (Courtesy John Sheldon and Charlie's, Brooklyn)

WINES

RED WINES

CALIFORNIA

		Full	Half
51.	**Napa Valley Red - Daniele** *Full, rich blend from several grape varieties*	6.00	
52.	**Gamay Beaujolais - Pedroncelli** *Light, fresh, with fruity character*	7.50	
53.	**Zinfandel - Pedroncelli** *Dry, clean, almost like a French chateau wine*	6.50	
54.	**Merlot - Rutherford Hill** *Rich, fine, from the style of Pomerol*	8.25	
55.	**Syrah - Joseph Phelps** *Big, dark, aggressive red from Napa Valley*	9.75	
56.	**Cabernet Sauvignon - Daniele** *Rich, youthful varietal from Lake County*	10.25	
57.	**Pinot Noir - Spring Mountain** *Full, big red made from several vintages*	13.00	
58.	**Cabernet Sauvignon - Clos du Val** *Classic Bordeaux style, rich Napa red*	18.50	

BORDEAUX

		Full	Half
61.	**Fonvillac - St. Emilion** *Soft, round, well-balanced young red wine*	12.50	6.50
62.	**Chateau Tourteran - Haut - Medoc** *Fine young red, very good with steaks and chops*	13.00	

BURGUNDY

		Full	Half
71.	**Beaujolais - Villages - Armand Roux** *Light, fresh, typical wine of this area*	8.75	
72.	**Cote de Beaune - Villages - Louis Jadot** *Young, fine French wine from a famous wine-house*	17.75	

RHONE

		Full	Half
81.	**Cotes du Rhone - Armand Roux** *Light, pleasant wine that is an all-purpose red*	8.50	4.50

ITALY

		Full	Half
85.	**Valpolicella - Bolla** *Light, clean, smooth red from Northern Italy*	7.50	
87.	**Chianti Classico Ducale - Ruffino** *A Tuscan delight, for many Italian dishes*	10.00	5.50
88.	**Barolo Riserua Special V - duca D'asti** *The king of Italian red wines*	14.00	

SPARKLING WINES

CALIFORNIA

		Full	Half
91.	**Brut - Korbel** *New York's favorite California sparkler*	12.50	
92.	**Napa Brut - Domaine Chandon** *Dry, French style from a French owned Napa winery*	17.00	
93.	**Cuvee de Pinot - Schramsberg** *The newest, French style with a blush of pink*	21.00	

CHAMPAGNE

		Full	Half
95.	**Brut Imperial - Moet & Chandon** *Fine dry Champagne from the world's leading producer*	26.00	14.00
96.	**Brut Vintage Gold - Veuve - Clicquot** *An exciting example from a great Champagne house*	34.00	18.00
	House Wines Available in White, Red or Rose	7.00	3.50

In arranging your wine list, group the wines according to type—reds, whites, rosés, sparkling wines, dessert wines. A short description of each wine is helpful to the novice—often a phrase will do. It should indicate whether the wine is sweet or dry, full-bodied or light, and perhaps what food it goes well with (Table 6.2). The list as a whole should show imagination and enthusiasm. It should make the whole subject of wine simple and inviting, not formidable and exclusive.

What your wine list says about your wines should be clear and honest and useful. It should include key information relevant to making a choice, such as the place of origin and the vintage year. If you are catering to connoisseurs, give all the essential information that will establish the wine's pedigree. Your wine expert should help you with this.

A number for each wine is helpful too. It will spare the customer embarrassment if the wine name is unfamiliar and hard to pronounce. The number should correspond to the bin number where the wine is kept. This system reduces errors on the part of serving personnel. Figure 6.14*b* is a nicely developed wine list, chosen, annotated, numbered, and priced with the customer in mind.

Before you send your list off to the printer, be sure everything is spelled correctly. Check everything carefully against the labels. Check the printer's proof carefully too. If a wine is misspelled, a knowledgeable customer may read it as an indication that you don't know anything about wines.

table 6.2 **WHICH WINE WITH WHICH FOOD?**

menu item	*wine suggestion*[a]
Appetizer	Champagne, dry white wine, dry sherry
Salad	No wine
Fish or seafood	Dry or medium-dry white
Beef	Hearty red
Lamb	Hearty red
Veal	Light red or full-bodied white
Ham or pork	Dry or medium-dry white or rosé
Turkey, duck, chicken	Full-bodied white or light red
Game (venison, pheasant, wild duck)	Hearty red
Lasagna, spaghetti, pizza	Hearty red
Cheeses, full-flavored	Hearty red; sweet white (with roquefort)
Cheeses, mild	Sherry, port, madeira, mild table wines of any type
Desserts, pastries, fruits, mousses	Semisweet sparkling wine, sweet white table wine

[a] The diner's choice takes precedence. If more than one wine is to be served, the general rules are white before red, light before hearty, and dry before sweet.

Train your servers to know the wines on your list. A chart like Table 6.2 with your menu items on the left and appropriate wines from your cellar on the right will help your servers to help your customers. Servers should know firsthand how each wine tastes.

Tasting wines . . .

When you taste wine, pour only a small amount. Hold the glass by the foot or the stem, never by the bowl, which would convey the heat of your hand to the wine. Swirl the wine around in the glass and breathe in the aroma before tasting. Notice whether the wine runs down the side of the glass quickly or in "legs"—this tells you about the wine's body.

Take a small sip and roll it all over the tongue. Breathe in air over it. You will find there are many different tastes and impressions that tell you much more about the wine than if you quaffed it like a beer. Notice the aftertaste too. You will begin to see that the vocabulary of the experts is in fact a technical language for identifying the many nuances they perceive in a wine.

Here are some of the words used to describe sensory qualities even a beginner can perceive.

- ***Degree of sweetness.*** If the grape sugar has been entirely consumed during fermentation and none has been added, a wine will be totally lacking in sweetness, or ***dry***. A dry wine or one with only a little sweetness is usually preferred with a meal, while a sweet wine is an appropriate finish to a meal. Many Americans brought up on carbonated soft drinks prefer a wine with some sweetness.
- ***Body.*** Body is the feel of the wine in the mouth; it comes from the amount of alcohol, sugar, and glycerine (one of the soluble substances formed during fermentation). A light-bodied wine is usually referred to as a light wine; it is low in one or more of the body components. A full-bodied wine is typically high in body components. It will cling to the sides of the glass if you swish it around, and run down slowly in legs. When you taste a full-bodied wine it will fill your mouth in a sensuous way.
- ***Aroma, bouquet, nose.*** One of wine's greatest pleasures, often overlooked by the beginning wine drinker, is the aroma or bouquet that rises as you swirl the half-filled glass. If it has the fresh-grape fragrance of a young wine, it is called an aroma; if it has the scent of a mature wine, changing as it develops in the glass, it is called the bouquet. Either one is referred to as the nose.
- ***Tartness.*** A sharp, acid taste, like green fruit, not to be confused with dryness. Some tartness may be desirable. However, a wine having the acid taste of vinegar has outlived its life span.

- ***Softness.*** Lacking in tartness; said of a mature wine. The term has also been adopted by makers of jug wines to apply to new wines that have less alcohol and more residual sugar than wines where fermentation has run its course.
- ***Astringency.*** A taste the beginner may mistake for dryness or acidity; this one puckers the mouth. You find it in young red wines, not quite ready to drink. It comes from the tannin derived from the grape skins and will disappear as the wine matures and mellows.
- ***Mellowness.*** The opposite of astringency; softened with age, ripe. Also used of jug wines that are "soft."
- ***Finish.*** Finish is the aftertaste. A good wine should have a pleasant aftertaste in keeping with the wine itself.
- ***Character.*** Positive, distinctive taste characteristics, as opposed to noncommittal taste.

As to taste preferences, the customer is always right. There are no longer rigid rules about what kind or color of wine to drink with what type of food. In the end, taste is a personal matter. The Romans sometimes mixed their wines with pitch or sea water.

Imported wines . . .

If you are reading about imported wines for the first time, you may want to pour yourself a glass of wine as you thread your way through this complicated subject. Within the scope of this book we can tell you what wines are well known and what to look for on their labels, and what more modest imports are worth investigating.

A bottle of imported wine *must* carry on it the name and type of wine, country of origin, name and address of the importer, net contents, and an indication of alcoholic strength. In addition, labels may state the vineyard name, vintage date, grapes, controlled area in which the grapes were grown, bottler and place of bottling, a quality designation, and the shipper's name and address (usually required by the exporting country). Most of this information is important to the buyer, but precisely what is important varies for different wines. The information may be carried on a single label or on a main label and one or more smaller ones (see Figure 6.18*b*, page 173). Often the net contents are stamped into the bottle itself.

A well-known wine is often expensive and may be hard to find. Nevertheless, in a restaurant with an expensive menu, customers who know wines may expect to see several well-known wines at high prices, even if they do not order any of them, and several more in the medium-price range which they will probably buy. As we have said before, choosing such wines is a job for an expert. But the more you know, the better you can judge the investment you are making and the more likely you will be to take proper care in storing and serving them.

Wines from France. French wines lead the list of imports. Look for the Appellation Contrôlée on the label; there are few wines worth buying without it.

The wines of the Bordeaux region have been famous since Roman times. They come in the high-shouldered bottle you see in Figure 6.15*a*. A wine spoken of as "a Bordeaux" means a red wine from Bordeaux. This is the claret after which American generics are named, but you never see the name claret on a Bordeaux label. Nor do you see the name of the principal grape —Cabernet Sauvignon—except on lesser wines catering to the American enthusiasm for varietals. Bordeaux is the connoisseur's wine; many uninitiated wine drinkers do not even like it.

Most prestigious are the Bordeaux wines named for the château (vineyard) producing them. The best wine from each vintage is usually bottled at the vineyard and carries the phrase *Mis en bouteille au château* (literally, placed in the bottle at the vineyard), or *Mise du château.* The rest of a vintage may be sold to a shipper *(negoçiant)* who blends wines from various vineyards to sell under the shipper's label or under a registered brand name (*monopole*). Figure 6.16*a* shows a typical château-bottled Bordeaux label.

The most famous Bordeaux reds are Château Lafite-Rothschild, Château Margaux, Château Latour, Château Mouton-Rothschild, and Château Haut-Brion. Many other château wines are almost their equal. Vineyards in certain districts have been classified according to excellence, and anything

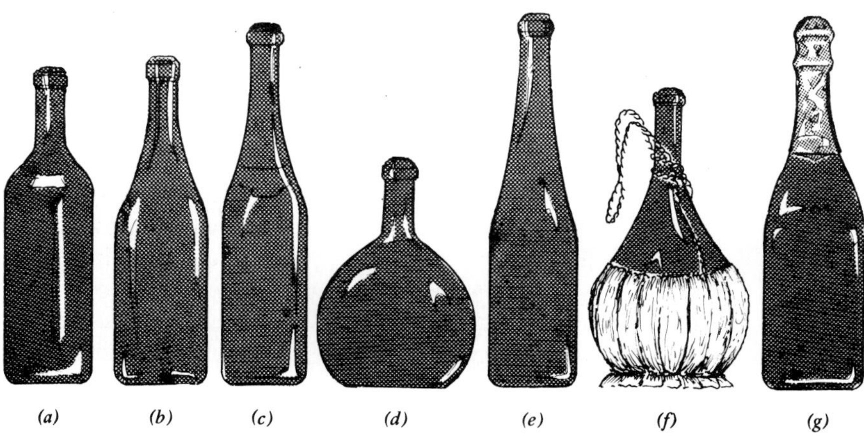

figure 6.15 Traditional bottle shapes. (*a*) Bordeaux. (*b*) Burgundy. (*c*) Rhine/Moselle. (*d*) Bocksbeutel (German Franken, Chile, Portugal). (*e*) Rosé. (*f*) *Fiasco* (chianti). (*g*) Champagne and other sparkling wines.

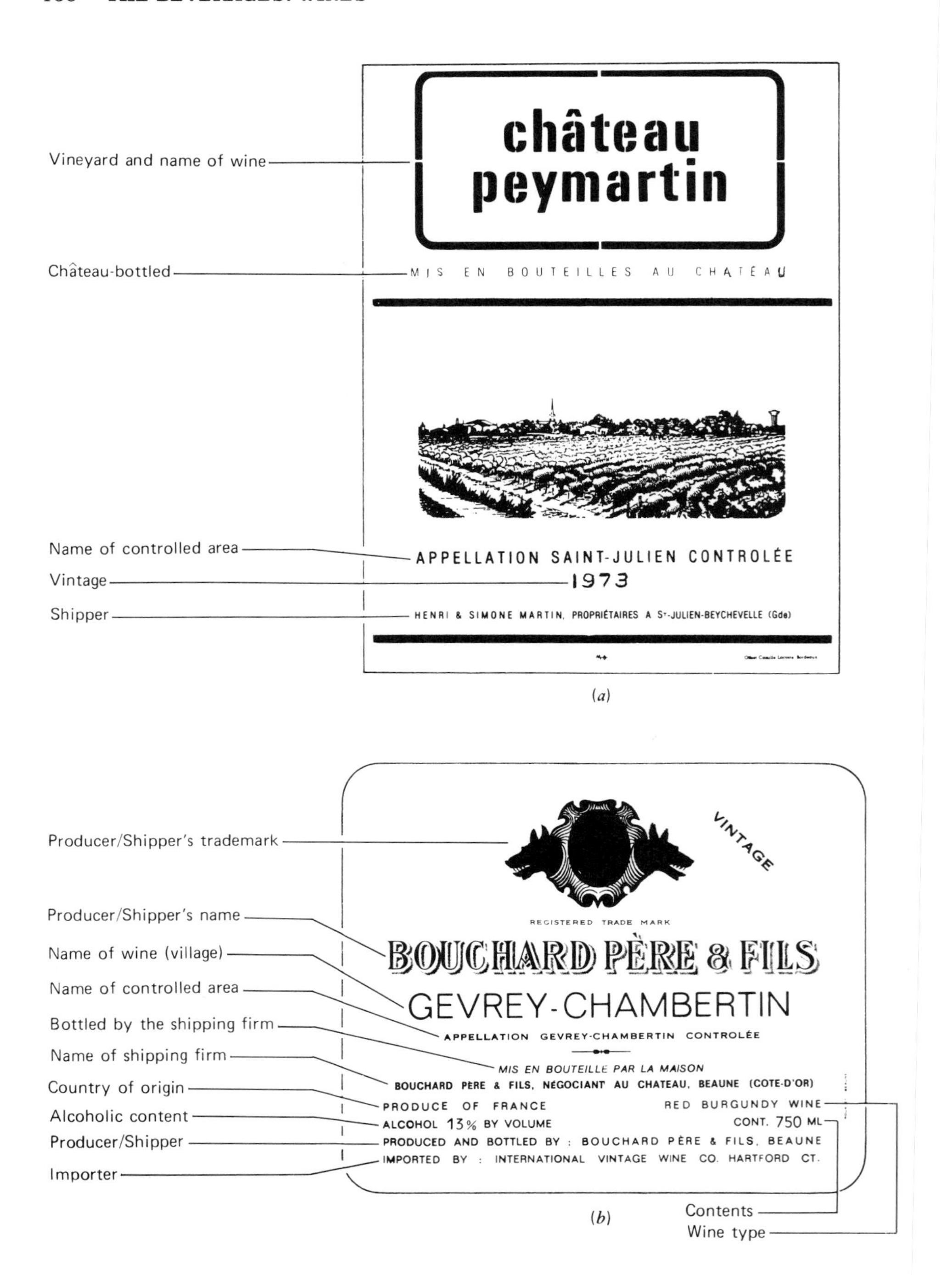

figure 6.16 Typical French wine labels. (*a*) Bordeaux. (*b*) Burgundy. (*c*) Beaujolais. (*d*) White burgundy.

Name of wine — Beaujolais

Name of controlled area — Appellation Beaujolais Contrôlée

Type of wine — RED BEAUJOLAIS WINE

Bottler/Shipper — Bottled & Shipped by : JEAN-JACQUES JOSEPH

NÉGOCIANT A F - 01800

Importer — Imported by : ACCENT WINE & SPIRITS, HOUSTON, TEXAS

Contents — NET CONTENTS 750 ML

Country of origin — PRODUCT OF FRANCE

Alcoholic content — ALCOHOL 12% BY VOL.

IN BERTHON LIBOURNE

(c)

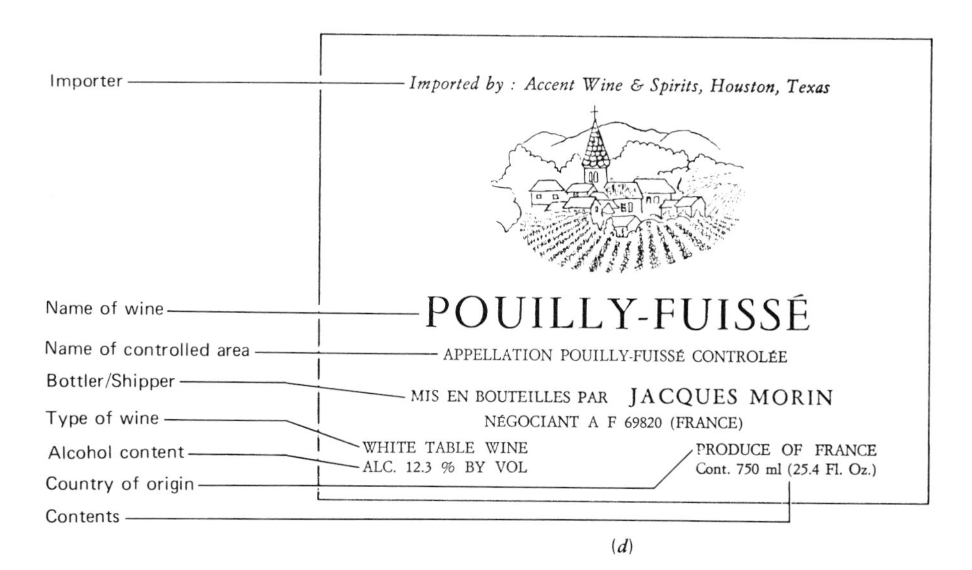

(d)

on a label indicating such classification—*Premier Cru* (first growth), *Cru Exceptionnel* (exceptional growth), and so on—will be impressive on a wine list.

Next in fame among the Bordeaux, and probably in quality, are wines with the Appellation of certain communes, such as Margaux, Pauillac, St. Julien, St. Estèphe, and of certain districts such as Médoc, St. Emilion, Pomerol.

In wines of this quality, vintage is also part of the pedigree. In buying, you must know that the vintage was a good year and that it is still within the life span of the wine. The vintage date belongs on the wine list. (If the supply is small, the word Vintage is sometimes used on the list instead of a date.)

Bordeaux produces two types of white wines, the dry wines of the Graves district and the sweet wines of Barsac and Sauternes, of which Château d'Yquem is the ultimate in fame and fortune.

The wines of the Sauternes district are extraordinarily sweet and rich. These qualities come from the warm climate and especially from the method of harvesting. The grapes are left on the vines until they are overripe and develop a special mold, *Botrytis cinerea*, known as "noble rot" (*pourriture noble* in French, *Edelfäule* in German). This mold dries the grapes, concentrating the sugar and the flavor. Each bunch is individually selected at its peak of desirability; a vineyard may be hand-picked in this way as many as eight times.

Burgundy wines also belong on a fine wine list. The French wines referred to as Burgundy are the classical red Burgundies made from the Pinot Noir grape in the Côte d'Or region. Like the Bordeaux, they are wines of great character and long life. They are shipped in the Burgundy bottle shown in Figure 6.15*b*.

Knowing what sort of wine lies beneath the label of a Burgundy is a far more complicated business than it is for a Bordeaux. Whereas in Bordeaux each vineyard (château) is owned by a single family or corporation, in Burgundy a single vineyard may be owned by many individuals, each with a few acres or perhaps just a few rows of vines. Each vineyard is an Appellation Contrôlée, and each owner/grower is therefore entitled to make an individual wine with the name of the vineyard on the label. Thus there may be great variation in quality depending on the skill of the grower, and conceivably as many as 5 to 50 different wines of the same name and vintage could come from a single vineyard. Likewise the phrase *Mis en bouteille au domaine* (bottled at the vineyard) does not guarantee quality or consistency; even when bottled at the vineyard there is no uniformity among growers' products.

For this reason it has become the practice for Burgundy shippers to buy wines from several growers in a vineyard and make a wine carrying the vineyard label on which they stake their own reputation. Thus the key to a good Burgundy is the shipper's name. Figure 6.16*b* shows a label for such a Burgundy.

Burgundy vineyards are also classified for quality; the magic phrase here is *Grand Cru* (great growth). Of the Grand Cru vineyards, the most famous are Romanée-Conti, Clos de Vougeot, and Chambertin, followed by certain hyphenated Chambertins and Musigny.

The district of Beaujolais is also part of the Burgundy region, but it produces a very different wine from the classical Burgundy. The wine called Beaujolais is a fresh, light wine with a short life span of two to three years.

It too is shipped in the Burgundy-shaped bottle shown in Figure 6.15. Unlike most red wines it is often served at a cool cellar temperature.

The best Beaujolais is identified as Cru Beaujolais plus the Appellation of the commune it comes from. Next in rank is the Appellation Beaujolais-Villages. Third rank goes to Beaujolais Supérieur, followed by just plain Beaujolais (Figure 6.16*c*). French Beaujolais wine should not be confused with the California varietal made from the grape called Gamay Beaujolais, which is a very different wine.

A red wine often mistakenly listed as a Burgundy is Châteauneuf-du-Pape. This Appellation Contrôlée area in the Rhone valley is said to have been planted originally by Pope Clement V. The wine comes in a Burgundy-shaped bottle, which may account for the mistakes in identity. Across the Rhone is the Tavel district, famous for its rosé.

The term white Burgundy encompasses several rather different wines. The vineyard of Montrachet and its neighbors are one group. Mâcon wines are another, and Chablis wines are a third.

Montrachet is a small vineyard that produces a white wine of great prestige and excellence. (It is made from 100% Chardonnay grapes, but the grapes are not cited on the label.) The vineyard is so famous that its neighbors have hyphenated its name to theirs. Other well-known white Burgundies from this region are Meursault and Corton-Charlemagne. It is said Charlemagne himself planted this vineyard. As with the red Burgundies there are many small growers in these vineyards, and a reliable producer/shipper rather than estate bottling is often the key to quality.

Of the wines of the Mâcon region of Burgundy, the best-known in this country is Pouilly-Fuissé (Figure 6.16*d*), another Chardonnay wine. Other Mâcon wines with a district or village Appellation are also imported and are often better value.

Chablis is a Burgundy region producing an entirely different white wine from the others, though made from the same Chardonnay grape. In fact it is seldom called a white Burgundy; it is just called Chablis. It is pale, greenish, light, and very dry, with a taste always described as flinty—very different from the typical American chablis. There are four classifications: Grand Cru, Premier Cru, Chablis, and Petit Chablis, in descending order. Quality varies greatly from one vintage to another because of the weather, which is often terrible.

Another group of fine French white wines comes from the Loire valley. The best-known are Pouilly-Fumé (often confused with Pouilly-Fuissé but very different), Sancerre, and Vouvray. This last is made in several versions from bone-dry and fruity to sweet to sparkling.

The Champagne district of France produces the sparkling white wine that bears its name. All French Champagnes are blends of wines from several vineyards and carry the name of the shipper/producer rather than a vineyard label. They are made from both white and dark grapes unless they are labeled *Blanc de blancs* (literally, White from whites). A French Champagne carries a vintage date only if it is from an exceptionally good year.

A Champagne label indicates its degree of sweetness. The French terms may confuse the novice. They range from *Brut* or *Nature* (the dryest) through *Extra Sec* (Extra Dry—having a small amount of added sugar), *Sec* (Dry—actually slightly sweet), to *Demi-Sec* (Semi-Dry—actually quite sweet), and *Doux* (downright sweet—seldom used). The French words should appear on a wine-list entry.

Wines from Germany. Germany is famous for its white wines. There are many young white wines of general appeal at moderate prices, and there are truly fine wines, beloved of connoisseurs, for the prestige wine list.

Germany has its own system of classification and quality control. (At this point you might want to pour yourself another glass of wine.) There are three categories of wine:

- *Tafelwein* (table wine).
- *Qualitätswein* (quality wine).
- *Qualitätswein mit Prädikat* (quality wine with special attributes).

Tafelwein, the first and lowest class, is seldom exported. Wines of the other two classes come from 1 of 11 regions delineated by the government and carry the name of the region on the label. Every wine of these two classes is chemically analyzed, tasted, and approved by government officials and is given a quality control number that appears on the label.

Wines labeled simply *Qualitätswein*, or Q.b.A. for short, are above average in quality. The label may state the vintage year, the variety of grape (if 85% or more), the village or vineyard (or both) from which the wine comes, and the name of the bottler (required). A wine bottled by the grower/producer will say *Erzeugerabfüllung* or *Aus eigenem lesegut*—the equivalent of château or estate bottling. There may also be a brand name and a shipper's name. Figure 6.17 shows typical German labels.

At this point the reader of a German wine label may come with relief to the familiar word Liebfraumilch. This is a Q.b.A. designation for wine blended from anywhere in the four regions of the Rhine valley and replaces the regional designation on the label. Liebfraumilch is a soft all-purpose Rhine wine—the real Rhine wine, not its California generic name borrower.

The other familiar German wine type is Moselle, a lighter, more delicate wine of only 8 to 10 percent alcohol. The collective name for a Moselle blend is Moselblümchen. The wines from the Rhine and Moselle valleys together account for most U.S. imports of German wines. You can tell them apart at a glance by the color of the bottle—brown for Rhine and green for Moselle. Both bottles have a typical tall tapered shape, like a stretched-out Burgundy bottle (Figure 6.15*c*). A green Bocksbeutel is used for wines from the Franken region (Figure 6.15*d*).

Most of the *Qualitätsweins* are best drunk when young, and there is no

(a)

*The term Table Wine is allowed as a statement of alcoholic content (between 7% and 14%).

(b)

figure 6.17 Typical German wine labels. (*a*) A *Qualitätswein*. (*b*) A *Spätlese*.

need to worry about vintage dates except to drink a wine soon enough (two to three years for Moselles, five to seven for Rhines).

Wines in the highest quality category—*Qualitätswein mit Prädikat*—belong to one of the following five groups:

- *Kabinett.* Light dry wines of some elegance, meeting specified technical standards (e.g., no sugar added for fermenting).

- *Spätlese* ("late picking"). Wines of more body and sweetness, made from grapes picked three weeks after the general harvest.
- *Auslese* ("selected picking"). Wines richer, fuller, and usually sweeter than *Spätlese*, made from grapes picked selectively by the bunch as they ripen.
- *Beerenauslese* ("selected berry picking"). Very rich top-quality wine made from perfectly ripened grapes chosen grape by grape.
- *Trockenbeerenauslese* ("selected dried-berry picking"). The ultimate in a rich, luscious wine, made from individually selected grapes that have dried and shriveled with *Botrytis cinerea*, noble rot—*Edelfäule*.

For wines of this category, vintage is important. It should be a good year for the specific region.

Here are a few famous villages and vineyards for your connoisseur's wine list:

- From the Moselle valley—Bernkasteler Doktor, Wehlener Sonnenuhr, Piesporter Goldtröpfchen, Ockfener Bockstein.
- From the Rhine valley—Steinberger, Schloss Johannisberg, Liebfrau-enstift-Kirchenstück, Bad Dürkheim.

Wines from Italy. Italian wines are becoming increasingly popular. Much less expensive on the whole than the French wines, they fill a need on the moderately priced wine list. Figure 6.18 shows you two Italian wines.

Of the Italian reds, Chianti is certainly the best known, its familiarity promoted by the gay straw-covered bottle (*fiasco*) that has been its hallmark (Figure 6.15*f*). For today's imported Chiantis the wicker container has largely given way to a more practical Bordeaux-shaped bottle, but the rough young wine is the same. A smoother and more mellow Chianti, known as Chianti Riserva, is aged at least three years. The best Chianti is from the Chianti Classico district, identified by a black cockerel on the neck of the bottle (Figure 6.18*b*). None of the Italian Chiantis should be confused with the generic chianti from California.

Experts will tell you that the king of Italian wines is Barolo. Other reds worth exploring are Gattinara, Barbaresco, Bardolini, Valpolicella, and Barbera, a varietal. A varietal of a different sort, and one very familiar in this country, is Lambrusco, a wine with some sweetness and a hint of sparkle. It accounts for up to half the wine imported from Italy. This is an unsophisticated wine for a young clientele, never a candidate for a connoisseur's list.

figure 6.18 (opposite) Typical Italian wine labels. (*a*) A white wine with superior credentials. (*b*) A well-known red wine with superior credentials. (Photo by Pat Kovach Roberts)

(*a*)

Italian seal showing wine meets all requirements for export to U.S.
Black cockerel identifying Chianti Classico
Vintage date
Wine from Classico district of Chianti zone
Controlled zone
Aged at least 3 years
Producer/Bottler
Country of origin
Producer/Bottler
Net contents
Alcohol content

(*b*)

It is served chilled—unusual for a red wine—and makes a good cocktail or party wine for a young crowd.

Italy exports some good white wines to the United States—Soave, Orvieto, Frascati, Est! Est!! Est!!! There are also good white varietals made from Verdicchio, Pinot Bianco (Pinot Blanc), and Pinot Grigio grapes.

Est! Est!! Est!!! is best known for its legend: A medieval bishop traveling to see the pope sent his servant ahead to seek inns with good wine and mark them with the word *Est* (This is it). The servant was so enthusiastic about the wine at Montefiascone that he chalked the inn door with Est! Est!! Est!!! The bishop agreed, settled down there, and never got to Rome at all.

All the Italian wines mentioned are from quality-controlled wine districts and will have the DOC label on the bottle indicating that district quality standards have been met.

Wines from other countries. Today imports are arriving from all over the world, most of them moderately priced and well worth looking into. Though they may be unfamiliar to your patrons, you can promote them as Specials or center a wine-tasting promotion around them—"Undiscovered Wines!!"

Spain is doing well with its Rioja, a red wine in the Bordeaux style. Spain is, of course, the homeland of the original Sherry, made in the Jerez district using time-honored methods and strict controls that yield a connoisseur's

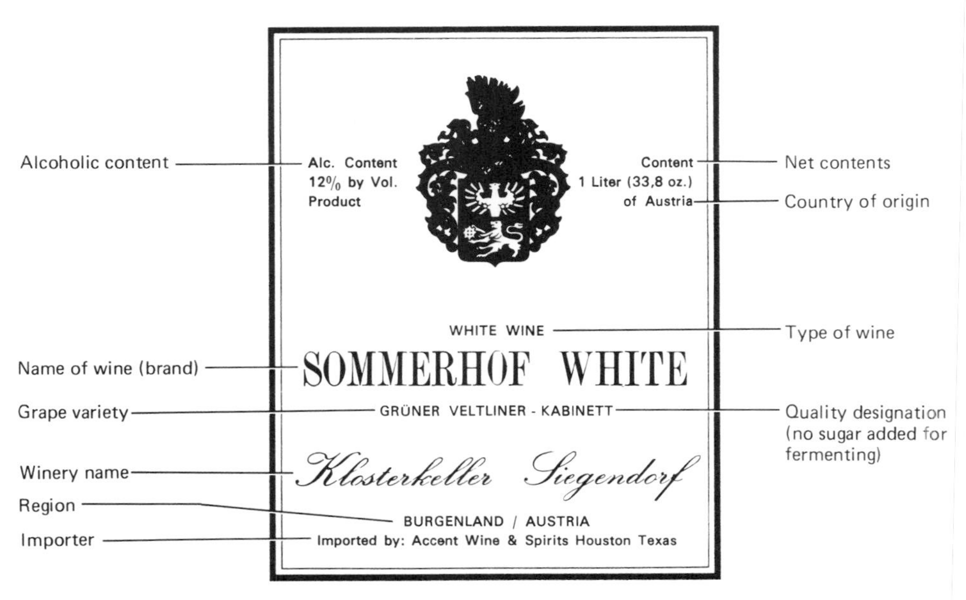

(*a*)

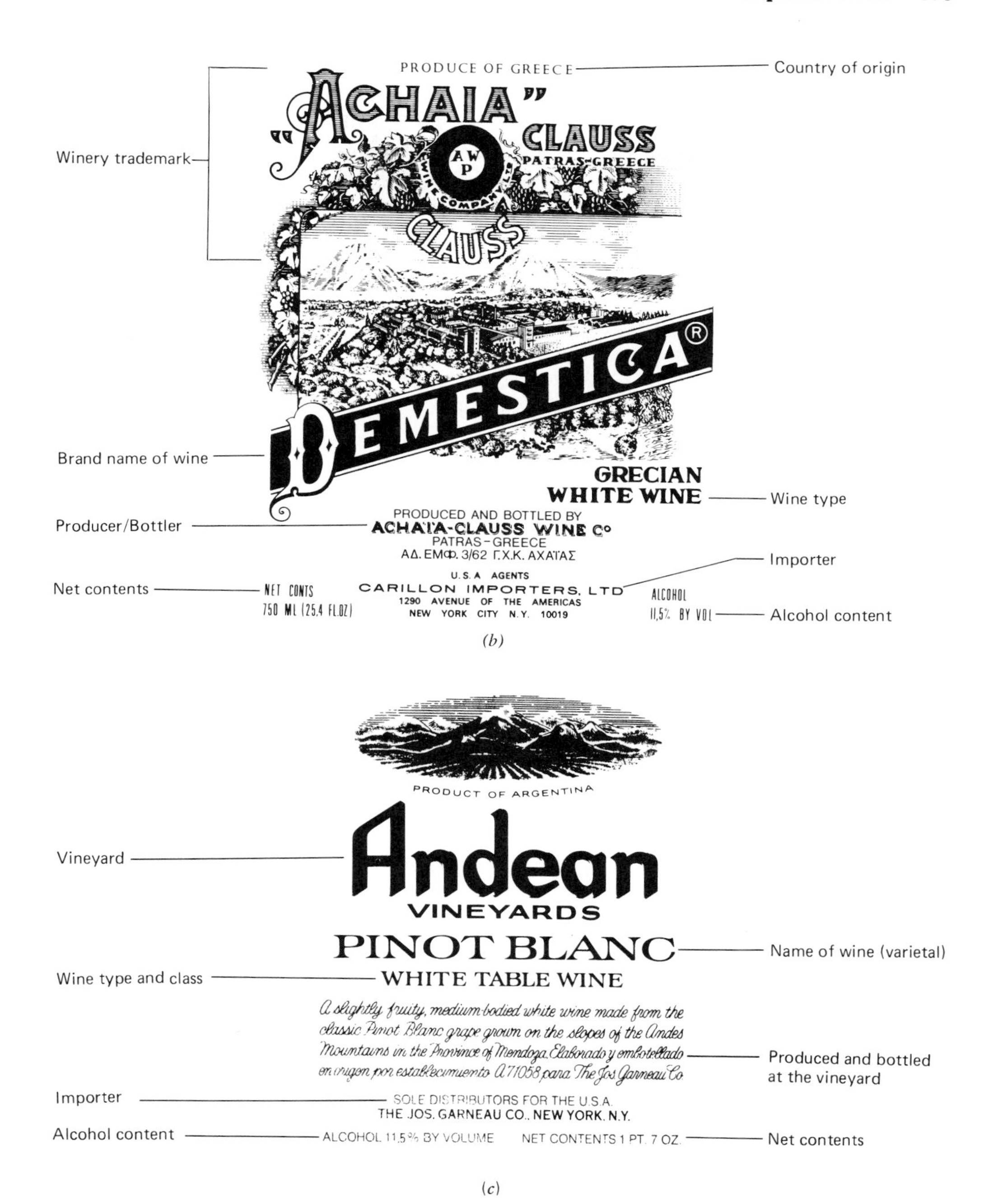

figure 6.19 Wines from other countries. (*a*) An Austrian brand-name varietal. (*b*) A well-known Greek wine from a well-known shipper. (*c*) A varietal from Argentina.

product. For a knowledgeable clientele you should consider imported Sherries—dry for aperitif use and cream for after dinner.

Portugal is to Port wine, or Porto, as Spain is to Sherry. Vintage is significant in its selection. Of broader interest are the white and rosé wines from Portugal, the most familiar being the brand-name rosés Mateus and Lancers (Figure 6.4).

Chile, Argentina, Austria, Greece, Romania, Hungary, and Australia are all contributing good, moderate-priced wines to the current American market. Figure 6.19 shows three such candidates. It is best to choose such wines by taste and introduce them gradually until you see what sells. Some excellent buys are available.

Summing up . . .

A book of this scope cannot hope to treat wines in any depth. By the same token a restaurant or bar serving wines for the first time should not plunge in and attempt to handle wines for a clientele of connoisseurs. But by all means serve wines! Wine by the glass or carafe sells easily and very profitably. A modest wine list of two or three wines of each type can also be very successful if the wines are of general appeal, the list is informative, the servers well informed and helpful, and the whole effort well organized.

A list of fine wines should be developed by an expert consultant, probably slowly and over several years in response to a given clientele. If you would like to broaden your own understanding, there are some excellent books on wine listed in the Bibliography.

chapter 7 . . .

THE BEVERAGES: BEERS

Beer has a large and loyal following, and every type of bar sells it in some form. Of the three types of alcoholic beverage, beer has the lowest percentage of alcohol and the highest food value—though food value is seldom the reason for its choice. Except for an opened bottle of wine, beer has the shortest life span of any alcoholic beverage. Draft beer needs the most space at the bar, the most attention, and the most care in pouring.

This chapter explains the nature of beer, how it is made, and how it is stored, cared for, and poured. You will become aware of beer's requirements at the bar and you will be able to produce a perfect glass of beer from either the tap or the bottle.

Noah took beer on the Ark. Human beings may have made grain into beer even before they learned to bake it into bread. Primitive peoples derived much of their body fuel from beer's carbohydrates and alcohol. Whichever came first, beer and bread were together the principal items of the ordinary family diet for centuries.

Columbus found Indians making a beer from corn and sap from the black birch tree. The English and Dutch colonists could not get along without beer and made their own. George Washington had his private brewhouse; Sam Adams and William Penn both operated commercial breweries. The soldiers of Washington's army each received a quart of beer a day.

Today's huge brewing industry really began in the nineteenth century, when German immigrants brought European know-how and beer-drinking customs to the United States. The brewer's art received an enormous stimulus from Pasteur's experiments with yeasts. Not only did he unravel the mysteries of fermentation; he developed the technique of sterilizing through pasteurization. The process was used to stabilize beers 22 years before it was applied to milk.

Beer shared the limelight with bourbon in the old-time saloon of the late nineteenth century. It was the excesses of both the beverage industry and the individual drinker during this time that finally brought about Prohibition.

Today beer has a totally different image and cultural function. It is a mainstay at many types of bars and restaurants. Once considered the drink of the male blue-collar worker, the college student, and the sports buff, it is acquiring a much more diverse clientele. The introduction of light beer has appealed to popular interest in health and fitness, while growing numbers of imported beers are intriguing people into tasting and exploring beers as they taste and explore wines.

Americans brew more beer than any other country in the world. Let's take a look at their products.

Beer types . . .

Beer is the generic term for all fermented alcoholic beverages that are made from malted grain and flavored with hops. In this broad usage the term beer includes not only familiar beer types but such distinctive beverages as malt liquors, bock beer, steam beer, ale, porter, and stout.

Classes of beer. These various malt beverages are classified into two groups, ***lager beers*** and ***ales***. The two types use different strains of yeast and somewhat different methods of fermentation. A lager beer is fermented by yeast at the bottom of a cold tank, a process called, quite logically, ***bottom fermentation***. An ale is fermented by a yeast that gathers on top of a somewhat warmer liquid—that is, by ***top fermentation***. Lager beer is stored, or ***lagered***, for several weeks or months before packaging. Ale is stored only a few days.

Lager beer tends to have a lower percentage of alcohol than ale, though some lagers have more than some ales. The maximum alcohol content for lager beer is 6% by weight (7.5% by volume). Maximus Super, a regional beer brewed in Utica, New York, maintains this percentage.

Alcoholic content of beers is not readily apparent. Federal regulations do not allow it to be mentioned on the label unless required by state law. Alcoholic content is usually spoken of as a percentage by weight, but not always. You may hear of a 5% beer and a 3.2% beer, but you cannot compare them unless you know they are both expressed in the same terms. Percentage by weight is about four-fifths the percentage by volume and four-tenths the proof figure. For example, 3.2% beer by weight is 4% alcohol by volume and 8 proof, and 5% beer by volume is 4% alcohol by weight and 10 proof. The differences are small and do not affect taste. But you *can* taste the difference between a 3.2% beer by weight and a 5% beer by weight (4% and 6.25% by volume).

Lager beers. There are several kinds of lager beer as well as innumerable brands, each with its own flavor. The major kinds are the pilsner-style beers, light beers, malt liquors, bock beers, and steam beer.

Pilsner is a descriptive term that is often applied to the kind of beer we call *beer* in the United States—a lively, mild, dry, amber-colored, thirst-quenching liquid that may or may not say pilsner on the label. It is a style of beer rather than a distinctive type. The term is borrowed from the classic Pilsner-Urquell made in Pilsen, Czechoslovakia, a beer with a great deal more flavor and body than its bland American reflection. Pilsner-style beers are by far the biggest sellers in the United States. They include the best-selling Budweiser, Miller's High Life, Pabst Blue Ribbon, Coors, Schlitz, and Michelob, as well as such regional beers as Stroh's and Genesee. Generally they contain 3.2 to 4.5% alcohol by weight. Since state laws control the amount of alcohol allowed in beers sold within state borders, the same brand of beer may vary from one state to another.

Light beers are variants of the pilsner style. Catering to the current vogue of "light" liquors and diet soft drinks, these beers typically have one-third to one-half fewer calories than the regular lager beers. Since most of the calories in beer come from alcohol and carbohydrates, light beers have less alcohol or less carbohydrate or less of both, and a higher proportion of water. Why do we pay more for less? It costs more to produce less and sell it, and apparently people are willing to pay the price, because its popularity has risen dramatically in the last few years.

Malt liquors are lager beers with a higher alcohol content than the pilsners—generally 5.5 to 6% by weight. This is frequently produced by adding extra enzymes to increase fermentation. Leading malt liquors are Schlitz Malt Liquor, Colt 45, and Champale. Malt liquors have a small market and are over strength for some states.

Bock beers are rich, heavy, dark lagers with a caramel taste. They are made from malt toasted to a very dark color. They are often regional products and are usually produced as springtime specialties, though they can be made anytime. They typically have an alcohol content of not less than 6% by volume.

Steam beer is a truly American invention—the only kind of beer not borrowed from Europe. Its method of production developed in California during Gold Rush days and combines the bottom fermentation of lager beer with the higher fermenting temperatures of ale—ice was hard to come by in California in those days. It makes a beer with a lively head and the body and taste of ale. The name has nothing to do with brewing but comes from the "steam" released when the casks are tapped. It is made by one of the smallest breweries in the country—Anchor of San Francisco—but it is beginning to attract national attention. Its alcohol content is 5% by volume—the same as Budweiser's.

Premium and superpremium lagers, including the lights, have almost taken over today's market. (These terms derive from price tags and contrast with popular-priced beers.) On the one hand the lights are offering "less"; on the other hand the superpremiums such as Michelob and Löwenbräu are offering more—more taste and more body. Imported lagers are also becoming popular—Heineken from the Netherlands, Dos Equis and Carta Blanca from Mexico, St. Pauli Girl from Germany, Kirin from Japan, several Canadian beers, and many others, including Pilsner-Urquell. Many bars and restaurants are offering a menu selection of imported beers. Although sales are relatively small, they are rising fast, and a beer menu tends to stimulate beer sales in general.

Ales. The typical American ***ale*** is a copper-colored liquid with more body, more hops flavor, and more alcohol than the pilsner-style lagers—typically 5 to 5.5% by weight and sometimes up to 6.5%. Connoisseurs say that ales should be served at higher temperatures than beer for full flavor enjoyment—55°F at least. However, your clientele may have something to say about this. We Americans tend to drink all beers as if they were soft drinks to quench thirst, instead of savoring the flavor as Europeans do.

Porter is a dark, bittersweet, top-fermented brew that was brought by the colonists from England, where it is no longer made. It has survived here as a specialty. America's oldest existing brewery—Yuengling's in Pottsville, Pennsylvania—makes a "Celebrated Pottsville Porter" among other specialties. McSorley's Old Ale House in New York City, one of the few surviving old-time saloons, serves a McSorley's Porter and a McSorley's Ale, specially brewed for them.

Stout is the successor to porter in England—a "stouter" porter, fuller-flavored, darker, more alcoholic. It ranges from 3 to 7.5% alcohol and from sweetish to bitter. Guinness, imported from Ireland, is the best-known bitter stout. In some circles it is familiar as an ingredient of Black Velvet, a mixture of equal parts champagne and bitter stout, served in the tallest possible glass.

How beer is made . . .

Though there are many kinds of beers, the method of manufacture is essentially the same for all. The differences come from variations of the ingredients and methods, and from the art of the brewmaster, who must take charge of this living, active substance as it develops.

The ingredients. There are four ingredients essential to the making of beer in all its forms: ***malt*** (nearly always barley malt), ***water***, ***hops***, and ***yeast***. Often there is a fifth ingredient, another cereal in addition to the malt, called a ***malt adjunct*** or ***grain adjunct***.

- Water is nine-tenths of the beer. It is the vehicle for everything else. It also contributes certain minerals to the taste.
- The malt and the adjuncts provide the sugars to be fermented, yielding alcohol and carbon dioxide. They also contribute flavor, head, body, and color—especially the malt. Translated into nutrients, the body elements are carbohydrates, proteins, and traces of the vitamins riboflavin, niacin, and thiamine.
- Hops provide a typical bitter flavor essential to all beers. They also have an antiseptic effect during the brewing.
- Yeast is the agent that acts on the sugars to break them down into alcohol and carbon dioxide. It may also affect taste and the hangover potential of the beer.

Variations in the character of each ingredient are important to the final product. Imported hops, for instance, are more flavorful than domestic hops. Different grain adjuncts give different tastes. Coors' emphasis on its Colorado mountain spring water is not all advertising art; it is essential to the taste of Coors. Some waters are suitable for ale but not for beer, and some are suitable for beer but not for ale. Many brewers modify their water—adding various salts, for example—to produce a successful, standardized product.

Most brewers buy their malt in the form of dried or roasted malt or even malt extract; many of the cheaper beers use malt extract. Anheuser-Busch, Coors, and some small regional breweries malt their own barley. You will remember from Chapter 5 that malt is sprouted grain, the sprouting being necessary to provide enzymes that will break down the grain's starch molecules into the simpler sugar molecules that will in turn break down into alcohol and carbon dioxide when attacked by yeast. Without malt little or nothing would happen.

Except for a few made from wheat, all beer malts are barley malts. Malt producers sprout the barley, then dry it in kilns to stop the sprouting at the proper moment. They then roast it according to the brewer's instructions. The time and temperature of roasting affect the dryness or sweetness of the final product as well as its color—the more roasting, the deeper the color.

The hops that give beer that characteristic suggestion of bitterness are the blossoms of the female hop vine. The best are Bohemian hops, imported from Czechoslovakia, but quality hops are grown in the American West, especially the Yakima Valley in Washington. The blossoms are picked, dried, and refrigerated until used.

The brewer's yeast is the special laboratory product of each brewer, and its behavior (it is constantly active) is closely watched. Erratic behavior or a stray yeast from the air getting into the brew could cause a disaster, requiring shutting down the plant, cleaning and sterilizing the equipment, and starting all over again.

Adjuncts are commonly used in American brewing. They give beer a lighter color and milder flavor, and they are cheaper than barley malt. The most commonly used adjuncts are rice and corn, but sugar, wheat flakes, soybean flakes, and potato starch are also used. Rice gives the lightest color. In general, the higher the proportion of barley to adjunct, the more flavor and body in the beer and the better the head. Superpremium beers typically use a higher proportion of barley malt: Anheuser-Busch's Michelob, for example, uses 95% barley malt and 5% rice while its premium beer, Budweiser, uses 65% barley malt and 35% rice. In some European countries, notably Germany, Switzerland, and Norway, adjuncts are prohibited by law and beers are made entirely with barley malt.

Another type of ingredient has become more common in recent years—the additive. Additives are used to stabilize beer foam, to prevent cloudiness, to facilitate conversion of starch to sugar, to prolong shelf life, to adjust color. All additives are substances approved by the Federal Food and Drug Administration. Many brewers, however, continue to produce beers without additives, relying on good ingredients and good production methods to prevent the problems the additives are intended to solve.

The process. Making beer is a four-step process: (1) mashing, (2) brewing, (3) fermenting, and (4) lagering. The first three steps are very similar to the first stages of making whisky. Figure 7.1 diagrams the whole sequence.

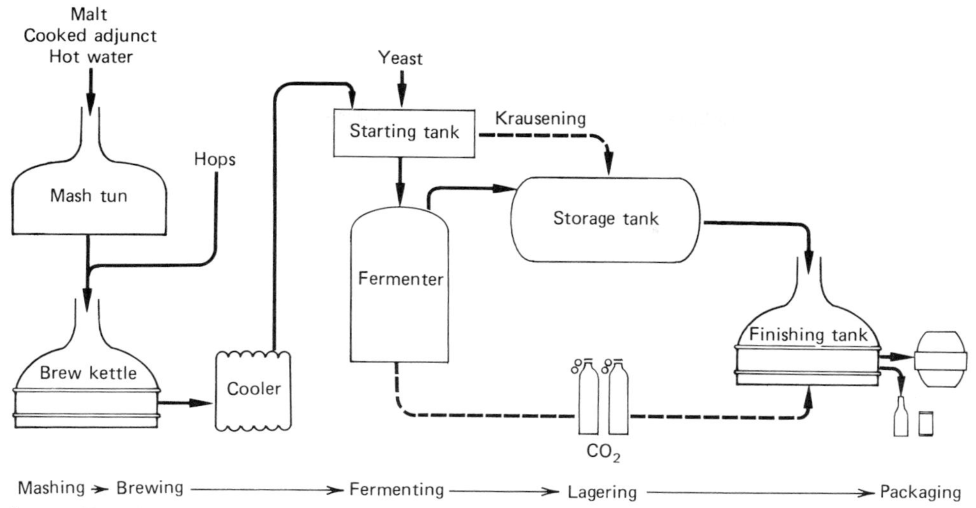

figure 7.1 Making beers and ales. Broken arrows show alternate ways of carbonating.

Mashing converts starches into sugars. The barley malt is ground into grist, which is fed into a container called a ***mash tun*** along with hot water. Adjuncts—usually corn or rice—are precooked and added to the mash tun. Everything is mixed and cooked together at low temperatures (up to 154°F) for one to six hours. During this process the malt enzymes are activated and turn starches to sugars. Then the grain residue is strained out and the remaining liquid—now called ***wort***—is conveyed to the ***brew kettle***.

Brewing—boiling the wort with hops—extracts the distinctively bitter hops flavor that makes beer taste like beer. In huge copper or stainless-steel brew kettles, the wort-plus-hops is kept at a rolling boil for 1 to 2½ hours. The boiling also sterilizes the wort and draws out antiseptic elements in the hops that protect beer from spoilage.

After brewing, the hops are strained out and the wort is cooled. At this point the techniques vary according to whether a lager beer or an ale is being made. For ales the wort is cooled to 50° to 70°F. For lager beers it is chilled to 37° to 49°F.

Fermenting—converting the sugars into alcohol and carbon dioxide—begins when yeast is added. If a lager beer is being made, the yeast settles to the bottom of the fermentation tank, and the action proceeds from the bottom. The yeasts for ales—different strains that work at warmer temperatures—rise to the surface and work from the top.

The usual fermentation time is a week or more—a bit less for ales, a bit more for beers. During fermentation the carbon dioxide being given off may be collected and stored under pressure, to be added again at a later stage.

Lagering, or storing (from the German *lagern*, to be stored), matures or ripens the beer, mellowing its flavor. Some further slow fermentation may also take place and impurities may settle out. Lagering of beer takes place at near-freezing temperatures. It may last for several weeks to several months. Ales are ripened too, for a much shorter time. Both beer and ale are matured in stainless-steel or glass-lined tanks, in contrast to the wood casks in which spirits are aged. Wood casks spoil the beer taste. (Anheuser-Busch uses beechwood chips in its lagering tanks to clarify the beer, but since these are specially cooked and sterilized they impart no taste.)

During this period some beers are given a small additional amount of newly fermenting wort, to add zest and carbonation through further fermentation. This process is called ***krausening*** (from the German *Kräusen*, referring to froth that forms in fermenting wort at a certain stage), and is sometimes referred to as ***natural carbonation***. If the beer is not krausened, it is carbonated by adding the stored carbon dioxide at the end of the lagering period.

After storage, the beer is filtered and then kegged, bottled, or canned. Kegs, or half-barrels containing 15½ gallons, provide bar supplies of draft beer. Bottles and cans provide the bar with individual servings.

Pasteurizing. Most canned and bottled beers are stabilized by ***pasteurizing***—exposing them in the container to 140°–150° heat for 20 minutes to an hour, to kill bacteria and any remaining yeast cells. Draft beers may be flash-pasteurized with steam, but most are not pasteurized at all. This is why they taste better. It is also why they are packaged in metal kegs that will withstand increased pressure that may come from slight continuing fermentation, and why they are kept refrigerated without interruption from brewery to bar. A constant temperature is essential to maintain the quality of the beer. A beer that has been warmed and cooled again is known as a ***bruised*** beer; it suffers a loss in quality.

Some canned and bottled beers are not pasteurized. Instead the beer is passed through ultrafine filters that remove yeast cells and other impurities. Thus it retains many of the characteristics of draft beers and may be labeled and advertised as draft beer. Coors beers are the best-known examples of unpasteurized beers. They are made under hospital-clean conditions, and not only kegs but cans and bottles are shipped and stored under refrigeration. They are intended to be kept cold right to the consumer (bar managers please note!).

Under federal regulations pasteurized beers in cans or bottles may refer in advertising to "draft flavor" or "on-tap taste" if the label states clearly that they have been pasteurized. They may not call themselves draft beer.

Storing and caring for beer . . .

Even pasteurization does not confer indefinite shelf life. All beers should be kept cool and used promptly. Beers kept too long will lose both flavor and aroma.

Although canned and bottled beers are either pasteurized or specially filtered, they still have a limited shelf life. They should be used within three to four months of the date of packaging. Some brewers mark each package with a ***pull date***, the date you should pull it off your shelves if you haven't served it yet. Others code their containers with the date of brewing. Your supplier can tell you which system applies to the beers you are buying. If you rotate your stock, using the oldest first and putting new supplies behind existing stocks, you will avoid serving over-age beers.

Cans and bottles will stay fresh when stored between 40° and 70°F. Unpasteurized beers should be refrigerated at all times. Temperatures above 70° will destroy flavor and aroma. Light will also cause deterioration. Direct sunlight will bring about change in a matter of minutes. This effect of light is the reason for putting beer in brown bottles. But don't depend on the bottle; keep it away from light and heat in all cases.

If beer gets cold enough to freeze it is likely to precipitate its solids and form flakes that will not dissolve when thawed. Beer kept too cold for a long time may gush when opened. This may also happen when cans or bottles are handled roughly.

Draft beer must be kept cold from brewer to distributor to storeroom to bar, preferably at 36° to 38°F. Since it has not been pasteurized in the way that bottled beers have been, it is much more susceptible to deterioration. Even at ideal temperatures its shelf life is 30 to 45 days in the untapped keg.

Much of the quality of a good glass of draft beer lies in the proper use and care of the beer system. A beer system includes one or more kegs of beer, a CO_2 cylinder, a tap (faucet), lines running from the CO_2 cylinder to the keg and from the keg to the tap, and a refrigerated beer box or remote cooler (Figure 7.2).

The CO_2 cylinder contains carbon dioxide gas under pressure of 1000 pounds per square inch (psi) at room temperature. It has a pressure regulator that reduces the pressure of the gas between cylinder and keg to 12 to 15 psi depending on the brand of beer. The carbon dioxide under pressure has two functions: it maintains the carbonation of the beer, and it moves the beer through the line to the tap when the tap is opened. The cylinder should be kept at room temperature. It should not be in the beer box with the keg of beer.

A pressure gauge indicates the pressure in the keg. A correct pressure constantly maintained will keep the beer in the keg lively and tasty. With too low a pressure the beer will lose carbonation and taste flat. With too high a pressure the beer will froth and "pour wild."

The carbon-dioxide and beer lines are connected to the beer keg by couplings that fit into valves in the keg. Connecting the lines to the keg is called ***tapping***. There are two principal types of connections. In one system the carbon-dioxide coupling fits into a valve in the top of the keg and the beer-line coupling fits into a valve at the bottom (Figure 7.3*a*). In the other system a single coupling with two branches connects both carbon-dioxide

figure 7.2 A self-contained beer system with two kinds of beer. Dark lines carry beer from keg to tap. Light lines carry carbon dioxide from cylinder (not shown) to keg. (Photo courtesy The Perlick Company, Milwaukee)

and beer lines through a valve in the top of the keg. Carbon dioxide pressure forces the beer to rise through a long hollow rod reaching up from the bottom of the keg to the beer line (Figure 7.3*b*).

The two kegs in Figure 7.2 also show the different systems. Note that the one on the left has a tilt plate under it to send the beer in the bottom of the keg toward the line so that it is not wasted.

In either system kegs may be connected in series, giving high-volume operations a continuous supply of beer without having to change kegs frequently (Figure 7.4).

Your beer supplier will determine which tapping system is best for you and will supply the couplings that go with the type of kegs being delivered. Usually you must get your carbon dioxide from a different supplier. If you are serving more than one kind of draft beer, each beer type must have its own carbon dioxide supply, since the pressure required may be different.

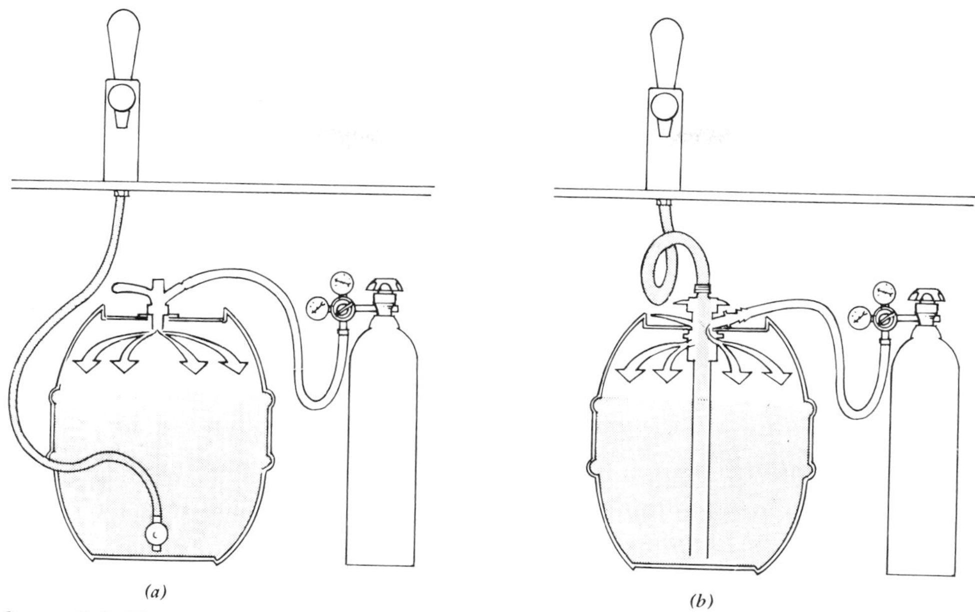

figure 7.3 Two tapping systems. (*a*) Carbon dioxide enters at top of keg. Pressure sends beer up through line connected to keg at bottom. Keg must be tilted toward connection. (*b*) Carbon dioxide enters at top of keg. Pressure sends beer up through internal rod and into beer line. Both lines are attached to keg at top with single tapping device.

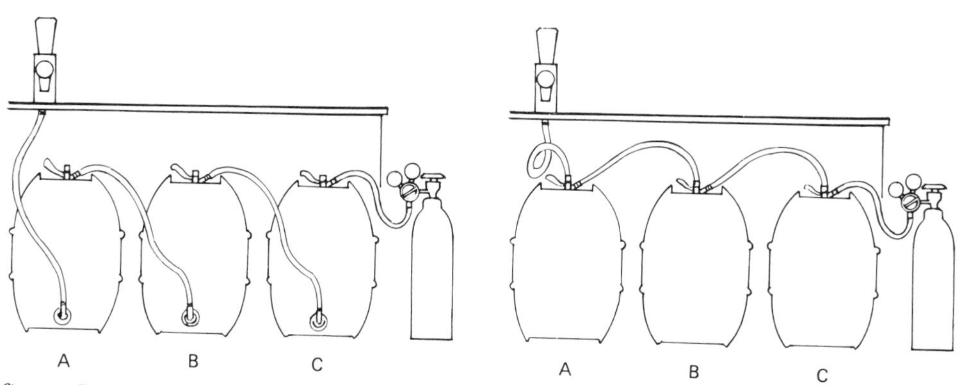

figure 7.4 Kegs in series. Beer is drawn from keg nearest tap (A). Carbon dioxide enters Keg C, pushing beer flow from keg to keg. When fresh supplies are added, partly empty keg must be placed nearest the CO_2 cylinder.

Where the size of a bar precludes the luxury of an underbar beer system, the kegs may be housed in a remote beer box or walk-in cooler. For example, the Jack Tar bar described in Chapter 3 has its beer in its own walk-in cooler in the garage below the bar. For a remote system the line from keg to tap must be insulated to maintain temperature, and a motor is necessary to pump the beer over the extra distance.

A remote beer supply should not share a walk-in cooler with food storage, since frequent opening and closing of the door will make it impossible to keep the beer at a constant temperature. In fact, remote systems should be avoided altogether if possible. For the same reason, reserve kegs for an underbar beer system should have their own refrigerated storage space.

The beer box or cooler should be maintained at a constant temperature between 36° and 38°F. Beer allowed to warm over 45°F may become cloudy or turn sour. Beer allowed to chill below freezing may become unsalable. At 32°F the water in the beer freezes and separates from the alcohol. If allowed to thaw at the normal temperature of 36° to 38°F it may possibly be blended back together by gently rotating the kegs. (This is best done by your distributor.) If the beer is cloudy after thawing and rotation it is beyond salvaging.

To care for your beer properly you have to care for the whole beer system. Your thermometer and your pressure gauge are your allies. If they are inaccurate they can be your enemies. Check them frequently. If the arrow on the pressure gauge moves up or down from the pressure at which you set the regulator, either the regulator or the gauge is not doing its job properly. Check the thermometer by placing another thermometer in the cooler.

The beer lines are another key to good beer taste. They should be flushed daily at the close of service and kept full of water overnight. To flush a beer line you disconnect the line from the keg and connect it to a water faucet, using a special connector supplied by your distributor. Open the water faucet; then open the beer faucet and let it run until the water runs clear. Then close the beer faucet and the valve on the connector, leaving the line full of water overnight. Do this with each beer line. *Never let the lines dry out.*

To reconnect the beer line to the keg, first let fresh water run through the line and the beer faucet. Then close both the beer faucet and the connector valve, disconnect the coupler, and reconnect the line to the valve on the keg. Open the beer tap and let it run until the water runs out and the beer appears.

Another housekeeping tip is to flush your drains with a pitcherful of water when you are closing for the night. Sometimes yeasts and bacteria can become active in the pipes overnight and clog your drains.

In addition to daily flushing, beer lines, valve tap, and faucet should also be cleaned thoroughly every two weeks. Usually this is part of a distributor's service. If not, your distributor will show you how to do it or give you printed instructions.

Since your distributor is the person you are going to call if your beer begins to taste funny, he or she may well give you a list of troubleshooting suggestions such as the one in Table 7.1. If you can figure out what is wrong you will be able to give the distributor a clue to the trouble so that it can be taken care of faster.

table 7.1 **TROUBLESHOOTING DRAFT BEER PROBLEMS**

trouble	*causes*
Flat beer	Greasy glasses Beer drawn too soon before serving Pressure: too low, leaky pressure line, sluggish regulator, pressure shut off overnight Obstruction in lines Loose connections (tap or vent) Long exposure to air instead of carbon dioxide pressure Precooler or coils too cold
Wild beer	Beer drawn improperly Too much pressure: faulty pressure valve or creeping gauge Beer too warm in keg or lines Lines: too long, poorly insulated, kinked or twisted Faucets in bad condition
Cloudy beer	Beer too warm at some time (storeroom or delivery) Beer frozen at some time Beer too cold Defective valves at keg Old beer (was stock properly rotated?) Lines: dirty, hot spots, poor condition
Bad taste	Keg too warm: 50° and over at some time can cause secondary fermentation and sour beer Glasses: not beer-clean, not wet Dirty lines, dirty faucets Failure to flush beer lines; failure to leave water in lines overnight Bad air in lines, oily air, greasy kitchen air Unsanitary conditions at bar
Unstable head	Beer drawn incorrectly (tilt of glass) Glasses not beer-clean Too short a collar Flat beer causes

Serving beer . . .

Serving a perfect glass of beer depends on three things: the condition of the glass, the way the beer is poured, and the temperature of the beer.

The glass, first key to perfection, must be "***beer-clean***"—that is, grease-free, film-free, and lint-free. Any grease, oil, fat, or foreign substance, visible or invisible, will spoil the head on the beer; a dense, firm head of foam will deflate and break up, leaving large bubbles. The body of the beer will also lose carbonation. A less-than-clean glass can also cause an off taste or an off odor. A beer-clean glass, on the other hand, will support the original head and, as the glass is emptied, will leave the foam in rings on the sides of the glass. The taste will stay fresh and zesty all the way down to the bottom.

To produce a beer-clean glass, wash and dry it using the following steps, as shown in Figure 7.5:

- Using a special fat-free washing agent and hot water in your wash sink (Figure 7.5*a*), submerge the whole glass and thoroughly brush the inside and rim. Empty the glass of all wash water.
- In the rinse sink (Figure 7.5*b*), in running water if possible, immerse the glass bottom end first, to be sure you fill it completely. Swish it around or rotate it, then empty.
- In the sanitizing sink (Figure 7.5*c*), repeat the rinsing procedure.
- Turn the glass upside down on a clean rack or corrugated drainboard (Figure 7.5*d*), so that there is free access of air. Do *not* place the glass on a towel or rubber mat or any flat surface, and do *not* dry it with a towel.

The bartender who polishes glasses until they sparkle isn't around any more, if such persons ever existed outside of books. There is a good reason for not polishing glasses: the towel and the fingers can transmit grease, lint, chemicals, and bacteria to the glass.

The same rules apply, of course, to washing any glass. However, most drinks won't tell you whether the glass is clean.

The second key to a perfect glass of beer is the way you pour it. A good head is a thing of beauty to a thirsty customer. A good ***head*** is a collar of firm, dense foam reaching slightly above the top of the glass, ½ to 1 inch thick depending on what pleases both you and your clientele. If the head is scant the beer looks flat and lifeless, even when it isn't. If the head is too

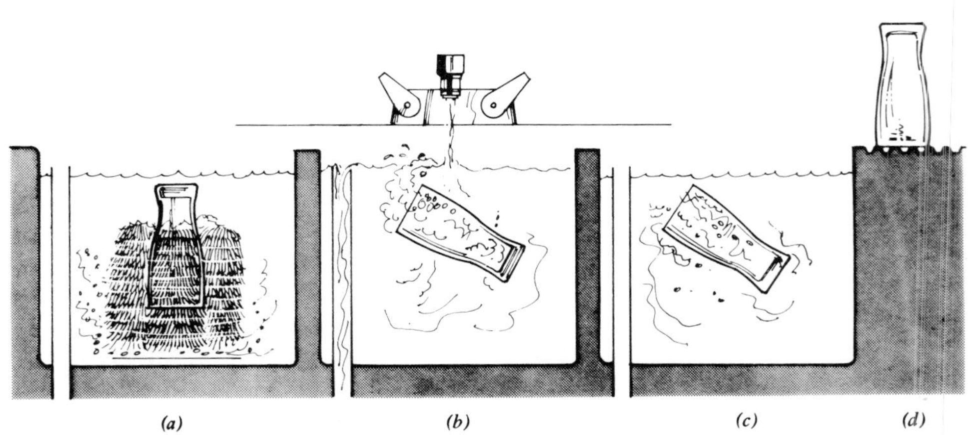

figure 7.5 How to wash a glass "beer-clean."

thick the customer may feel cheated—and rightly so. The more foam, the less beer in the glass.

The size of the head depends on two things—the angle at which you hold the glass while pouring and how long you hold the angle. Figure 7.6 shows you how. Rinse a beer-clean glass in fresh cold water and follow these steps:

- Hold the glass at a 45° angle about an inch below the tap (Figure 7.6*a*) and open the tap all the way.
- When it is about half full, straighten the glass upright and let the beer pour right down the middle (Figure 7.6*b*), still keeping the tap wide open.
- When the head has risen a little higher than the rim of the glass (Figure 7.6*c*), close the tap.

Notice that the draw is a single motion from beginning to end, not little spurts. A few practice draws will establish for you the angle of the glass and the time you should hold the angle to produce the head you want.

How big a head do you want? Sales representatives are fond of pointing out that your profit is in your head. Figure 7.7 shows what they mean. But customer taste is your real criterion, and customers don't come back for little glasses with big heads. You have to measure your profit-by-the-glass against repeat business; then choose your glass, your head, and your price accordingly and see that your bartenders hit your policy on the head.

In serving beer from a can or bottle you can produce a good head by reversing the tilting procedure: you place the glass upright and tilt the can or

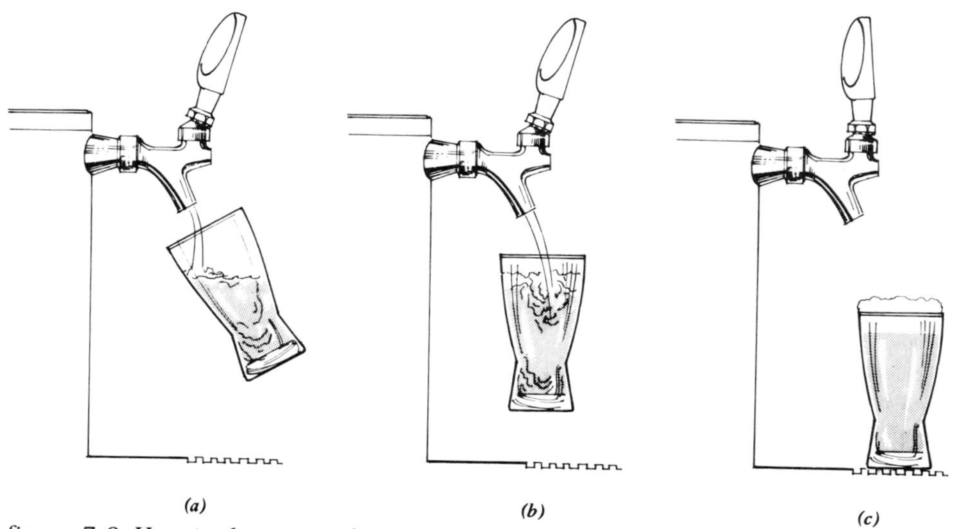

figure 7.6 How to draw a perfect beer.

TYPE OF GLASS	SIZE	1" HEAD	¾" HEAD	½" HEAD
		APPROXIMATE GLASSES PER ½ BBL.	APPROXIMATE GLASSES PER ½ BBL.	APPROXIMATE GLASSES PER ½ BBL.
SHAM PILSNER	8 oz.	343	325	283
	9 oz.	292	279	260
	10 oz.	265	245	223
	12 oz.	221	204	186
TULIP GOBLET	8 oz.	305	292	275
	10 oz.	248	230	207
	11 oz.	227	209	185
	12 oz.	210	191	167
FOOTED PILSNER	8 oz.	325	292	280
	9 oz.	282	259	245
	10 oz.	250	233	215
SHELL	7 oz.	360	336	315
	8 oz.	315	292	275
	9 oz.	270	255	243
	10 oz.	245	236	220

TYPE OF GLASS	SIZE	1" HEAD	¾" HEAD	½" HEAD
		APPROXIMATE GLASSES PER ½ BBL.	APPROXIMATE GLASSES PER ½ BBL.	APPROXIMATE GLASSES PER ½ BBL.
HOURGLASS	10 oz.	264	248	233
	11 oz.	235	220	205
	12 oz.	220	204	189
	13 oz.	198	184	173
MUG STEIN	10 oz.	248	233	223
	12 oz.	203	189	176
	14 oz.	169	158	153
	16 oz.	149	140	134
HEAVY GOBLET	9 oz.	378	331	294
	10 oz.	330	296	264
	12 oz.	248	220	204
	14 oz.	209	194	172

PITCHER	SIZE	1" HEAD	1½" HEAD
		APPROXIMATE PITCHER PER ½ BBL.	APPROXIMATE PITCHER PER ½ BBL.
	54 oz.	47	50
	60 oz.	39	42
	64 oz.	35	38

figure 7.7 Servings per keg in relation to glass and head. (Courtesy Miller Brewing Company, Copyright © 1980)

bottle instead, as shown in Figure 7.8. You should open the can or bottle in the customer's presence, to show you are serving what the customer ordered, and proceed thus:

- Pour the beer straight into the center of the wet glass with the can or bottle at a steep angle (Figure 7.8*a*) so that the beer gurgles out.
- When it creates a fine-textured head of some substance, lower the angle (Figure 7.8*b*) and fill the glass slowly until the foam rises to just above the lip.
- Wipe the container and set it beside the glass (Figure 7.8*c*).

Temperature is the third key to a perfect glass of beer. A beer that tastes right to the typical American customer is served at 40°F. Ales may be served at 45°F. Guinness and bock beers are usually chilled only lightly, and other imports may have different serving temperatures.

If beer is served in a thin glass at room temperature, the beer temperature will rise two degrees. If served in a heavy glass such as a thick mug, goblet, or schooner, it may rise as much as five degrees. Thus a 40° beer served in a mug at room temperature may be 45° when the customer begins to drink it.

Frosted glasses or mugs for beer are a recent fashion, especially in warm climates. To some extent the freezer-frosted glasses are a merchandising gimmick; they spark interest and underscore beer's cooling, thirst-quenching character. They also keep the beer cold longer. But some people don't like them. If it is served too cold, too fast, the glass can stick to the lip with the effect of frostbite. And if it makes the beer too cold, the taste buds will perceive less flavor. Refrigerated glasses may be the happy medium.

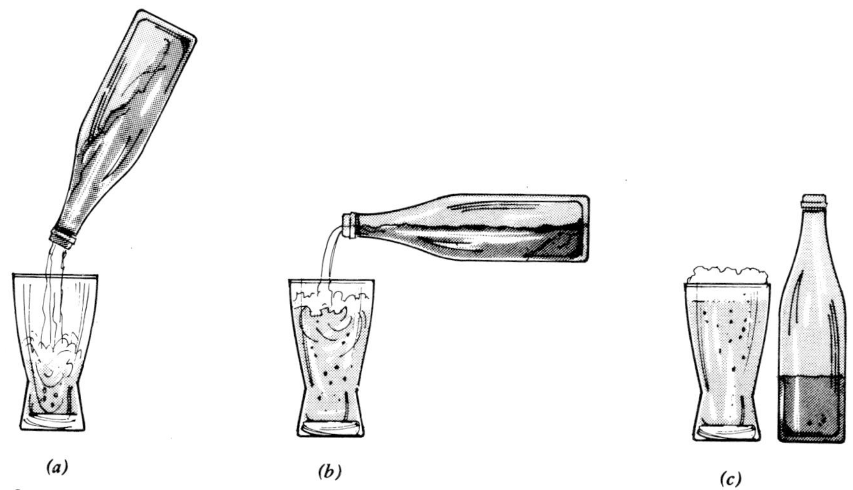

figure 7.8 How to serve beer from a can or bottle.

A beer that has been cared for properly, served with a modest head at the proper temperature, will probably call for another.

Summing up . . .

Beer rounds out the trio of alcoholic beverages usually found at every bar. For certain clienteles draft beer is a big attraction and is worth the space it takes and the care required in keeping it at its best. Employees should be well trained in the care of the beer system, the washing of glasses, and the pouring of the beer in order to serve a consistent, top-quality product.

Until recently most beer drinkers stuck with a single brand, but people have begun to experiment. Premium and superpremium beers, light beers, and imported beers are growing in popularity. Some enterprises have used beer tastings as promotional devices. Changing trends in beer preferences offer good merchandising opportunities.

chapter 8 . . .

SETTING UP THE BAR

Setting up the bar means organizing it for smooth operation each day. Organization is a management responsibility even though bar personnel do the setting up. The manager should see to it that every bartender and bar back understands clearly the routines of setting up and carries them out with precision and care. The routines of closing the bar are equally important. At both ends of the day, and all day in between, sanitation is a primary concern.

This chapter stresses the importance of organization and sanitation while detailing every aspect of opening and closing routines. There are several reasons why management should know this detail: for setting standards, for training employees, for follow-up, and for substituting at the bar in emergencies. As for sanitation, it is the manager who has the responsibility, no matter who is careless.

Usually your bartender sets up the bar. You will have scheduled his or her shift to begin half an hour or so before you open your doors so that everything will be ready.

What is there to be ready? There is nothing very complicated about it —a series of routines, a few rules, good organization. Good organization is the ingredient you supply—the way you want things done, the way you have trained your employees. If you have things organized, the day will flow smoothly. If something is overlooked, if things are left undone, customers will wait for their drinks and your bartender will be playing catch-up all evening.

The essentials of setting up are few. Everything must be superclean. The day's supplies of everything must be on hand and in position—liquor, mixes, beer, wine, ice, glasses, garnishes, condiments, utensils, bar towels, napkins, snack foods, ashtrays and matches, money in the register. That's *it*. How to get them ready is the meat of this chapter.

Sanitation . . .

Supercleanliness is essential for two reasons: customer appeal and customer health. Your local health department comes into the picture too, as guardian of customer health. You must meet its standards or lose your health permit.

A superclean bar is attractive—sparkling glassware, gleaming countertop, clean ashtrays, fresh-looking garnishes ready to go, everything neatly arranged. The underbar should be in the same condition—shining stainless steel, bottles all in order, ice bins full to the brim with fresh ice. It is all visible to someone—if not directly, then in the mirror. Even though the underbar functions as the kitchen area of the bar, the bartender plays to the public all the time, whereas the cook can make a mess behind scenes and clean up later. Train your bartenders to start clean and work clean.

Bacterial hazards. You are lucky that the bar is not so hazardous a place as a kitchen. There are few things you serve from the bar that are potential vehicles for disease. In fact, it is mostly the other way around: alcohol kills bacteria. There are a few things, however, that may carry bacterial hazards, and these your personnel should know about and guard against. They should also be aware of the nature and habits of bacteria, so that both they and you take cleanliness very seriously.

Bacteria that cause disease in humans have two characteristics that are of peculiar importance to food and beverage enterprises: they multiply at room and body temperatures, and they multiply very fast. They do this by splitting in half—doubling themselves—every 20 minutes or so, so that a single bacterium can become 4000 bacteria in four hours. To do this they need moisture, warmth, and food.

Fortunately few foods at the bar make good bacterial nourishment. Hazardous foods include eggs, milk, and cream, plus hors d'oeuvres containing any of these foods or meat, fish, poultry, stock, or sauce. So few of these items are typical of today's bars that it is easy to forget the whole subject and become careless. But don't. A single episode of food-borne illness can be devastating. The way to avoid it is to keep your hot hors d'oeuvres hot—above 165°F (74°C)—and your cold foods cold—40°F (4.4°C) or below —until they are about to be served. The danger range is 45° to 140°F (7° to 60°C), as shown in Figure 8.1. Check your refrigerator temperatures daily, and keep them at 40°F.

Preventing the spread of bacteria. More likely to happen is that germs brought in by people will be transmitted to other people. Your health department may keep your employees from bringing in chronic diseases by requiring an examination for each employee hired. But everyone carries bacteria and viruses of various sorts in the nose, mouth, and throat and on the hands, skin, and hair. Many germs can be transmitted through the air by breathing, sneezing, coughing, or directly by touching, or indirectly via glasses, towels, napkins.

To counter all this undercover bacterial action, not only must all bar equipment be kept superclean but scrupulous personal cleanliness is necessary. Each staff member must have clean hands at all times, clean nails, clean clothing, clean hair. Hands should be washed as a matter of habit before beginning work, before handling equipment and supplies, after using the bathroom, after blowing the nose or covering a sneeze, after smoking, and as necessary during work. Many health codes require a separate hand sink at the bar for this purpose. Soap and hot water and thoroughness are essentials. Towels used for the hands should be used for nothing else.

Employees should not come to work if they have colds or other illnesses. A cut, sore, or lesion of any kind should be securely covered.

Spills should be wiped up promptly; warm and moist, they can be ideal breeding grounds for bacteria. Used bar towels should be relegated to the laundry, never washed in the bar sinks and reused. Towels are among the worst offenders in spreading disease. A common habit among service personnel is to carry a towel on the shoulder or tucked in the belt. It isn't sanitary and shouldn't be allowed.

Washing and handling glassware and utensils. Glassware and utensils should be washed as soon as possible after use; dirty glasses are breeding grounds. The same method that produces a beer-clean glass (Chapter 7) produces a bacteria-free glass.

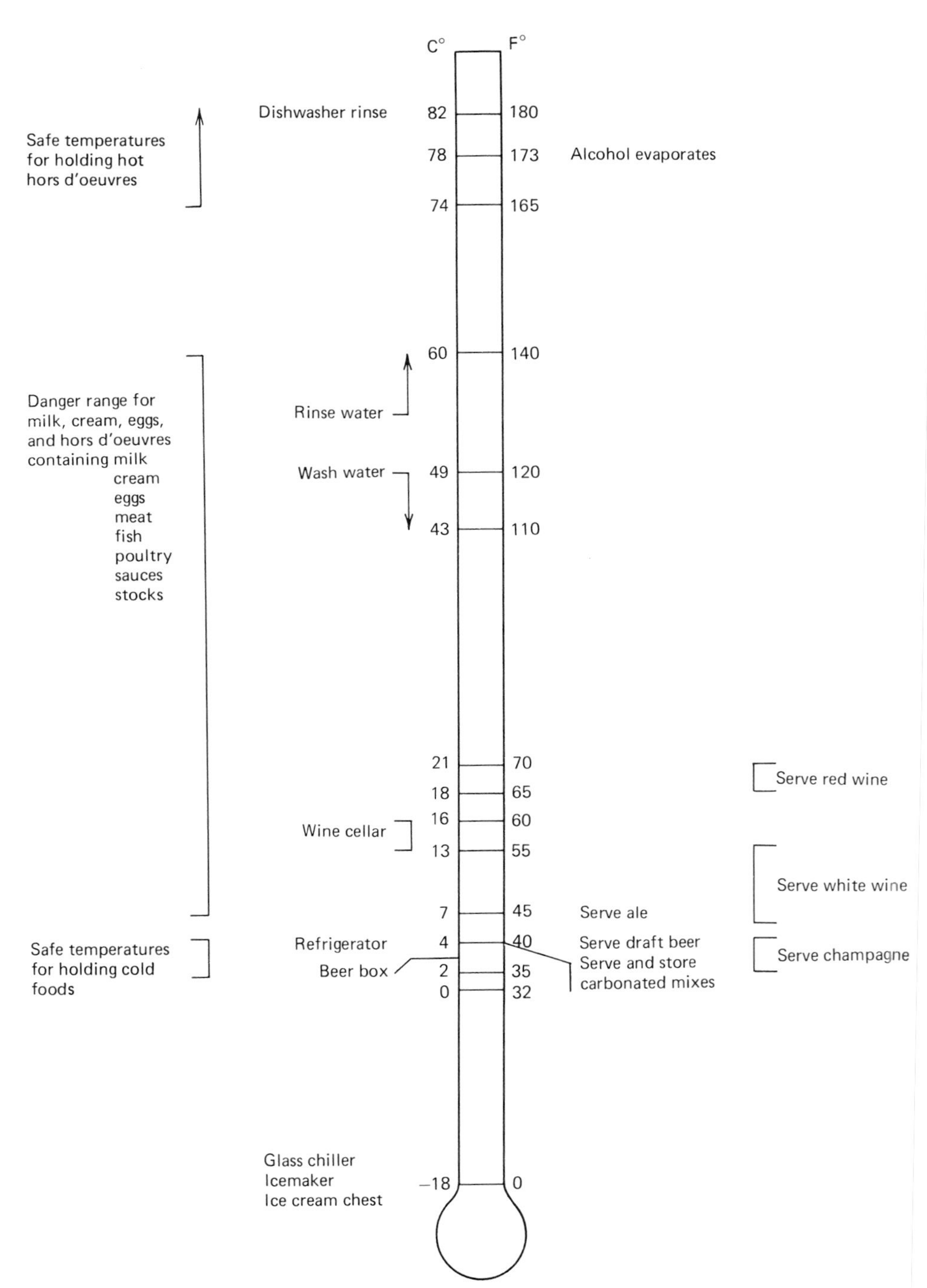

figure 8.1 Bar temperatures.

The water in the wash sink should be 120°F (49°C); do not let it cool below 110°F (43.3°C). It is important to use the right amount of detergent. The package instructions will indicate X amount to the gallon of water, so you must know how many gallons your sink holds. The nonfat detergent you use for your beer-clean glasses is suitable for all the others. Brushing the glass thoroughly is particularly important, with special attention to the rim for traces of lipstick as well as the invisible residues of use.

The rinse water in your middle sink should be 120° (49°C). It should be changed often, or the faucet can run slowly and the overflow drain can take away sudsy or cloudy water.

The third sink is for sanitizing, required in most health codes. Again, the right amount of sanitizer per gallon of water must be carefully observed. Use too much and it may linger on the glass; use too little and it won't do the job. Glasses should be submerged for at least 60 seconds. This solution too should be changed if it becomes cloudy.

All glasses should be air-dried by inverting them onto a deeply corrugated drainboard, a wire rack, or thick plastic netting made for the purpose. These surfaces should of course be clean. Air-drying avoids recontamination by fingers and towels. The rack, netting, or corrugated surface allows air to remove the moisture that germs need for multiplying—if any have survived the rigors of the thorough washing routine.

As part of their initial training all your service personnel should be coached in handling glassware. The fingers should never touch the inside of the glass, nor the rim, nor the outside of the glass below the rim as far down as lips may touch it. This rule holds whether the glasses are clean or dirty. Likewise the insides of mixing glasses and cups and plastic containers, the bowls of spoons, and the business end of any utensil should not be touched. In short, anything that touches food or drink should be untouched by human hands, including food and drink themselves and including ice.

The National Sanitation Foundation says that sanitation should be a way of life. If this is to be true in your enterprise, your own attitude plus the training you give must set the tone.

Liquor supplies . . .

One of the first things the bartender does in setting up is to replenish the supplies of liquor used the day before.

Each bar should have a standard brand-by-brand list of liquors, beers, and wines that should be in stock *at the bar* to begin the working day, with specified quantities of each brand. This is known as ***par stock***. The primary purpose of par stock is to ensure an adequate working supply, one day at a time. In setting up the bar, the bartender checks the bottles in stock against the list and adds from the storeroom whatever is needed to bring the stock up to par. Figure 8.2 is a typical par stock form.

Getting the liquor from the storeroom typically involves a requisition slip or some other form of record of what is being withdrawn, with the signature of the person withdrawing it. In a small enterprise only the manager and perhaps one other person—the assistant manager or head bartender—have access to the storeroom. In a large operation there is a storeroom staff and no one else is allowed in. A requisition slip requesting the liquor is presented to the person in charge, who issues the liquor and signs the slip. In this way, the amount is recorded as the liquor leaves the storeroom, and responsibility is pinpointed for issuing it. It now becomes the responsibility of the person withdrawing it, who has also signed the slip. Requisitioning and issuing are described in greater detail in Chapter 11.

Next, the beverages must be arranged for efficient use. Bottled beer, white wine, and sparkling wine must be chilled, rotating the stock in the refrigerator to bring forward the cold bottles from the day before. The speed rail at each station must be checked to be sure there is a starting supply in every bottle, and reserves must be set up, with pourers in place, for those that are almost empty. The arrangement of bottles in the well must also be checked. Most bartenders arrange their stations with the bottles in a certain order according to frequency of use. They know where each one is and they can reach the right bottle with speed and accuracy. If one is out of place, another will be too, and the guest may end up with tequila in a Vodka Martini.

Pourers should be checked daily. If corks on stainless-steel pourers are wearing loose, now is the time to replace them.

Draft beer must be readied for the first draw. Fresh water must be run through each line and faucet; then the line must be disconnected from the water tap and reconnected to the keg. The beer faucet is then opened and allowed to run until the water is drained off and the beer appears. The pressure gauge should be checked at the keg.

Mixes . . .

The supply of mixes must be checked and replenished, and any you make up fresh must be prepared. It is advisable to have a list of quantities needed daily—a kind of par stock for mixes—based on records of past sales. Amounts may vary according to the day of the week, with the quantity rising on weekends. Fresh-made mixes are one thing you don't want an oversupply of, since they won't stay fresh indefinitely.

Carbonated mixes. The indispensable carbonated mixes are club soda, tonic, ginger ale, cola, and 7-Up. In addition you may use collins mix, diet

SKY CHEFS

STOREROOM: ______________________

FACILITY: ______________________

ALCOHOLIC BEVERAGE
PAR STOCK RECORD

Date Prepared: ______________________

Name of Preparer: ______________________

SCOTCH	Par Stock	Bottle Size
Pouring Brand		
BOURBON		
Pouring Brand		
WHISKEYS		
Pouring Brand		
RUM		
Pouring Brand		
TEQUILA		
Pouring Brand		

GIN	Par Stock	Bottle Size
Pouring Brand		
VODKA		
Pouring Brand		
BRANDIES		
Pouring Brand		
CORDIALS		

APERITIFS	Par Stock	Bottle Size
ALMADEN		
Burgundy		
Burgundy		
Chablis		
Chablis		
Rosé		
Rosé		
BULK WINE		
Burgundy		
Chablis		
Rosé		
CORKED WINES		
BEER		
Local – Draft		
Premium – Draft		
Heinekens–Bottled		

figure 8.2 Par stock form used by Sky Chefs in its airport bars and restaurants. Each bar has its own par stock. (Courtesy Sky Chefs)

drinks, root beer, Sprite, and Dr. Pepper, depending on your part of the country, your type of operation, and the preferences of your clientele.

In setting up carbonated beverages there are two criteria: an *adequate* supply and a *cold* supply. A carbonated beverage at room temperature will lose all its bubbles as soon as you pour it, filling the glass with fizz and then leaving a flat and scanty drink. It is critical to have your mixes as cold as possible.

There are three kinds of carbonated mixers: bottled, premix, and postmix. As explained in Chapter 3, both premix and postmix come in 5-gallon cylinders and are chilled and carbonated automatically at the time they are dispensed. Setting these systems up simply consists of having the right number of containers—at least one in reserve for each mix—and checking the pressure gauge on the CO_2 cylinder. This gauge should read 60 psi, which is the amount of pressure needed to carbonate the beverage and deliver it to the dispenser. If the indicator gives another reading, pressure should be adjusted accordingly.

If you use bottled mixes, the small bottles are the only way to go. They must be thoroughly chilled and opened only as needed. If an opened bottle sits more than half an hour it should be discarded; your next customer wants a sparkling drink. A 12-ounce bottle will make three highballs.

Juices and juice-based mixes. Another group of mixes are fruit juices and mixes having a fruit-juice base. The most common of these are orange, tomato, V-8, pineapple, grapefruit, cranberry, lemon, lime, and sweet-sour mixes. Which of these you need will depend on your type of enterprise, your drink menu, regional preferences, and your own clientele. You may also want to develop a house specialty based on your own freshly made mix.

Orange juice may be bought fresh in cartons, fresh in the orange, frozen in concentrated form, and canned. Fresh orange juice is variable in taste depending on the seasonal availability of different types of oranges. Canned orange juice, even the best of it, tastes unmistakably canned. Frozen unsweetened orange-juice concentrate is the most consistent in flavor, keeps for months unopened in the freezer, and is quickly prepared.

Some proprietors find that drinks made with fresh orange juice squeezed to order are a great house specialty even when they carry the high price tag of a labor-intensive product. The juice should be strained through a coarse strainer to exclude pith and seeds, but a little fine pulp gives authenticity. Florida and Texas oranges give the most juice.

Grapefruit juice comes in cans or as frozen concentrate as well as fresh in the grapefruit itself. Each form has the same advantages and drawbacks as the comparable forms of orange juice, except that there may be less flavor variation in fresh grapefruit juice.

Tomato, V-8, and pineapple juices come only in cans. Cranberry juice comes in bottles or as a frozen concentrate.

Lemon and lime juices may be freshly squeezed in quantity and kept in glass or plastic containers. Lemon and lime granules are also available, both sweetened and unsweetened. They leave much to be desired. Frozen concentrated unsweetened lemon and lime juices make the best drinks.

A sweet-sour mix of fresh lemon juice and sugar can be made ahead. Sweet-sour mixes can also be bought bottled or as frozen concentrate or in powdered form. Some have a foaming agent—frothee—to simulate egg white. Some mixes are better than others. Try them out.

There are several mixes for specific drinks—Bloody Mary, Daiquiri, Margarita, Mai Tai, Piña Colada. All you do is add the liquor. Of these, the frozen concentrates are likely to be best.

All the juices and mixes keep for at least a day or two. Some keep a week or more—cranberry, tomato, and pineapple, for example. As soon as they are opened, canned juices should be transferred to glass or plastic containers. A new batch of juice should never be added to an old batch. It is a good idea to tape the date of preparation on the container.

Setting up juices and mixes is a matter of checking supplies, tasting leftover supplies for freshness, making new batches as necessary, and arranging them in place for efficiency and speed. All should be refrigerated until just before you open your doors. Although they are not subject to bacterial contamination, they will lose their flavor at room temperature and they will melt the ice in the drink too fast, diluting its fresh tart flavor.

If you have a frozen drink machine, making a recipe for the day ahead is part of setting up. However, use first anything left from the day before.

Other liquids for mixing. Another mix you will find convenient to have in constant supply is ***simple syrup***. This is a laborsaving substitute for sugar in other forms, since it blends more quickly than bar sugar or the traditional lump of sugar muddled with bitters in an Old-Fashioned. To make simple syrup you use equal parts by volume of sugar and water and boil one minute, or blend 30 seconds, or shake until thoroughly mixed. For 1 teaspoon of sugar in a recipe use ½ ounce of simple syrup (1½ teaspoons). Keep it in a bottle or plastic container. It is safe at room temperature.

Some other nonalcoholic liquid ingredients are Rose's lime juice, grenadine, beef bouillon, and cream of coconut. There is no substitute for Rose's lime juice, a sweet and pungent syrup of lime and sugar. Grenadine is a sweet red syrup flavored with pomegranates, used as much for color and sweetness as for its special flavor. Beef bouillon is an ingredient of the Bullshot and is best purchased in 1-serving cans. Cream of coconut comes in cans and is used in tropical drinks. Passion fruit syrup is a bottled mix of tropical juices, sugars, and additives.

Less common are orgeat (pronounced *oar*-zha), a sweet almond-flavored syrup; falernum, a slightly alcoholic (6%) sweet syrup with an almond-ginger-lime flavor and a milky color, and orange flower water, a flavoring extract. All these syrups are okay to store at room temperature.

Milk and cream are often called for in drink recipes. Cream is usually half and half but may be heavy cream for a well-to-do clientele. Whipped cream is especially popular for dessert/coffee drinks. Milk and cream must be kept refrigerated both for health reasons and because they quickly turn sour at room temperature. Cream substitutes are sometimes used, but the drink is not as good and the substitutes spoil easily. Keep them cold in the refrigerator too. When setting up, you can whip cream ahead and it will hold well for the evening if you whip in a little fine sugar at the end. Consistency should be just short of stiff for most drinks.

If coffee drinks are on your menu, the coffee must be made. It is well to keep in mind, though, that coffee held for longer than an hour changes flavor and will spoil the finest recipe. Fresh coffee is essential!

Bottled waters such as Perrier are currently in vogue, though the vogue is waning. They should be kept in the refrigerator and served in the opened bottle with an empty glass. Ice cubes and a wedge of lime or lemon may be included if the customer wishes.

Garnishes . . .

Preparing fruits for garnishing the cocktails, highballs, and other drinks is one of the most important parts of setting up. The standard items include lemon wedges and lemon twists, lime wedges, orange and lemon slices, cherries, olives, and cocktail onions (not a fruit but replacing a fruit in some cocktails). Other fruits and vegetables sometimes used for eye and taste appeal are pineapple spears or chunks, cucumber spears or celery sticks, fresh mint, stick cinnamon for hot drinks, and anything of your own inspiration.

Preparing citrus fruits. Lemon wedges are used for appearance and for squeezing juice into individual drinks. Lemon twists are used for the flavor of the rind; they are rubbed along the rim of the glass and twisted to squeeze the oil into the drink. Whole lemons are also squeezed for fresh lemon juice in quantity.

All citrus fruits should be washed thoroughly before cutting, and so should the hands. For cutting, use a sharp knife and cut on a cutting board, not in midair. Always cut down and away from yourself, keeping the fingers and thumb of your other hand curled out of the way (Figure 8.3).

The best lemons are medium in size, with medium-thick skin (too thick is wasteful; too thin is hard to work with and not as nice to look at). A good size of lemon is 165 ***count*** (165 lemons to the case). You can increase juice yield of lemons by soaking them in warm water and rolling them back and forth on a hard surface while exerting pressure with the flat of the hand.

To cut lemon wedges (Figure 8.4), first cut a small piece off each end—just skin, not pulp (*a*). Cut the lemon in half lengthwise (*b*) and, cut side

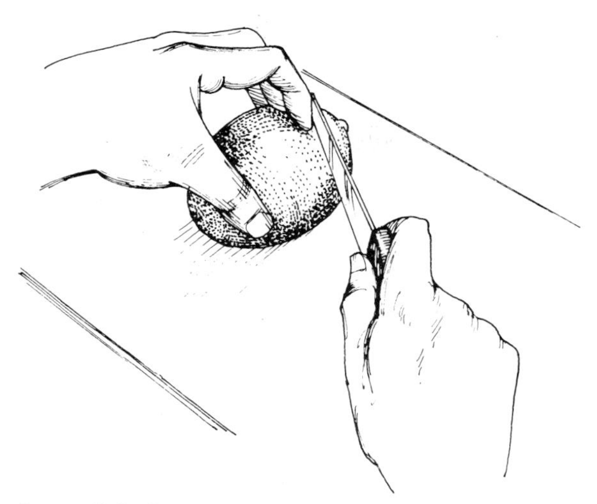

figure 8.3 Cutting technique. Hold fruit firmly on board by placing fingers on top as shown, fingertips curled under slightly, thumb well back. Cut down toward board. Knuckles can guide knife.

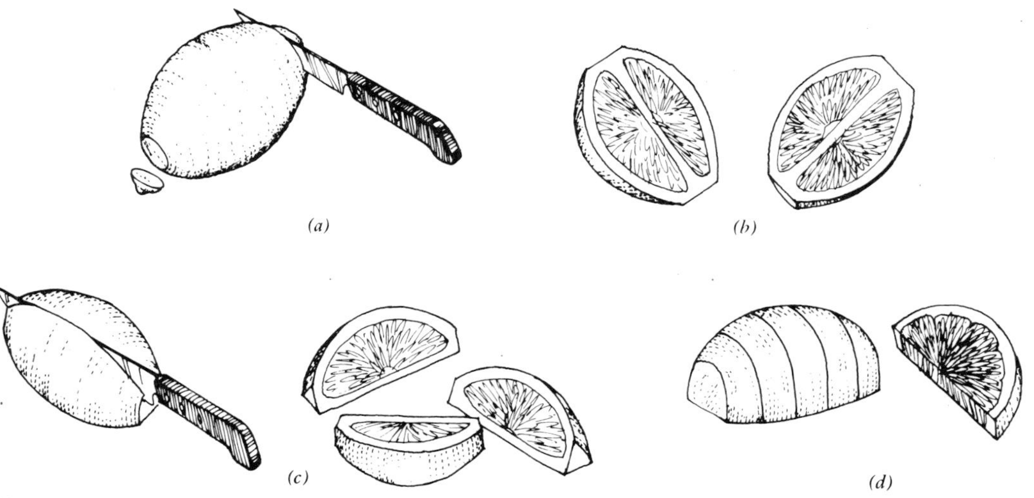

figure 8.4 How to cut lemon wedges.

down, cut each half lengthwise into wedges of the size you want (*c*). An alternate way of cutting wedges is to cut the lemon in half lengthwise and then, with the cut side down, slice each half crosswise into half-inch slices (*d*). These smaller wedges fit nicely into the hand squeezer or are easy to squeeze between the fingers. If you want a wedge that will hook onto the rim of the glass, make a cut lengthwise down the middle of the half-lemon before slicing.

To make lemon twists use a zester or stripper. These tools strip off just the yellow part of the skin—the ***zest***. Cut pieces about 1½ inches long. To produce twists without a special tool (Figure 8.5), first cut off both ends of the lemon (*a*), then scoop out the pulp with a barspoon (*b*), saving it for juice if you want it. Then cut through the rind (*c*) and lay it flat. Scrape away the white pith (*d*) and discard it, leaving about ⅛-inch thickness of yellow skin. Slice this in half-inch-wide strips (*e*).

If you want lemon ***wheels*** for garnishes (Figure 8.6), simply cut crosswise slices beginning at one end of the lemon (*a*). Discard end pieces having only skin or pith on one side. Slices should be thin yet thick enough to stand up on the edge of the glass. Make slits halfway across slices for this purpose (*b*).

The best limes are deep-green, seedless, and on the small to medium side (54 count—the cases are smaller than lemon and orange cases). But the bar manager is often at the mercy of the market, taking what is seasonally available.

The ideal size lime will make 8 neat wedges (Figure 8.7). First you cut off the tips, then cut the lime crosswise (*a*). Then put the cut sides down (*b*) and cut each half into four equal wedges (*c*). Lime wheels are made the same way as lemon wheels.

Orange slices are made by slicing the orange crosswise as you do the lemon. Make slices ¼ inch thick; if they are any thinner they don't handle

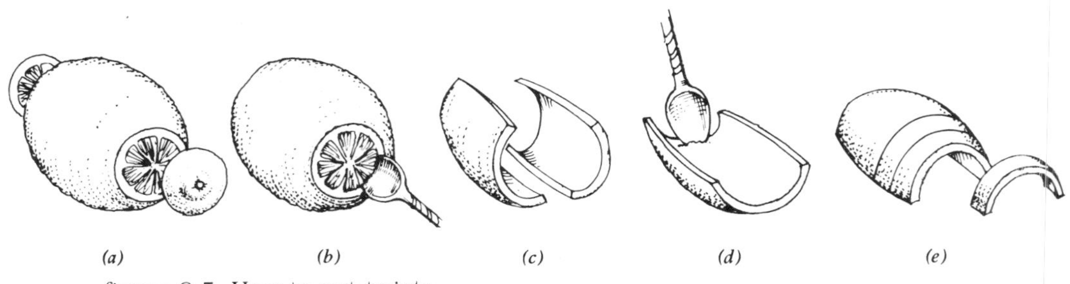

figure 8.5 How to cut twists.

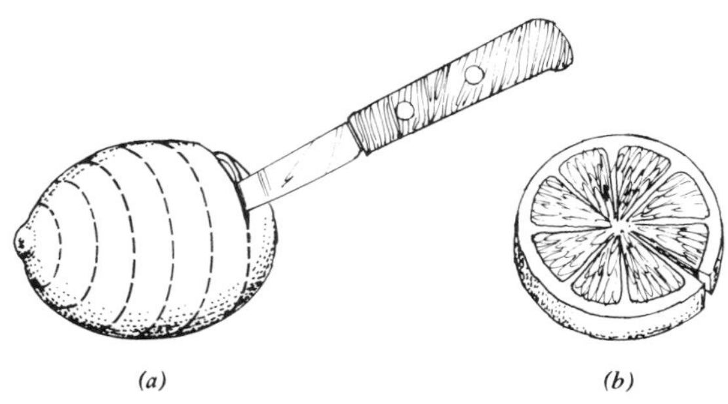

figure 8.6 How to cut wheels.

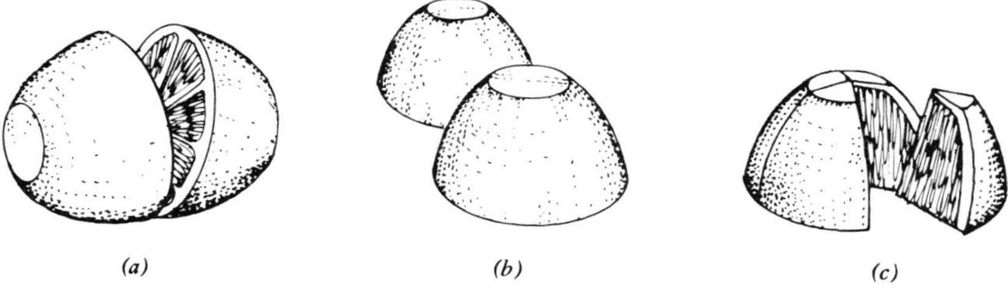

figure 8.7 One way to cut lime wedges.

well and tend to dry out. Orange slices can be used whole, as wheels, or quartered and impaled on a pick—***flagged***—with or without a cherry as in Figure 8.8 (*a*) and (*b*). The best-looking oranges are the California varieties; the navel oranges are ideal because they have no seeds. A case count of 80 gives you a good size.

All citrus garnishes should be kept moist. They keep best if you can form each fruit back together again, but often there are too many pieces. Covering them with a damp bar towel helps to retain moisture and appearance. So does refrigeration; you can bring them out in small batches. Often you can prolong life by spraying them with 7-Up. Twists dry out especially quickly and should not be made too far ahead.

Other garnishes. Cherries, olives, onions, and pineapple chunks come in jars or cans and need no special preparation. Cherries used as garnish are maraschino cherries, pitted, both with and without stems depending on the drink. Cocktail olives are small pitted green olives of the manzanilla type. Although they are available stuffed with anchovies, nuts, or pimientos, your customers may not like your choice, and it is best to go with the traditional empty olive they expect. Cocktail onions are little onions pickled in brine, and pineapple chunks are chunks of canned pineapple.

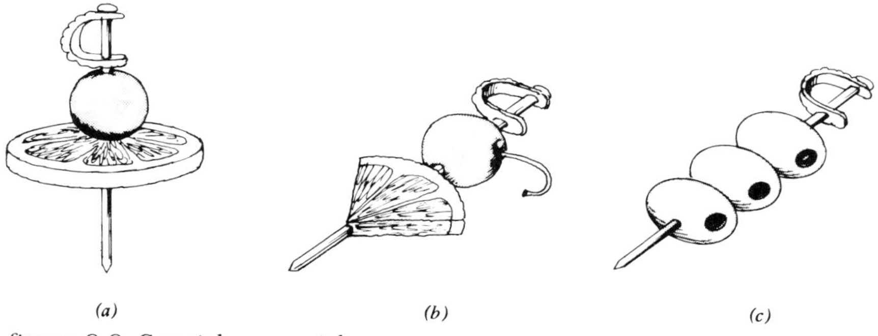

figure 8.8 Garnishes on picks.

These garnishes are removed from their juices, set up in glasses, cups, or a condiment tray, and kept moist until time to serve. The damp-towel covering is good for them too.

Other fresh-cut garnishes such as fresh pineapple spears, cucumber sticks, and celery sticks should be cut to size and shape with an eye to appearance in the drink. They too are kept chilled and moist. For added crispness, celery and cucumbers can be kept in ice water in the refrigerator until needed.

Just before serving time all perishable garnishes are set up on the bar in an arrangement that is both efficient and attractive (Figure 8.9). If the bartender does the garnishing, everything should be within easy reach—a separate setup for each station. If servers garnish the drinks, the garnishes should be at the pickup station. Each set of garnishes must have a supply of picks for spearing the garnishes to go into the glass. If you don't want picks in your drinks, have tongs handy for placing the garnish. Don't allow fingers for this ceremony; that is both unsanitary and unsightly.

Condiments for flavor . . .

For some reason the term ***condiments*** when used in the context of the bar often refers to the garnishes prepared for eye appeal. In more general usage the term condiments applies to pungent or spicy foods used to add a special flavor or enhance the total flavor. These too we have at the bar—Tabasco, bitters, Worcestershire, sugar, spices.

Tabasco is liquid hot pepper. There are other brands too, but Tabasco is the one your customers expect to see go into their drinks. It goes in drop by drop, through a built-in drop-size dispenser.

Bar bitters are actually nonbeverage spirits (undrinkable by themselves). They are like liqueurs without sugar, made by distilling or infusing alcohol with secret formulas of bitter herbs, spices, and other flavoring agents. You will need two kinds at the bar: Angostura bitters from Trinidad and orange bitters. A bitter bitters from New Orleans, Peychaud's, is also called for in some drinks. Bitters are dispensed by the dash, which is the equivalent of $^1/_6$ teaspoon. A dasher is built into the neck of the bottle. If you want only a drop, you can hold the bottle level and tap it with one finger.

Worcestershire sauce is a nonalcoholic kitchen condiment that found its way to the bar to season some versions of the Bloody Mary. It comes with a built-in dasher/pourer: turn it horizontally and it is a dasher; turn it vertically and you can pour from the bottle.

You will need salt at the bar in two forms: ordinary table salt and coarse salt, called kosher salt or Margarita salt (probably at a higher price). You'll use table salt in Bloody Mary mixes. The coarse salt is for rimming the Margarita or Salty Dog glass.

figure 8.9 Garnishes and serving accessories set up for service. Hinged top on condiment tray can be closed to keep garnishes moist and fresh. (Photo by Pat Kovach Roberts)

You will need sugar in several forms. Simple syrup blends best in drinks. Superfine—often called bar sugar—is the best granulated type, but you can use ordinary sugar if you mix longer. Cubes are not altogether obsolete; they are used in bars that make Old-Fashioneds in the old-fashioned way. They are also an excellent means of making flambéed drinks: soak a cube in 151° rum and set it aflame. Even honey figures in some recipes. You have to decide how far you want to go.

Certain ground spices should be on hand: nutmeg, cinnamon, pepper, celery salt. And everything at each station, no matter how small or how seldom used, must be in its appointed place.

Ice . . .

Shortly before opening time, the ice bin at each station must be filled with fresh ice, with clean ice scoops ready in the bins.

Some precautions are essential to keep the ice in the bins clean and fresh. All your service people should observe them:

- Never touch ice with the hands. Use a scoop to fill the glass or shaker.
- Never use a glass as a scoop. This is a common practice because it is fast, but it is a very dangerous one. You can easily chip the glass or break it outright; then you will have glass mingled with your ice. *You* won't find it; some customer will. As soon as a glass breaks, empty the bin and wash it out.
- Never put anything in the ice bin to cool—no wine bottle, no warm glass. It may transmit dirt and germs. It may also damage the warm glass.
- Do not position condiment trays over an ice bin; something may drop in.
- Never reuse ice, even if you wash it. Throw out all ice from used glasses. Start each drink with fresh ice in a fresh glass, even if it is the same kind of drink for the same person.
- Do not use your ice scoops for anything but ice. Keep them in the bins.

Accessories to service . . .

Near the condiments at each station go the accessories to garnishing—the picks, straws, sip sticks, stir sticks, and the cocktail napkins. You will have chosen them all carefully as you planned your drink menu and your garnishes, so that the visual impression of each drink served carries out the total image of your enterprise.

Picks may be either the colored plastic kind—sword-shaped—or round wood toothpicks. They are used to spear the olive or the onion or the cherry-plus-orange, and both spear and garnish go into the glass. Some places fill up the pick with three olives or onions, adding flare and an impression of largesse (Figure 8.8c).

Straws are useful in two lengths—the 5-inch length for drinks in stem or rocks glasses, the 8-inch length for highballs and collins-size drinks. Straws are essential for sipping frozen drinks, while in highballs and rocks drinks customers use them as stirrers.

You may prefer to use plastic stir sticks for highballs. These can be custom-made as souvenirs of your place for patrons to take home as mementos. Sip sticks are somewhere between straws and stir sticks; they are hollow but firmer than a straw and smaller in diameter, and usually only one is used per drink.

The final essential is a supply of cocktail napkins, stacked with the folded edge toward the bartender or server for easy pickup, or sprayed into a fan so they don't stick together. These napkins should be 2-ply; anything thinner disintegrates in no time. There should be a good stack at each station and plenty in reserve.

Ashtrays head the list of top-of-the-bar accessories. They should be sparkling clean—and kept emptied and wiped clean each time they are emptied. Smoking tends to make people thirsty. Quite often the emptying and wiping of an ashtray will trigger a second drink or another glass of wine.

Matches—with your logo and address and phone number—must be set up along with the ashtrays. Again, there must be a plentiful reserve supply; these too go home with customers.

The snacks you serve are placed on the bar just before you open. These can be part of your image (are you famous for your hot hors d'oeuvres?) or they may be subtle thirst promotion—peanuts, pretzels, popcorn, cheese crackers.

All the essentials for table service must be clean and in readiness at the pickup station:

- Drink trays.
- Tip trays.
- Guest checks for servers.
- Wine lists.
- Wine coolers.
- Bottle openers.
- Bar towels or white table napkins.
- Cocktail napkins.
- Coasters for beer.
- Extra pens or pencils.
- Matches.

Opening the cash register . . .

Most bar operations work with a standard amount of starting cash for the cash register, called the ***bank***. The purpose of the bank is to have ample change in coins and small bills such as ones and fives. The amount of the bank will vary according to the sales volume and the policy of the bar. When the bar is closed the bank is put into the safe, usually in a compartmentalized cashbox or a cash-register drawer with a locked lid.

In opening the register the first task is to count the bank to make sure there is adequate change. In some systems the person closing the bar the night before has left a ***bank count slip*** (Figure 8.10) listing the amounts of the various coins and bills. The opening bartender checks the opening count against this slip. If everything checks out, the money goes into the register drawer, ready for business.

The next step is to check the register tape to see that there is an ample

BANK COUNT SLIP

Total in drawer: $ 300.00				Name $ C. Smith	
Bills: $ 250.00		Coins: 50.00		Date 7/4/81	
$1 Ones	$ 100.00	Pennies	$ 3.00	Shift: 11-3 3-7 (7-close) (Circle one)	
$5 Fives	$ 30.00	Nickels	$ 6.00		
$10 Tens	$ 60.00	Dimes	$ 20.00		
$20 Twenties	$ 60.00	Quarters	$ 20.00		
		Halves	$ 1.00		

This form is to be used by cashier turning in the bank after sales have been accounted for. This slip should be left with the bank.

figure 8.10 Bank count slip.

supply for the day and that the printing on the tape is clear. A clear tape is essential to the recordkeeping of the bar. Next, the register is cleared to make sure no transactions have been recorded since the last shift. The next step is to ring up a *No Sales* transaction to obtain the first transaction number on the tape. This number must be recorded on the cashier's check-out slip (discussed later). This is a control procedure enabling management to have an audit trail when reconciling the register and the moneys at the end of the shift.

The person opening the bar should make sure that the usual materials are available at the register—pencils, pens, paper clips, rubber bands, pay-out vouchers, credit-card forms.

Mise en place . . .

There is a fine tradition in the restaurant industry expressed in the phrase *mise en place*—putting in place. It means that the setting up is complete and everything is in position ready to go, right down to the last olive. The first customer has only to cross the threshold to set it all in motion.

A perfection of mise en place brings a moment of equilibrium between preparation and action that is important in starting the action off right. Not only is everything ready, but the bartender *knows it.* The resulting confidence influences all the bartending that follows; the readiness is psychological as well as physical. A relaxed, confident bartender is a better host, makes fewer mistakes, and can cope with emergencies far better than one who wasn't quite ready for that first customer and is still trying to catch up. A bartender who takes pride in mise en place is a real pro, and one to cherish.

A good mise en place is all the things we have been talking about plus a few more:

- Glasses of all the necessary kinds and sizes clean and in place in the numbers needed.
- Bar implements clean and in place at every station—jiggers, mixing glasses and cups, shakers, barspoons, strainers, squeezers, openers, zesters, scoops and tongs, blenders and mixers and ice crushers clean and in working order.
- Supply of guest checks in place; credit-card slips in place in good supply; credit-card machine in position.
- Coffee made, if coffee served.
- Money in cash register sorted, counted, ready to go.

Figure 8.11 is a suggested checklist for a complete mise en place. Such a list can be very useful in keeping procedures standardized and orienting substitute bartenders and new personnel.

Behind-the-bar behavior . . .

Your bartender is the key person in conveying the image of your enterprise. You may have special requirements in personality, dress, and behavior if you are after a certain specialty image. But certain basic rules apply across the board.

Prompt, friendly, courteous service is the overriding requirement. Greeting a new person immediately conveys a sense of welcome and belonging. Remembering an old customer's favorite drink makes that person feel appreciated. One warning applies here, however: if a regular customer comes in with a new companion it's best not to give away that "regular" status. A genuine smile for everyone is a great sales asset—and a safe one.

If uniforms are not required, dress should be conservatively appropriate

CHECKLIST FOR SETTING UP THE BAR

Liquor: Bring stock up to par. Turn in empties.

Well: Check supplies. Set up reserves. Check pourers. Line up bottles.

Bottled drinks: Replenish and rotate beers, wines, mixers, bottled waters in cooler. Check cooler temperature (40°).

Draft beer: Check supply. Wash lines, connect to keg, drain water, taste beer. Check pressure. Check beer-box temperature (36°–38° F).

Soda system: Clean gun. Check soda cylinders. Check CO_2 pressure.

Glasses: Wash used glasses. Check supplies. Arrange in order.

Implements: Check and set up blenders, mixers, mixing glasses, shakers, jiggers, barspoons, strainers, squeezers, openers, scoops, tongs.

Juices and mixes: Taste leftovers for acceptability. Prepare fresh juices, mixes; refrigerate. Prepare frozen-drink mix; start machine. Check and replenish bottled-mix supplies. Make simple syrup.

Ice: Check and clean bins. Fill with ice.

Garnishes: Prepare and set up lemon wheels, wedges, twists, lime wheels, wedges, orange wheels, flags, cherries, olives, onions, cucumber sticks, celery sticks, special garnishes.

Condiments: Set up salt and sugar for rimming, bar and cube sugar, bitters, Tabasco, Worcestershire, pepper, nutmeg, cinnamon, celery salt.

Serving accessories: Set up picks, straws, stir sticks, sip sticks, cocktail napkins, bar towels, coasters, bottle openers, wine carafes, wine coolers, ashtrays, matches, serving and tip trays.

Cash register: Count bank and set up register. Check tape. Check and set up cashier supplies: guest checks, credit-card forms, payout vouchers, pencils, pens, paper clips, rubber bands, stapler and staples.

Coffee: Make fresh. Set out sugar and cream.

Snacks: Set out.

figure 8.11

to the general atmosphere. Blue jeans and T shirt are out of place behind a hotel bar, just as a tuxedo is at a truck stop. White shirt, dark pants, and conservative tie are safe dress anywhere. A bartender should never outdress the clientele. It goes without saying that clothes should be as neat and clean as the personal cleanliness you require.

Bartenders should not drink while working. Smoking, eating, and chewing gum are distasteful to many customers and should be confined to breaks away from the bar.

The legendary bartender you tell your troubles to does not belong in most bars. It is best for bar personnel not to get too involved in conversations, lest they neglect other customers or seem to play favorites. This is especially true of personal conversations, and most especially remarks about the bar, the boss, the other help, and the other customers. It makes a bad impression all around. It is bad, too, to butt into an ongoing conversation, whether it be football, politics, men, women, or religion, and it's disastrous for the bartender to get involved in an argument.

Discretion is an absolute must. The bartender's lips should be sealed when a customer reappears after getting drunk the night before, or brings a different date, or brags about something the bartender knows isn't true. If a phone call comes for a customer, the latter's presence should not be acknowledged without checking with the customer.

Courtesy goes beyond the initial greeting; it extends to keeping one's cool under duress. There will always be that customer who challenges everything—the drinks, the prices, the right change. Only the cool calm bartender can handle this one. And be sure the change is right every time.

If a customer begins to have too many, it is almost universal practice to shorten the drinks—less liquor, more ice, more mix. Just a touch of spirit on the top of the next drink without stirring is a common way of dealing with the problem. The bartender has no choice but to stop serving a customer who is really drunk; it is against the law. If there is trouble, it is up to management to deal with it as tactfully as possible: an escort to the door and off the premises, a taxi home, even the police if necessary. Your reputation is on the line, and maybe your license too, and maybe even more: some states have passed legislation holding that if someone becomes drunk in your establishment and later causes damage to property or person, your establishment and/or the person who served the alcohol can be held liable.

Closing the bar . . .

After standing on one's feet for six or eight hours, maybe more, the prospect of cleaning up the bar is not thrilling. But clean he/she/you must.

A good place to begin is by putting up all the perishables, such as cream and juices. These go into the refrigerator. Juices should be in covered plastic containers. If you have any opened cans of Bullshot, Coco-Lopez, or other mixes, empty them into glass or plastic containers and refrigerate.

Along with the juices, look after the cut-up fruits. Generally cut limes will last for 24 hours. Place them in a jar with a lid, add some 7-Up or collins mix to keep them moist, put the lid on the jar, and place it in the refrigerator. Cut-up oranges do not keep well, and it really does not pay to keep them. Whole fruits go into the refrigerator.

Olives and onions should be put back into their jars, submerged in their original brine, with the lid of the jar closed. The same goes for cherries: be sure they are covered with syrup, otherwise they will dry out and wrinkle. Everything goes into the refrigerator. Its 40° moist air will preserve the quality of the product and will help assure the health of your patrons.

Next to be put away are the bar snacks—pretzels, chips, peanuts, crackers, popcorn. Empty all leftovers into the trash. Never put anything touched by customers' hands back into a container with fresh items, unless each one is individually wrapped. This is a health precaution to protect your clientele. Close all the lids tight to keep everything fresh.

At this point wash all the glasses and put them on their assigned shelves. Wash all tools and equipment—blender and mixer cups, utensils. Leave them upended on the drainboard to air-dry.

Next will be the cleaning of the lines in your beer box, as described in Chapter 7. Carbon dioxide pressures should be checked and adjusted. If any wines were on temporary display, they must be locked safely away.

If you have a frozen-drink machine, drain the contents into a plastic container and store them in the refrigerator. Turn off the machine and clean it out according to the manufacturer's instructions.

The next step is to remove and soak all the pourers. This will prevent the sugars found in most liqueurs from building up and slowing the flow. While the pourers are soaking, clean the soda gun and the ring that holds it. Remove the nozzle parts and wash, rinse, and sanitize them. Wipe down the flexhose with a damp cloth. After the pourers have soaked for about 10 minutes, dry them and replace them on the bottles. Wipe each bottle with a damp cloth to remove any spills.

When all the supplies have been put away, the glasses washed, and the liquor taken care of, it is time to clean out the ice bin. First you scoop out all the ice into the nearest compartment of the bar sink. Then you run hot water into the bin to melt all the remaining ice and wash down any debris left in the bin. Most of this debris will be trapped at the drain. With a towel, remove this and shake it off the towel into the trash can. Then with a clean towel scour the walls and bottom of the bin. If your bin has an old-style cold plate at the bottom (part of your dispensing system), pay particular attention to cleaning thoroughly under and around it. This is a great place for collecting lots of crud. It can make your bins smell stale and affect the taste of your ice, as well as being unsanitary.

After cleaning the ice bin, proceed to the bar top. Remove everything and wipe down the surface with a damp cloth. A bar with a top of brass or other metal should be polished at this time. Empty the ashtrays, making sure all cigarettes are out before you dump them into the trash. Wipe the ashtrays clean. Replenish the supply of matches, ready for a new day.

Now that everything has been washed, it is time to empty the sinks and scour them with a mild abrasive and very fine steel wool.

After the sinks are cleaned the bar floor must be cleaned—swept and mopped. Remove the floor mats and hose them down outside at least every other day.

Emptying the trash cans is the next closing ritual. Hose them down and give each one a new plastic trash liner.

Up to this point the cleanup has dealt with the sanitation and safety needs of the bar. The next step is to get the supplies replenished for tomorrow's business. If the storeroom is closed at this hour, the stock can't actually be brought to the bar, but the accumulated empties can be counted and listed out and a requisition completed for their replacement. Beer and wine supplies should be checked along with spirits, and necessary replacements added to the list.

Supplies of bottled mixes, fruits, and condiments should also be counted and necessary replacements ordered. The syrup cylinders should be checked too, as well as the pressure at the CO_2 cylinder.

The entire closing-down procedure is designed to do two things for you:

- It ensures that the sanitary practices essential to successful operation are carried out.
- It ensures that, if anything happens to the person who opens up tomorrow, the bar is ready to go with very little effort by a substitute. This is crucial—to be able to open at a moment's notice if you must.

In some enterprises, closing the bar also includes closing down the register. This will differ from bar to bar, but here is an order of procedure:

1. If tickets/checks are used, be sure that all have been rung up.
2. If a tape is to be used for checking out the register, remove it, sign it, and date it.
3. Read the register and record the readings on the ***cashier's checkout slip*** (Figure 8.12). (This will depend on the system used.)
4. Remove the cash drawer with the cash and all supporting papers such as credit-card charges, checks, payout vouchers.
5. Reconcile the total register sales with the actual cash plus house-credit slips plus credit-card slips plus checks plus payout vouchers. Record everything on the cashier's checkout slip.
6. Count out the bank and place this money in the cash-register drawer or cashbox. Write on a new bank count slip (Figure 8.10) the amounts of the various coins and bills. This procedure is a continuous activity. It helps the person who opens the next shift to check the bank.
7. Turn the money over to the manager or lock it in the safe.

Leave the empty register drawer open. In case of a break-in, if the drawer is empty and open, it won't be pried open and ruined.

Now the closing down is complete. Check your refrigerators to be sure

CASHIER'S CHECKOUT SLIP

Date______________________ Opening transaction #__________

Shift | 11-3 | 3-7 | 7-2 | (Circle appropriate shift)

Cashier____________________ Checked by____________________

Beginning bank ____________

Bills ____________

Coins ____________

Checks ____________ Number of checks________

Charges:

Amer Express ____________

Visa/MasterCard ____________

Other ____________

House credit ____________

Payouts ____________ (Itemize below)

TOTAL ____________

Less bank ____________

CASH ____________ (Include payouts and charges)

Total sales ____________ (From register tape)

OVER / SHORT ____________ (Circle one)

Overring/Underring____________ Transaction #__________Amount_____

ACTUAL OVER/SHORT____________

ANALYSIS OF PAYOUTS

Name	Amount	Name	Amount

figure 8.12

they are running and cold, with all their doors tightly closed. Check your sinks to be sure they are clean. Make sure your coffee maker is turned off. Check to see that no lit cigarette butts, your own included, are left on the bar or anywhere else. Lock everything that should be locked . . . turn off the lights . . . check to see that the Exit lights are on . . . and you are ready to go home.

Summing up . . .

The bar is the heart of each beverage service, and its smooth operation affects sales and profits in many ways. It influences the quality of the drinks, the quality of the service, the number of drinks that can be poured, and the number of people needed to serve them. Only if everything is in order at the start can the bar run smoothly and bartenders keep calm and serve customers with speed and good humor.

Management plays an important role in efficient bar setup—by setting the standards and requiring their observance, by thoroughly training employees in the routines, and by frequent follow-up on employee performance.

chapter 9 . . .

MIXOLOGY ONE

The word mixology usually refers to a bartender's knowledge and skill, but a manager must know everything a bartender knows and more. A manager may not have a bartender's dexterity, nor know as many different drink recipes, but it is the manager who decides how drinks are to be made in the enterprise he or she manages, and it is the manager who trains the bartenders in the ways of the house.

This chapter and the next are aimed at giving the manager a thorough understanding of (1) the structure of a good drink, (2) the structure and essential ingredients of each different type of drink, and (3) the basic mixing methods. Illustrated step-by-step instructions facilitate detailed understanding and provide good training tools.

The biggest mixed drink on record was a rum punch made in the West Indies in the eighteenth century. The ingredients included 1200 bottles of rum, 1200 bottles of Málaga wine, 400 quarts of boiling water, 600 pounds of the best cane sugar, 200 powdered nutmegs, and the juice of 2600 lemons. It was made in a marble basin built in a garden especially for the occasion, and served by a 12-year-old boy who rowed about the basin in a mahogany boat serving 600 guests, who "gradually drank up the ocean upon which he floated."[1] Did they drink it all—four bottles of rum-plus-wine per person? Maybe the bottles were small, or some of it evaporated in the West Indian sun. Or maybe there was a collective hangover as big as the recipe.

Today's mixology concerns not big recipes but single drinks made to the customer's order. It is a skill dating from the nineteenth-century days of the grand hotel and the fashionable cocktail bar. Before that there were a few mixed drinks, mostly hot, dating from colonial days—flips, toddies, slings, and a colonial brew called Whistle-Belly Vengeance, made with sour beer, molasses, and "Injun bread," simmered together.

The mixed drink as we know it was developed to cater to the well-to-do in the big cities and the fashionable resorts that opened up with the building of the railroads. (Ordinary folk in taverns and saloons drank straight liquor or beer.) One of the essentials—ice—was just becoming available. It came in huge blocks which were chipped or whacked to drink-size lumps or shaved for fine ice.

A new breed of bartender developed to man the elegant bars. He was splendidly dressed, wearing diamonds and gold, and poured with such flourish that one contemporary proclaimed, "Such dexterity and sleight-of-hand is seldom seen off the conjurer's stage." It was these artists who began the mixed drink as we know it today. They were the first mixologists.

The term ***mixology*** is typically defined as the art or skill of mixing drinks containing alcohol. It includes the techniques of the bartender, which do indeed require skill and sometimes art, and it also includes the knowledge that backs up the skill. The bartender must know the drinks by name, the ingredients, the mixing methods, and the way they are served. The manager must know even more. The manager must be able to set the house standards for the drinks—drink size, glass size, type of ingredients (premix or fresh), proportions—and develop drink menus and specialty drinks.

About mixed drinks in general . . .

The term ***mixed drink*** includes any drink in which one alcoholic beverage is mixed with another or others, or with one or more nonalcoholic ingredients. This includes cocktails, highballs, tall drinks, frozen drinks, coffee

[1] This recipe is quoted in Alexis Lichine's *New Encyclopedia of Wine and Spirits*, Knopf, 1974.

drinks, and almost every other bar product except a glass of beer or wine or a straight shot of whisky or brandy.

Structure and components of a mixed drink. Mixed drinks of all kinds have certain characteristics in common. One of these is a structure that is loosely typical of all drinks.

Each drink has (1) a major alcoholic ingredient, or base, usually a spirit, which determines its character and usually its predominant flavor, and (2) one or more complementary ingredients, which modify or enhance that flavor. A Manhattan, for example, has whisky as the major ingredient and sweet vermouth as the modifier or enhancer, while a highball has a carbonated mixer or water as the modifier. A drink may also have (3) one or more minor ingredients that add a flavor or color accent, and (4) a garnish. Thus a Manhattan sometimes has a dash of bitters or a drop of oil from a lemon twist added for a flavor accent, and it has a stemmed maraschino cherry as its standard garnish.

The major ingredient is the base of the drink. The modifiers and flavor accents make each drink different from all others having the same base.

Some highly flavored mixes manage to reverse flavor roles with the major ingredients, as in a Bloody Mary or a Cuba Libre. Such drinks are often ordered by people who do not really like the taste of the spirit and want the mix to cover it up. In this case, from the drinker's perspective, the mix is the major flavor ingredient and the liquor gives it the desired kick. But from the bar point of view the spirit is still the major ingredient.

Most drinks contain 1 jigger of the major ingredient, the jigger size being a policy decision of the management. If the modifier is another liquor, it is typically a smaller amount—anywhere from one-half to one-eighth the amount—from half a jigger to the splash of vermouth in today's Dry Martini. Even when several modifiers are added, the major ingredient typically comprises at least half the liquor in any drink. Accent ingredients are nearly always added in drops or dashes.

Many drinks have standard garnishes that customers expect and want; these are as much a part of the drink as the liquid ingredients. Change the garnish on one of these, and you have to change the name of the drink as well. Thus, add an onion to a Martini instead of an olive, and the drink becomes a Gibson.

Some drinks have no prescribed garnish, but today's trend is toward the showmanship of dressing them up. It would be a great mistake, however, to garnish such sacred standards as a Scotch and Soda, or indeed any drink that has been ordered by call brand, unless there is a standard garnish. Patrons who order such drinks usually want the unadulterated taste of the liquor itself.

Developing your drink recipes. A successful mixed drink is based on carefully calculated relationships of ingredients, and on a carefully calculated relationship between glass, ice, and drink ingredients. You should make these calculations when you plan your drink menu and before you buy your glasses and choose your size of ice cubes. If you write down your calculations for each drink you serve, you will have a set of standardized recipes for your bar. Then if you train your bartenders to follow the recipes consistently, you will have a consistent product no matter who is tending bar.

For each drink you establish the following:

- The amount of major ingredient to be poured (1 ounce, 1¼ ounces, 1½ ounces, ⅞ ounce, or whatever). This becomes your jigger size.
- The other ingredients and their proportions to the major ingredient.
- The size glass to be used.
- The amount of ice in the glass.
- The garnish and its arrangement.

The ice in the glass is a key ingredient in the taste of any drink made with a carbonated mix or fruit juice. While its primary function is to chill the drink, it also controls the proportion of liquor to mix by taking the place of liquid in the glass. The ice goes into the glass before the mix, and the more ice, the less mix.

Suppose, for example, you want a highball to have 3 ounces of mix to 1 ounce of liquor in an 8-ounce glass. You put enough ice in the glass to take the place of 3 ounces of liquid, which will bring the finished drink up to a volume of 7 ounces, about half an inch below the rim. An 8-ounce glass filled three-quarters full with small rectangular cubes will displace 3 to 4 ounces of liquid. (Different sizes and shapes of cubes will make a difference; with large square cubes you have to fill the glass fuller with ice because the cubes leave big spaces between them.) If you want a strong proportion of mix in relation to liquor you use less ice or a larger glass. If you want a stronger liquor taste you use more ice or a smaller glass.

Measuring. The only way to pour a drink that follows a recipe is to measure every ingredient. There are various ways of measuring liquor. There is the metered pour (discussed at length in Chapter 3), in which at least the major ingredients are measured and dispensed through a handgun or through pourers that shut off at the proper measure. A second way is for the bartender to pour into a lined jigger of your chosen size and to stop pouring at the line.

A third way is to free-pour. Free-pouring is a subjective form of measurement that involves turning the bottle—with pourer in place—upside down for full-force flow while counting in one's head. To pour an ounce you count "one-twenty-three" or "ten-twenty-thirty." "One-two-three" will give you ⅞ ounce. "One-two-three-four" will yield 1¼ ounces. And so on. Each

table 9.1 **BAR MEASURES**

dash	= $^1/_6$ teaspoon or 10 drops
teaspoon (tsp)	= $^1/_6$ ounce (oz, fluidounce) or 5 milliliters (ml)
barspoon	= 1 teaspoon
standard jigger	= $1^1/_2$ ounces or 45 milliliters (or whatever amount you set as your basic drink)
pony	= 1 ounce or 30 milliliters
wineglass	= 4 ounces or 120 milliliters
1 fluidounce	= 30 milliliters
1 ounce by weight	= 28 grams

person develops an individual way of counting that ensures the greatest accuracy for that person.

Free-pouring takes practice, experience, confidence, and good reliable pourers. It is usually the least accurate way to pour, since it is only as consistent as the individual pouring and it is likely to vary from person to person and day to day. Even the best bartender should check his or her pour every few days to see if it is still on target.

If the free-pour is accurate and consistent it can have the advantages of speed and showmanship. But few bartenders can rival the accuracy and consistency of an objectively measured drink.

The typical manager tends to think of measured pour in terms of achieving the full value of each bottle in sales, and that is certainly a major reason for measuring. But stop and think about this other aspect of measurement —the best drink, consistently, every time. It is the proportions of the ingredients to one another that make a drink what it should be. If you pour an extra quarter-ounce of gin in a Martini but only the usual amount of vermouth, that Martini is going to be drier than usual. Perhaps the customer doesn't like that as well. Or perhaps that customer thinks you made a great drink. Fine—but when the next one is your standard Martini that customer is disappointed.

So measurement is important to keep the proportions right. We measure liquors in terms of jiggers or ounces. We measure ice in terms of how full to fill the glass. We measure condiments by drops or dashes, sugar by teaspoonfuls. Table 9.1 gives you the standard measurements and their relationships.

Mixing methods. The way you want a given drink made in your enterprise is another aspect of mixology related to quality and consistency, and to speed and service as well. Many drinks are always made the same way; for others you have a choice.

There are four basic mixing methods: ***Build***, ***Stir***, ***Shake***, and ***Blend***.

- To ***build*** a drink is to mix it step by step in the glass in which it will be served, adding ingredients one at a time. You typically build highballs, fruit-juice drinks, tall drinks, hot drinks, and drinks in which ingredients are floated one on another.
- To ***stir*** a drink is to mix the ingredients together by stirring them with ice in a mixing glass and then straining the mixture into a chilled serving glass. You stir a cocktail made of two or more spirits, or spirits plus wine—ingredients that blend together easily. The purpose of stirring is to mix and cool the ingredients quickly with a minimum of dilution.
- To ***shake*** a drink is to mix it by shaking it by hand in a shaker or by mixing it on a mechanical mixer (shake mixer). You shake a drink having an ingredient that does not readily mix with spirits, such as sugar, cream, egg, and sometimes fruit juice.
- To ***blend*** a drink is to mix it in an electric blender. You *can* blend any drink you would shake, and you *must* blend any drink that incorporates solid food or ice, such as a Strawberry Daiquiri or a Frozen Margarita. Some bars use a blender in place of a shaker or mixer, but it is not nearly so fast and easy as a mechanical mixer and doesn't make as good a drink as the hand shaker.

Before the days of electrical mixing equipment, all drinks were either built in the glass, or stirred, or shaken in a hand shaker. In some establishments today, where the pace is leisurely and the emphasis is on excellence, they are still mixed by hand from scratch—no premixed ingredients, shortcuts, substitutes. The blender is used only for the newer drinks that can't be made without it, and the shake mixer is never used. There is no question that the drinks are better.

At the other end of the spectrum is the high-speed, high-volume operation—the bar at the airport or the fairground or the baseball stadium—that blends or shakes everything mechanically and stirs only in the serving glass if it stirs at all. It uses premixed products, shortcuts, and substitutes, and has a limited selection of drinks. The drinks are not perfect, but they have the right ingredients and they are cold and they are fast and they are what the customer expects.

Between the two extremes are many enterprises that use some elements of both, with the majority near the simplified, mechanized end of the scale. Most bars have eliminated the mixing glass and hand shaker altogether, and use at least some premixed products. There are valid reasons—speed, volume, cost, suitability to the enterprise and the clientele. Many types of customers are not connoisseurs of premium drink quality and don't appreciate premium prices.

But as a bar manager you need to know both ends of the scale—how to make drinks from scratch as well as all the shortcut options. You need to know what quality compromise a given shortcut will make in order to judge its suitability for your enterprise. Then you can tailor your drink menu to your type of enterprise and your clientele, and choose your equipment and supplies and train your personnel accordingly.

To this end, in describing how to make the various types of drinks, we will often give you both the original method of mixing (which is what most bartender manuals give you even though three-fourths of today's bars don't use it) and the speed method most commonly used today. It will enable you to understand each drink type and see how the shortcuts fit into it, so that you can work out what best suits your goals.

Drink families. Though mixed drinks, like human beings, have a structure, they also, like human beings, have countless variations. Pick up any bartender's manual and unless you know your way around you will find it mind-boggling; one of them has some 4000 drink recipes and the New York Bartenders Union lists 10,000. And new drinks are appearing all the time, proliferated by distillers, magazine writers, bartenders, and beverage managers.

Fortunately for the manager planning the drink menu, and for the bartender learning dozens or even hundreds of drinks, these drinks evolved in families—again like human beings. The number of families is fairly small, and if you know the family characteristics you have a handle on every family member. There are two keys to family character—the ingredients and the method of mixing the drink. A third element often comes into play—the size and type of glass. Whether the glass determines character or character determines glass is an interesting point to ponder.

In the rest of this chapter and in the next one, we will look at different drink families, with their characteristic ingredients and the mixing method that applies to each. Table 9.2 summarizes the drink families.

The highball family . . .

A ***highball*** is a mixture of a spirit and a carbonated mixer or water, served with ice in a highball glass. It is said to have got its name from the railroad signal for full speed ahead used in the late 1800s—a ball raised high on a pole. Originally it was probably a speedy drink of whisky-and-water tossed down at a brief train stop. Figure 9.1 gives you the whole story of the highball.

In the original method of building a highball, you would use a small bottle of mixer and go through all the steps as they are given. In the speed method you would add the mixer from a handgun and your stirring would probably be limited to a couple of swirls with the stir stick, counting on the customer to finish the job.

Now let us look more closely at the whole procedure. First, the sequence.

table 9.2 DRINK FAMILIES

drink type	ingredients[a]	method	glass
Buck	Liquor, lemon, ginger ale, cube ice	Build	Highball
Coffee	Liquor, (sugar), coffee, (whipped cream), (brandy float)	Build	Mug or wine glass
Collins	Liquor, lemon, sugar, soda, ice (cube or crushed), cherry	Shake/Build	Collins
Cooler	Liquor or wine, carbonated mix, (sweet, sour), (bitters), cube ice	Build	Collins
Cream	Cream, liquor, liqueur (or 2 liqueurs)	Shake	Cocktail or champagne
Daisy	Liquor, lemon, grenadine, crushed ice, (soda), fruit garnish	Shake	Mug, tankard, tall glass
Eggnog	Liquor, sugar, egg, milk, nutmeg	Shake	Mug
Fizz	Liquor, lemon, sugar, soda, cube ice	Shake/Build	Highball or wine
Flip	Liquor or fortified wine, sugar, egg, nutmeg	Shake	Wine
Flip, hot	Liquor, sugar, egg, hot milk, nutmeg	Shake	Mug
Frozen drink	Liquor, crushed ice, (any others)	Blend	8–12 oz, chilled
Highball	Liquor, carbonated mix or water, cube ice	Build	Highball
Hot Buttered Rum	Rum, sugar, hot water, butter, spices	Build	Mug
Hot lemonade	Liquor, sugar, lemon, hot water	Build	Mug
Hot toddy	Liquor, sugar, hot water	Build	Mug
Ice cream drink	Liquor, ice cream, (any others)	Shake-Mix or Blend	8–12 oz, chilled
Juice drink	Liquor, juice, cube ice	Build	Highball
Liquor on rocks	Liquor, cube ice	Build	Rocks
Martini/Manhattan	Liquor, vermouth, garnish, (cube ice)	Stir	Cocktail (rocks)
Milk punch	Liquor, sugar, milk, cube ice, nutmeg	Shake	Collins
Milk punch, hot	Liquor, sugar, hot milk, nutmeg	Build	Mug
Mint Julep	Bourbon, mint, sugar, crushed ice	Build	Tall glass or mug
Mist	Liquor, crushed ice	Build	Rocks
Old-Fashioned	Bourbon (or other), sugar, bitters, cherry, orange, cube ice	Build	Old-fashioned
Pousse-café	Liqueurs, (cream), (brandy), floated	Build	Straight-sided cordial
Rickey	Liquor, lime, soda, cube ice	Build	Highball
Sling	Liquor, liqueur, lemon or lime juice, soda, garnish, cube ice	Shake/Build	Highball or collins
Sling, hot	Liquor, sugar, lemon, hot water	Build	Mug
Smash (rocks)	Liquor, mint, sugar, cube ice	Build	Rocks
Smash (tall)	Liquor, mint, sugar, soda, cube ice	Build	Highball or collins
Sour	Liquor, lemon or lime juice, sugar	Shake	Sour or cocktail
Spritzer	Half wine/half soda, cube ice, twist	Build	Highball or wine
Sweet-sour cocktail	Liquor, lemon or lime juice, sweetener, (cube ice)	Shake	Cocktail (rocks)
Swizzle	Liquor, sweet, sour, (soda), (bitters), crushed ice	Build	Highball or specialty
Tom and Jerry	Rum, whisky, or brandy, egg-sugar-spice batter, hot milk or water, nutmeg	Build	Mug
Two-liquor	Base liquor, liqueur, cube ice	Build	Rocks
Tropical	Liquor (usually rum), fruit juices, liqueurs, syrups, ice, fruits, (mint), (flowers)	Shake	Specialty

[a] Ingredients in parentheses are optional.

HOW TO BUILD A HIGHBALL

Ingredients

Liquor
Carbonated mix or plain water
Cube ice
Garnish, varying with the drink, sometimes none

Glass

Highball (6 to 10 ounces)

Mixing Method

Build

Equipment and Accessories

Jigger (standard house size)
Barspoon
Ice scoop
Fruit squeezer (for some drinks)
Stir stick or straws
Pick (sometimes)
Cocktail napkin

step 1: Using the ice scoop, fill the glass with the required amount of ice and place it on the rail.

step 2: Add 1 jigger of the liquor ordered.

step 3: Fill the glass with mix to within ½ to 1 inch of the rim.

step 4: Stir with two or three strokes of the barspoon.

step 5: Add the garnish, if any, and a stir stick or straw. Serve on a cocktail napkin.

figure 9.1

Some people argue in favor of reversing Steps 2 and 3, using an ice-mix-liquor sequence. The rationale for this sequence is that if you pour the wrong mix by mistake you have wasted only the mix and not the liquor. If you use this sequence you must leave room for the liquor when you pour the mix, the exact depth depending on width and shape of the glass.

The rationale for pouring the liquor first is that most mixes, being heavier than liquor, will filter down through the liquor and you will probably need less stirring, which means less loss of sparkle from the carbonated mix. Also it is more natural to "think" the drink this way: ice, liquor, mixer. Another point: If you free-pour, you are more likely to pour a consistent amount of spirit if you are not looking at how much room there is left for it.

A point to note about working with carbonated mixes comes up in Step 4. Notice that the stirring with the barspoon is very brief—just enough to spread the liquor around in the mix. Carbonated liquids should always be handled gently (remember how it was with beer and champagne). A lot of stirring dissipates the bubbles. Vigorous stirring melts the ice too, diluting the drink. The customer can use the straw or stir stick for further mixing.

The finished drink should always be the same size—it should always reach an imaginary line half an inch (or some other distance you set) below the rim of the glass. Each drink will then be the same as the one before it, and will have the same proportions and the same taste. A glass should never be full to the brim, to avoid spillage.

As to what is the "right" glass size and jigger size and amount of ice, that is up to you. Highball glasses should be no less than 6 ounces nor more than 10. An 8-ounce glass is a good all-round glass. It will make an excellent highball using 1 to 1½ ounces of base liquor. Any smaller glass is likely to look stingy and certainly won't take care of anything stronger than a 1-ounce drink.

Now let us look at some of the highballs most in demand. In this book we use an abbreviated recipe format, which is all you need. We give you the method, glass, ingredients, and garnish. Anything in parentheses is optional. You do not have to follow our choice of glass size, but it will give you a proportion of glass to ice to ingredients that you can adapt to your needs.

Scotch and Soda Build

8-oz glass
¾ glass cube ice
1 jigger scotch
Soda to fill

Gin and Tonic Build

8-oz glass
¾ glass cube ice
1 jigger gin
Tonic to fill
Wedge of lime, squeezed

You can use these prototypes for all similar drinks.

Any liquor called for with soda or water is prepared like a Scotch and Soda. A garnish may be added if the customer wishes—usually a twist of lemon.

Many highballs, like the Gin and Tonic, are served with a garnish of lemon or lime—a wheel on the side of the glass or a wedge with juice squeezed in and the squeezed hull added to the drink.

In making the Gin and Tonic and similar drinks, allow room for the garnish when pouring the mix—say an extra quarter-inch of glass rim. Squeeze the lime wedge with the hand squeezer (original method) directly into the glass. Then drop the squeezed wedge, minus seeds, into the drink. Use your two or three strokes of the barspoon at this point. In the speed method you would probably squeeze the lime by hand or dispense with it altogether. When using a garnish whose flavor is added to the drink, standardize the garnish size so that your drinks will always have the same taste.

The procedures given in these two recipes apply to any liquor-mixer combination, for example, Rum and Coke, Brandy and Coke, Campari and Soda (a California favorite), Bourbon and 7-Up, Vodka and Bitter Lemon, Rye and Ginger. Some combinations have special names:

- Seven and Seven, 7-Crown whisky and 7-Up.
- Presbyterian, customer's choice of liquor with half ginger ale and half club soda.
- Cuba Libre, rum and cola with squeezed lime wedge.
- Mamie Taylor, scotch and ginger ale with a lemon twist or squeeze of lime.
- Moscow Mule, vodka and ginger beer with squeezed lime half, served in an 8-ounce copper mug.

Fruit-juice drinks . . .

Fruit-juice drinks are first cousins to the highball family; in fact, many people consider them highballs. They are made in a similar way in the same type of glass. The major difference is that fruit juice takes the place of the carbonated mix as the body of the drink. Figure 9.2 tells the story.

The original method and the speed method are identical here in most cases. One notable exception occurs in making a Bloody Mary: in the original method it is made from scratch, ingredient by ingredient, and is sometimes shaken, whereas in the speed method preprepared mix is poured from a bottle and stirred in the glass. We'll discuss it shortly.

Two points are worth noting in the basic method:

- In Step 1 the amount of ice is often less than in the highball, to give a higher proportion of juice. The added juice is enough to retain the full flavor to the last drop even though the melting ice dilutes the drink somewhat.
- In Step 4 the stirring is vigorous, since juice and liquor do not blend as readily as mixer and liquor, and there are no bubbles to worry about.

HOW TO BUILD A JUICE DRINK

Ingredients
Liquor
Fruit juice (sometimes premix)
Accent ingredients (sometimes)
Cube ice
Garnish (sometimes)

Glass
Highball (6 to 10 ounces)

Mixing Method
Build

Equipment and Accessories
Jigger (standard house size)
Barspoon
Ice scoop
Fruit squeezer (sometimes)
Stir stick or straws
Pick (sometimes)
Cocktail napkin

step 1: Using the ice scoop, fill the glass with the required amount of ice and place it on the rail.

step 2: Add 1 jigger of the liquor ordered.

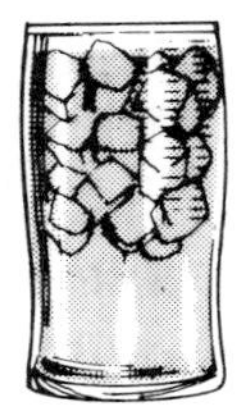

step 3: Fill the glass with juice to within ½ to 1 inch of the rim.

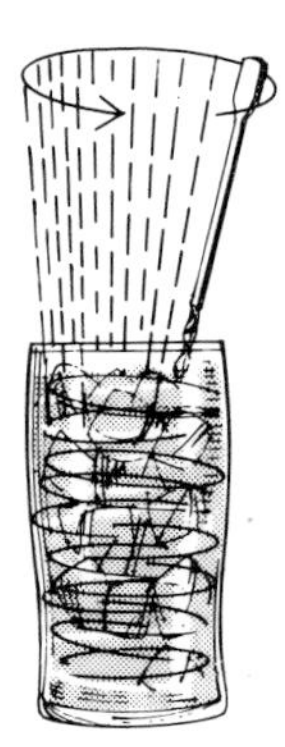

step 4: Stir vigorously with the barspoon.

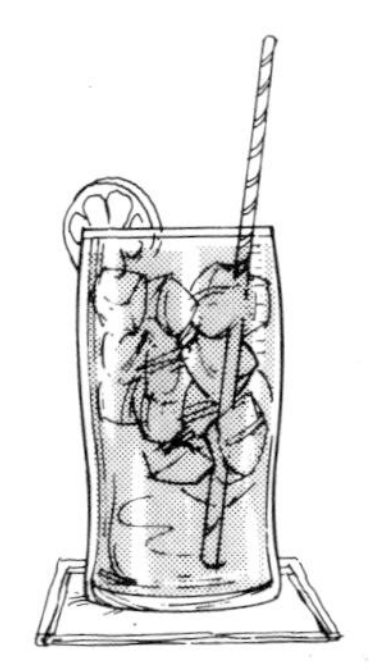

step 5: Add the garnish, if any, and a stir stick or straw. Serve on a cocktail napkin.

figure 9.2

Juice drinks are very popular and there are many of them. Here is a sample pair:

Screwdriver Build

8-oz glass
½ glass cube ice
1 jigger vodka
Orange juice to fill

Cape Codder Build

8-oz glass
½ glass cube ice
1 jigger vodka
Cranberry juice to fill
Lime wedge

Most juice drinks have special names. The only real trick in making them is in knowing what name goes with what juice and what liquor. Here are some variations of the Screwdriver:

- Left-handed Screwdriver, gin and orange juice.
- Kentucky Screwdriver or Yellow Jacket, bourbon and orange juice.
- Boccie Ball, amaretto and orange juice.
- Persuader, half amaretto/half brandy with orange juice.
- Cobra or Hammer or Sloe Screw, sloe gin and orange juice.
- Southern Screwdriver or Comfortable Screw, Southern Comfort and orange juice.
- Golden Screw, Galliano and orange juice.
- Madras, vodka and half cranberry/half orange juice.

Then there is a whole series of Screwdrivers with national names and the appropriate national liquor: Mexican Screwdriver (tequila and orange juice), Italian (Galliano), French (brandy), Greek (ouzo), Cuban (rum), Irish (Irish whiskey), Scotch Driver (scotch), Canadian Driver (Canadian whisky).

In other drinks the liquor is the same as in the Screwdriver but the juice changes:

- Greyhound, vodka and grapefruit juice.
- Salty Dog, Greyhound served in a salt-rimmed glass (usually a collins glass, and sometimes made with gin).

Rimming a glass with salt becomes the first step in making a Salty Dog. To rim the glass, run a cut lemon or lime evenly around the rim (half a lemon will give you a firm, even surface). Dip the rim in a shallow dish of salt; then build the drink as usual, taking care to keep the rim intact. A crisp, even rim enhances the drink, but if the lemon is applied unevenly the rim will be uneven. Too much salt will spoil the taste of the drink. There is a gadget called a ***glass rimmer*** that is made up of three trays. One contains a sponge that you saturate with lime juice, the second contains salt, and the third sugar. You press the glass rim on the sponge, then dip it in salt or sugar as the drink requires.

Another series of drinks takes off from the Screwdriver by adding another ingredient as an accent and traveling under a fanciful name that is sometimes better than the drink. The pair below contributed handsomely to the rise of vodka and tequila in national popularity:

Harvey Wallbanger Build

8-oz glass
½ glass cube ice
1 jigger vodka
Orange juice to fill
Top with ½ jigger Galliano

Tequila Sunrise Build

8-oz glass
½ glass cube ice
1 jigger tequila
Orange juice to fill
Top with ½ jigger grenadine

Harvey the Wallbanger was a surfer. Depending on who tells the story, he either won a surfing contest and celebrated with this drink, or lost a surfing contest and drowned his sorrows in it. Either way he had too many, and bounced from wall to wall as he staggered out of the bar.

In both these drinks, when you pour the orange juice you leave room at the top for the last ingredient, and you do your stirring *before* you add it. The Galliano, poured carefully with a circular motion, will float on top of the Harvey Wallbanger. The grenadine in the Tequila Sunrise will sink to the bottom, since it is heavier than everything else. There is one literalist school of thought that feels a true Sunrise should have the red color rising from the bottom. To create this you can pour the grenadine first and disperse it with a splash of soda, then add the ice, the tequila, and the orange juice, and forget the stirring.

A number of other drinks are variations of these two. Made with tequila, a Harvey Wallbanger becomes a Freddy Fudpucker or a Charley Goodleg (or Goodlay). Made from rum it is a Jonkanov or Joe Canoe. If you keep the vodka but change the juice from orange to grapefruit you have Henrietta Wallbanger. Or use half orange juice and half heavy cream and you have Jenny Wallbanger.

Variations on the Tequila Sunrise include Russian Sunrise with vodka replacing tequila, and Tijuana Sunrise using a dash of Angostura bitters instead of the grenadine.

The Bloody Mary is a fruit-juice drink and a classic in its own right. There are many versions. The essentials are vodka and tomato juice with accents of lemon or lime and spices. There are many different stories of its origin. One version has it that comedian George Jessel was mixing himself a heterogeneous early-morning pick-me-up in a Palm Springs bar when along came a woman named Mary, upon whom he spilled the drink. Anyway . . . here are two versions:

Bloody Mary *(from scratch)* Build

8-oz glass
½ glass cube ice
1 jigger vodka
3 oz tomato juice
Juice of ½ fresh lemon (½ oz)
2–3 dashes Worcestershire
2 drops Tabasco
Salt, pepper
Lemon wheel

Bloody Mary *(speed)* Build

8-oz glass
½ glass cube ice
1 jigger vodka
Bloody Mary mix to fill
Serve with celery stick as stirrer

The Bloody Mary has a host of close relatives that are made by substituting another liquor for the vodka or changing the mixer:

- Red Snapper, a Bloody Mary made with gin.
- Bloody Maria, substituting tequila for vodka.
- Danish Mary, with aquavit in place of vodka.
- Virgin Mary, everything but the liquor, also called a Bloody Shame.
- Bleeding Clam, or Clamdigger, with 1 jigger clam juice added.
- Bloody Bull, substituting half beef bouillon for half the tomato juice.
- Bullshot, substituting beef bouillon for tomato juice (with the same condiments).
- Bloody Caesar, substituting Clamato juice for tomato—a house specialty at Caesar's in Calgary, Canada.

Sometimes the Bloody Mary glass is dressed up with a rim of celery salt. Sometimes the glass is special—a tulip glass or a balloon wine glass. And many are the garnishes, the favorite alternates to the lemon slice being a lime wedge, celery stick, or cherry tomato.

Any of the highballs or juice drinks we have discussed can become tall drinks if the customer so specifies. To mix these, you use a larger glass, such as a collins or zombie glass, and you increase the amounts of everything in the same proportions, except for the liquor, which remains the same. What this customer wants is a long cool drink that is not as strong as a highball.

Many fruit-juice drinks are easily dressed up to become house specialties, using a specialty glass or a larger size and garnishing imaginatively. Sometimes a larger full-strength drink is offered at a come-on price for promotional and volume-building purposes.

Liquor on ice . . .

Another type of drink built in the glass consists of a liquor served over ice without adding anything else. Such drinks are typically served in a rocks or old-fashioned glass, usually of 5 to 7 ounces. Strictly speaking, they are not mixed drinks, but they are generally thought of as being related to the highball family.

Here is a duo of liquor-on-ice drinks:

Scotch on the Rocks Build

Old-fashioned glass
¾ glass cube ice
1 jigger scotch poured over ice

Scotch Mist Build

Old-fashioned glass
Full glass cracked or crushed ice
1 jigger scotch poured over ice
Lemon twist

In each case the method is so simple it needs no explanation. You do not even stir the drink. But a few comments may be useful.

A glass three-quarters full of cube ice will leave plenty of room for the liquor without any danger of spilling. It will also make the glass look more full of drink than if you fill it totally with ice. Do not worry if the ice stands higher in the glass than the drink does. People who order this type of drink do not expect a glass filled with liquor.

If your rocks drinks look too scant, you might try a different combination of glass size and ice cube. A 5-ounce footed glass with small cubes is often a good choice. Large square cubes do not fit well in small glasses; they leave spaces too large for the liquor to fill. Some establishments increase the jigger size for spirits served alone on the rocks.

Any kind of spirit can be served either on the rocks or as a mist. Those most commonly requested are whiskies, brandy, and a few liqueurs ordered as after-dinner drinks, such as amaretto and the coffee and fruit liqueurs. If these are served over crushed ice in a cocktail glass or snifter instead of a rocks glass they are known as frappés.

If you do not have crushed ice or an ice crusher you can wrap enough cubes for a mist or frappé in a towel and crack them with a mallet. You can also crack the wrapped ice by hitting the whole towelful on a hard surface such as a stainless-steel countertop.

Wines, especially white wines and the aperitif wines, are often ordered on the rocks. They will be served in 4- to 6-ounce portions poured over cube ice in an 8- or 9-ounce wine glass or footed highball glass.

Two-liquor drinks on ice . . .

Short, sweet drinks on the rocks have become very popular in the last decade. They appeal especially to younger drinkers and make good drinks to sip after dinner. They are too sweet to be true aperitifs, though people do order such drinks as a Black Russian or Stinger before dinner.

Two-liquor drinks typically combine a jigger of a major spirit (whisky, gin, brandy, vodka, tequila) with a smaller amount of a flavorful liqueur such as coffee, mint, chocolate, almond, anise, licorice. Proportions vary from 3:1 to 1:1, depending on the drink and the house recipe. Even with 3:1 the liqueur flavor often takes over the drink. Equal parts make a *very* sweet drink, though this varies with the particular liqueur.

Because the two liquors blend easily, these drinks are built in the glass and are among the easiest and fastest drinks to make. Together the two liquors usually add up to a 2- to 3-ounce drink. Anything less would be noticeably scant in the glass. Figure 9.3 shows the method.

Notice that in Steps 2 and 3 you add the base liquor first and the liqueur after. This is because the liqueur, having a higher sugar content, is heavier and will head for the bottom of the glass, filtering through the liquor and making it easier to blend the two.

Few of these drinks call for a garnish (see Step 5). If there is one it is an occasional lemon twist.

Step 6 is necessary because the liqueur, owing to its sugar content, clings to the sides of the jigger and will flavor the next drink.

Here are a couple of typical two-liquor drinks:

Black Russian Build

Rocks glass
Full glass cube ice
1 jigger vodka
½ jigger Kahlúa

Stinger Build

Rocks glass
Full glass cube ice
1 jigger brandy
½ jigger white crème de menthe

Each of these has many variations and spinoffs. The Black Russian spawns the following:

- White Russian, a Black Russian with a cream float (another version is made in a blender with ice cream).
- Black Magic, a Black Russian with lemon juice and a twist.
- Black Jamaican, rum substituted for vodka.
- Black Watch, scotch substituted for vodka, with a twist.
- Brave Bull, tequila substituted for vodka.
- Dirty Mother, brandy substituted for vodka.
- Siberian, a Black Russian with a brandy float.

In similar fashion there are many variations on the Stinger:

- Cossack, vodka substituted for brandy, also called White Spider.
- Irish Stinger, green crème de menthe substituted for white.

HOW TO BUILD A TWO-LIQUOR DRINK

Ingredients
A base liquor
A liqueur
Cube ice

Glass
Rocks (5- to 7-ounce)

Mixing Method
Build

Equipment and Accessories
Jigger
Barspoon
Ice scoop
Stir stick or straw
Cocktail napkin

step 1: Using the ice scoop, fill glass with ice to within ½ inch of rim and place on rail.

step 2: Add the base liquor.

step 3: Add the liqueur.

step 4: Stir with the barspoon.

step 5: Add the garnish, if any, and a stir stick or straw. Serve on a cocktail napkin.

step 6: Rinse the jigger.

figure 9.3

- International Stinger, a Stinger with Cognac as the brandy.
- Greek Stinger, a Stinger with Metaxa as the brandy.
- White Way, gin substituted for brandy.
- Smoothy, bourbon substituted for brandy.
- Galliano Stinger, Galliano in place of crème de menthe.

Other popular two-liquor drinks include:

- Rusty Nail, scotch and Drambuie (also called Knucklehead).
- Godfather, scotch or bourbon and amaretto (the bourbon version is also called The Boss).
- Godmother, vodka and amaretto.
- Spanish Fly, tequila and amaretto.

All these two-liquor drinks may also be served straight up in a cocktail glass. In this case they are stirred with ice in a mixing glass and strained into a chilled glass, in the manner we explore in the next chapter.

Collinses, rickeys, bucks, coolers, spritzers . . .

Several other drink families are also built in the glass. Some of them take off from the highball and juice drinks by adding other characteristic ingredients. The best known of these is the collins family.

Family Characteristics

- ***Ingredients.*** Liquor, lemon juice, sugar, soda, cube ice, maraschino cherry garnish, optional orange slice. Today the lemon juice, sugar, and soda are typically combined in the collins mix.
- ***Glass.*** Collins (12 to 14 ounces).
- ***Mixing method*** (speed version). Build with collins mix.

The basic steps are those for making a highball. We discuss making a collins from scratch in the next chapter.

In the collins family the name of the drink changes with the liquor—Tom Collins for gin, John Collins for bourbon, Mike Collins for Irish whiskey, Jack for applejack, Pierre for Cognac, Pedro for rum. There are as many other collinses as there are spirits—vodka, scotch, rye, tequila, and so on.

Cousins to the collinses are the rickeys. They use lime instead of lemon and are a shorter, drier drink—that is, they are served in a smaller glass and have little or no sugar.

Family Characteristics

- ***Ingredients.*** Liquor, fresh lime, soda, cube ice.

- ***Glass.*** Highball or old-fashioned.
- ***Mixing method.*** Build.

The basic steps are those for making a highball, but starting off with half a fresh lime squeezed over ice. Do not use Rose's lime juice; the unsweetened juice is essential to the drink.

A rickey may be made with any liquor or liqueur. A liqueur makes a sweeter drink, cutting the extreme dryness of the lime. Rickeys made with nonsweet spirits may have a small amount of simple syrup or even grenadine added.

Change the soda of the rickey to ginger ale and the lime to lemon and you have a buck. It is made like a highball, usually in a highball glass. It is not so dry as a rickey because the ginger ale is sweet. Today's buck usually goes by another name, such as:

- Mamie Taylor, made with scotch, with lemon wedge or lime juice.
- Mamie's Sister, made with gin, also called Fog Horn.
- Mamie's Southern Sister, made with bourbon.
- Susie Taylor, made with rum.

Yesterday's cooler was typically a long drink made with liquor and soda or ginger ale and served over ice in a collins glass decorated with a long spiral of lemon peel curling around inside the glass from bottom to top. One of the best-known coolers was the Horse's Neck With a Kick, which consisted of 2 ounces of liquor with ginger ale and the long lemon spiral. A plain Horse's Neck was a Prohibition drink without the liquor.

Today's cooler is likely to be half wine and half soda, iced, served in a collins or highball glass. If it is made with white wine it is called a spritzer. The long spiral of lemon peel has disappeared. These drinks are becoming more and more popular as part of the trend to wine and lighter mixed drinks. Here are two basic recipes:

Wine Cooler	Build
8-oz or 10-oz glass	
¾ glass cube or crushed ice	
Half fill with red wine	
7-Up to fill	
Twist or flag	

Spritzer	Build
8-oz or 10-oz glass	
¾ glass cube or crushed ice	
Half fill with white wine	
Soda to fill	
Twist, lemon slice, or lime wedge	

Another similar drink is the Vermouth Cassis, which couples 1½ to 2 ounces of dry vermouth with crème de cassis in a 4:1 to 6:1 ratio over cube ice, with soda to fill and a twist or slice of lemon. A different but related drink is Kir (pronounced Keer), a glass of chilled white wine with half an ounce or less of cassis—no soda and sometimes no ice.

Some old-fashioned drinks . . .

Some mixed drinks that are still around go back to colonial days; more originated in the era of the great hotel bars of the later 1800s or during Prohibition when new ways were devised to mask the awful taste of homemade gin. Some that we include here you may never encounter directly, but again they may have potential as specialty drinks, dressed up with modern techniques, intriguing names, and the romance of the past.

Two venerable drinks are still very much alive—*the* Old-Fashioned and the Mint Julep. In structure they are simply liquor-over-ice drinks that are sweetened, accented, and garnished. However, both involve more mixing than many other mixed drinks.

The Old-Fashioned is a cocktail that is never served in a cocktail glass. It is always built in the glass, like a highball, but it isn't a highball because there is little or no mixer. It is among the most venerable of mixed drinks, giving its name to its glass—a sturdy, matter-of-fact, all-business tumbler of 5 to 7 ounces, just the right size to make a full drink without adding water.

Family Characteristics

- ***Ingredients.*** Liquor, sugar, bitters, fruit (cherry, orange), cube ice.
- ***Glass.*** Old-fashioned.
- ***Mixing method.*** Build.

For recipes we give you the ritual method on the left and a modern version on the right:

Old-Fashioned *(traditional)* Build

Old-fashioned glass
1 lump sugar
Splash of soda on sugar
1-3 dashes Angostura bitters
Crush sugar with muddler; stir till dissolved
Full glass cube ice
Up to 2 ounces whisky
Stir briskly
Cherry, orange slice
Lemon twist (optional)

Old-Fashioned *(current)* Build

Old-fashioned glass
Full glass cube ice
Up to 2 ounces whisky
1-2 tsp simple syrup
1-3 dashes Angostura bitters
Stir briskly
(Soda or water to fill)
Cherry, orange slice, (pineapple chunk)
(Lemon twist)

This drink came out of Old Kentucky, and the tradition is part of the drink—the glass, the garnishes, and the tender loving care with which it is made. Older customers prefer it as it used to be; they want the rich taste of sweetened bourbon accented with bitters and fruit. If the undiluted drink does not fill the glass, use smaller cubes, more liquor, or a smaller glass.

Today's younger customer with a taste for lighter drinks often prefers it made with soda or water added. The customer should be queried as to preference.

You can make an Old-Fashioned with any whisky or with other liquors —brandy, applejack, rum, even gin—but unless the customer orders one of these, make it with bourbon in the South, a blended whisky in the East, and probably brandy in Wisconsin. Simple syrup really makes a better drink than lump sugar, unless you are emphasizing ritual for the customer. Stir it very well in any case, and garnish it handsomely. If you want to gild the lily you can add a touch of curaçao or a dash of sweet vermouth, or even a bit of the juice from the maraschino-cherry bottle. If you add these, use less sugar. You can change the bitters—such as orange bitters with rum or gin—or you can float a teaspoonful of 151° rum on a Rum Old-Fashioned. You can elaborate on the fruit garnishes—pineapple is often used.

Someone in Old New Orleans elaborated on the Old-Fashioned to create the Sazarac by coating the inside of the glass with absinthe. Today you use a substitute such as pernod, since absinthe is illegal. You roll a splash of it around inside the glass until the coating is complete, discard what is left, and proceed to make an Old-Fashioned using Peychaud bitters, often with straight rye as the liquor. The Sazarac is served in New Orleans as a prebrunch drink.

The Mint Julep, sipped on the porch of an old southern mansion, has always represented the quintessence of southern elegance and leisure. There are said to be at least 32 different recipes for this classic, but they all have certain things in common:

Family Characteristics

- ***Ingredients.*** Liquor (traditionally bourbon), fresh mint, sugar, crushed or shaved ice.
- ***Glass.*** 12- to 16-oz chilled glass or silver mug.
- ***Mixing method.*** Build (ritual methods stir first, then build).

Here is one of the 32 recipes:

Mint Julep Build

16-oz glass, chilled

10–12 fresh mint leaves, bruised gently

1 tsp bar sugar with splash of soda, or 1½ tsp simple syrup

Muddle sugar and mint

Add ½ glass ice

1 jigger bourbon

Stir up and down until well mixed

Fill glass with ice

Add another jigger bourbon

Stir contents up and down until glass/mug is completely frosted

Insert straws

Garnish with mint sprigs dipped in bar sugar or with fruit

You won't find this drink at a speed bar. Even this version, less elaborate than some, takes several minutes to make.

If the customer specifies, you can make a Mint Julep with rye, rum, gin, brandy, Southern Comfort, but the classical version is made with 100-proof, bottled-in-bond bourbon. Why else would you take all this trouble? It makes a good premium-price specialty drink for some enterprises.

A smash is a cross between a Mint Julep and an Old-Fashioned. It is made with Julep ingredients by the Old-Fashioned method in an old-fashioned glass. You muddle a cube of sugar and some mint with a splash of water, add cube ice and a jigger of liquor, and garnish with mint sprigs. Or you can make a smash as a long drink in a tall glass by adding club soda.

Another old-timer is the swizzle, a sweet-sour-liquor-soda drink served over crushed ice in a tall glass with a swizzle stick. This is a special stirrer (now rarely seen) with multiple arms at the bottom which are whirled about by rolling the top of the stick between the palms of the hands until the drink froths and the outside of the glass frosts. The old Caribbean recipe for such drinks is "one of sour, two of sweet, three of strong, and four of weak." The sour in a swizzle is lemon or lime. The sweet may be sugar, simple syrup, a flavored syrup, or a liqueur. The strong can be any spirit, and the weak in this case is melted ice or sometimes soda. Often a dash of bitters is added for accent. Fruit or mint garnishes are in order. This would make an interesting specialty drink if you could find some old-fashioned swizzle sticks.

Pousse-cafés . . .

"Coffee-pusher" is the literal translation of the term pousse-café. In France it is a sweet liqueur drunk with or after coffee at the end of a meal. In America the drink has been elaborated into a ribbon of different colors in a pony glass, made by pouring layers of different liqueurs in such a way that they remain separate.

Family Characteristics

- ***Ingredients.*** Liqueurs of different densities, sometimes nonalcoholic syrups or brandy or cream or all these things.
- ***Glass.*** Straight-sided pony or brandy glass.
- ***Mixing method.*** Build (float).

Many different liqueur combinations can go into a pousse-café; there is no one formula. The secret of layering is to choose liqueurs of differing density and to float them in sequence from heaviest to lightest. Density depends a good deal on the sugar content of a liqueur, but since there is no indication of sugar content on the bottle we can only infer it from the alcohol content, or proof. In general, the lower the proof, the higher the sugar

figure 9.4 How to build a pousse-café.

content and the density. Thus a 36° crème de cassis is usually heavier than a 50° triple sec, which is usually heavier than a curaçao at 60 proof. At the top of the scale is green Chartreuse at 110 proof. Grenadine, which has no alcohol, is often used as a pousse-café base, while brandy, which has no sugar, is often used as the top layer. Cream will also float atop most liqueurs.

But proof is not an infallible guide. Nor is the generic name of a liqueur: products of different manufacturers often have different densities. The best system is to work out your own recipes, by trial and error if necessary, and to use the same brands of liqueurs every time.

The pousse-café is mainly a drink for show rather than taste, the object being to create a handsome sequence of colors and to show off the bartender's skill. The technique is to pour each layer over the back of a spoon in or over the glass so that the liqueur spreads evenly on the layer below (Figure 9.4). If you hold the tip of the spoon against the side of the glass, the liqueur will run slowly down the side and thence onto the layer below.

This is no drink for a speed bar. Because of the time and skill required and the cost of the liqueurs, a pousse-café commands a good price, even though there may be only 1 ounce of liqueur in it altogether.

The pousse-café is never stirred. The customer drinks the rainbow of colors layer by layer—as nearly as possible. Since the ingredients are chosen for color and density rather than for complementary tastes, some sequences of flavor can be rather weird. Keep this in mind as you develop your own pousse-café specialty.

Here are some specific pousse-café recipes:

Pousse-café Build

Straight-sided liqueur glass
1/6 grenadine (red)
1/6 dark crème de cacao (brown)
1/6 white crème de menthe (white)
1/6 apricot-flavored brandy (orange)
1/6 green Chartreuse (green)
1/6 brandy (amber)

Angel's Kiss Build

Straight-sided liqueur glass
¼ dark crème de cacao
¼ crème d'Yvette
¼ brandy
¼ cream

Coffee drinks and some other hot ones . . .

Mixed drinks are not limited to the chilled glass, as the ski lodges and the bars and restaurants in cold-weather regions know well. Many of the mixed drinks in the early American colonies were warmer-uppers heated in the tankard by thrusting a red-hot poker or loggerhead into the liquid. Today's hot drinks are not limited to cold climes; coffee drinks are served all over.

Many dinner restaurants have developed specialty coffee drinks that double as dessert and coffee and end the meal on a note of excitement. Such a climax often lingers in memory and brings the diner back. Since these drinks are usually high-profit items, they are to the restaurateur's advantage in every way.

No doubt people have been spiking their coffee with spirits for generations and finding it delicious. It's what you do to dramatize it that makes it memorable. The Buena Vista Café in San Francisco started an Irish Coffee craze some 40 years ago when they put Irish whiskey and coffee and sugar in a goblet and floated whipped cream on top. People came from all over and fought their way through the crowds for a glass of it—and still do. Another way of adding drama is to flame a brandy float or a liquor-soaked sugar cube as the coffee is served. Tony's Restaurant in Houston has made Café Diablo into a dramatic tableside brewing ceremony involving a long flaming spiral of orange peel that can be seen all over the dining room.

The basic hot coffee drink is very simple. Figure 9.5 shows you how to make it.

For a hot drink the customer must have something to hold that is not too hot to grasp firmly—a cup handle or the stem of a glass. If you serve in a stem glass, it should be of tempered glass, able to withstand the heat without cracking. If your glass is not heat-treated, preheat it by rinsing in hot tap water. Thin glass is better than thick; it heats evenly and quickly.

For garnishes you can sprinkle nutmeg, cinnamon, shaved chocolate, or finely chopped nuts on top of the whipped cream—anything appropriate to the drink. A cinnamon stick can substitute for the stir stick or spoon in a shallow cup.

HOW TO BUILD A HOT COFFEE DRINK

Ingredients

Hot coffee
Liquor
Sweetener
Whipped cream topping (or brandy float)

Glass

Coffee cup or mug or preheated stem glass

Mixing Method

Build

Equipment And Accessories

Jigger
Barspoon
Coffee spoon or stir stick
Straws
Cocktail napkin

step 1: Add 1 spoonful sugar or other sweetener to cup, mug, or glass.

step 2: Add 1 jigger of the appropriate liquor.

step 3: Fill with hot coffee to within 1 inch of the rim.

step 4: Stir well with barspoon until sugar is dissolved.

step 5: Swirl whipped cream on top (or float brandy).

step 6: Add garnish if any, stirrer, and straws. Serve on a saucer or cocktail napkin.

figure 9.5

You can make a decaffeinated version by putting an individual portion of instant decaf in the cup along with the sugar in Step 1 and filling the cup with hot water in Step 3.

Some other variations include sweet liqueurs in place of sugar or a liqueur float in place of whipped cream. Or you can float a little additional high-proof spirit and serve it flaming. Or soak a sugar lump in spirit, put it on a spoon across the cup, and flame it, dropping it into the drink when the flame dies.

Here are some cream-topped hot coffee drinks:

- Irish Coffee, made with Irish whiskey.
- Café Royale, with bourbon or brandy; another version is made with half Metaxa/half Galliano.
- Dutch Coffee, with Vandermint, no sugar.
- Mexican Coffee, with tequila, sweetened with Kahlúa.
- Café Calypso, with rum and brown sugar. A Deluxe version substitutes dark crème de cacao for sugar, and a Supreme version uses Tia Maria in place of sugar. Also called Jamaican Coffee.
- Café Pucci, with half Trinidad rum, half amaretto.
- Kioki (or Keoke) Coffee, with brandy and coffee liqueur for sugar. Another version includes Irish whiskey.
- Royal Street Coffee, with amaretto, Kahlúa, nutmeg, no sugar.

In all these drinks the quality of the coffee is a major factor. Use fresh coffee, made within the hour.

Two other popular coffee specialties are Cappuccino and Café Diablo. Each of these has a number of versions depending on who is making it.

Cappuccino is an American elaboration of an ancient version of coffee-with-milk drunk by the Capuchin monks in Italy. The modern drink is made with espresso and has rum, brandy, crème de cacao, cream, and probably Galliano, all mixed together with the coffee and topped with whipped cream.

Café Diablo is a highly spiced brew. It is made with brandy as a base liquor, and includes Grand Marnier or Cointreau, orange peel or grated rind, and various sweet spices such as cinnamon, cloves, allspice, and sometimes coriander. And coffee. It may also include an anise-flavored liqueur, additional sugar, and even chocolate syrup in some versions. It can be built in the cup, stirring all the liquors and flavor accents together first, then adding hot coffee and flaming additional brandy on top. Or it can be flamed while mixing several servings tableside, then served in small cups with whipped-cream topping. Either way it makes an unforgettable drink, worthy of a high price tag.

Many hot drinks come to us down the centuries from days when they were the only kind of central heating around.

A hot toddy is made by mixing a jigger of liquor with sugar and hot water in a mug or old-fashioned glass. If the liquor is rum you may call the drink grog. If you use dark rum and add butter and spices you call it a Hot Buttered Rum. This drink is best if you premix the butter, sugar, and spices—cinnamon, nutmeg, cloves, salt—in quantity and stir in a teaspoonful per drink. Or use a packaged premix. In any case, serve with a cinnamon stick for stirring.

You can also add spices to any toddy and still call it a hot toddy. Or you can call it a hot sangaree.

Add an egg to a toddy, use hot milk instead of water, and it becomes a hot flip or an eggnog or a hot milk punch. Sprinkle nutmeg on the top.

Add lemon juice to a toddy and it becomes a hot lemonade or a hot sling or a hot scotch, gin, rum, rye, or whatever.

Feel free to make your own hot concoctions. Any flavorful spirit makes a good base (vodka is not distinctive enough). Build your drink in the glass or mug, one ingredient at a time. Make it festive, and make it smell good—aroma is important as a come-on. Orange and lemon make good flavor, aroma, and appearance.

There is one more hot drink that makes a good specialty for the Christmas season—the Tom and Jerry. It was invented by Jerry Thomas, bartender extraordinaire of the last century, who also invented the original Martini and something called the Blue Blazer.

The Tom and Jerry involves premixing a bowl of batter made of eggs, sugar, spices (allspice, cinnamon, cloves), and a little Jamaica rum (there is also a package mix on the market). You put a ladleful of batter in a mug,

figure 9.6 Professor Jerry Thomas preparing a Blue Blazer, a contemporary drawing. (Courtesy Alfred A. Knopf, Inc., and The New York Public Library, Astor, Lenox and Tilden Foundations)

add bourbon or brandy and hot milk or water, and stir vigorously until everything foams. Dust it with nutmeg and serve it forth.

As for Jerry Thomas's Blue Blazer, you usually find some version of his recipe in every bartending manual, but who has ever seen one made? You put whisky in one silver mug and hot water in another, set fire to the whisky and fling it with unerring accuracy into the hot water. Then you toss the flaming beverage back and forth between the two mugs to make a long streak of flame. "The novice in mixing this beverage should be careful not to scald himself," wrote Thomas in his 1862 treatise on mixing drinks.[2]

Figure 9.6 is a contemporary rendering of Jerry Thomas executing one of his Blue Blazers. It seems a fitting ending to this first installment of mixology.

Summing up . . .

Understanding drink structures and family relationships makes it easy to learn a great many drinks using just a few fundamentals. It also makes it possible to recognize a drink type from a list of its ingredients and to make it even when no instructions are given. Beyond that, a grasp of drink types and ingredient relationships makes it easy to invent drinks by substituting or adding appropriate ingredients to familiar drinks.

Drinks built in the glass are the easiest and fastest kinds to make. The next chapter presents drinks made by other methods that require mixing equipment beyond the barspoon and the handgun.

[2] Professor Jerry Thomas, *The Bon Vivant's Companion, or How to Mix Drinks*, edited by Herbert Asbury, Knopf, 1928.

chapter 10 . . .

MIXOLOGY TWO

Continuing the story of mixology, this chapter presents the remaining methods of mixing drinks, including the original method of shaking by hand and the current methods of blending and mechanical mixing that make frozen drinks and ice cream drinks possible. It explores additional drink families as well as current methods and techniques of preparation and the filling of drink orders for speed and quality. The story then moves on to discuss how the manager can use all this knowledge to plan drink menus and create specialty drinks—a most profitable endeavor these days.

Picking up the historical note on which we ended Mixology One, we find that Jerry Thomas of Blue Blazer fame was also a key figure in developing the mixology of cocktails. The name cocktail is said to have originated well before Thomas's time, in a tavern somewhere north of Manhattan where French and American officers of George Washington's army met to dine and drink the beverage concoctions of tavernkeeper Betsy Flanagan. One night, so the story goes, Betsy served them chickens patriotically stolen from a nearby farm owned by enemy Tories, and she decorated her drinks with feathers from the cocks' tails. The drinks inspired a toast—"*Vive le cock tail!*" Whether or not the story is true (there are many others), Betsy had a good merchandising gimmick.

When Jerry Thomas wrote his book in 1862 the word cocktail referred to "composite beverages" which were generally bottled to take on picnics or hunting trips. As the world's most prestigious bartender, Thomas turned the cocktail into the most fashionable bar drink of his era. He became known as The Professor because of his dedicated researches and experiments in mixology.

Which brings us to those contemporary "composite beverages"—many of them cocktails—that are not built in the glass but are stirred, shaken, blended, or mechanically mixed with the shake machine. Again the drinks can be grouped in families, with common ingredients and mixing methods as the family ties.

The Martini/Manhattan family . . .

It is hard to know whether to call this group of drinks one family or two. One half of the family is descended from the original Martini cocktail invented by our friend Jerry Thomas, and the other from the original Manhattan cocktail introduced at New York's Manhattan Club by Winston Churchill's mother. Today's Martinis and Manhattans are distinctly different from one another, but when you come to second-generation variations and refinements, lines cross and distinctions blur.

At any rate, the basic characteristics of both branches of this drink family are very much the same:

Family Characteristics

- ***Ingredients.*** Liquor, vermouth (in a 4:1 to 8:1 ratio), garnish.
- ***Glass.*** Stemmed cocktail, chilled.
- ***Mixing method.*** Stir.

In a Martini the liquor is gin, the vermouth is dry, and the garnish is an olive or a lemon twist. In a Manhattan the liquor is whisky, the vermouth is sweet, and the garnish is a cherry.

The mixing method spelled out in Figure 10.1 is for a ***straight-up*** drink

THE STIR METHOD: HOW TO MAKE A MARTINI OR MANHATTAN

Ingredients
Liquor, 4 to 8 parts
Vermouth, 1 part
Garnish

Glass
Stemmed cocktail glass, chilled

Mixing Method
Stir

Equipment and Accessories
16-ounce mixing glass with strainer
Jiggers
Barspoon
Ice scoop
Tongs, pick, or condiment fork
Cocktail napkin

step 1: Place a chilled cocktail glass on the rail, handling it by the stem.

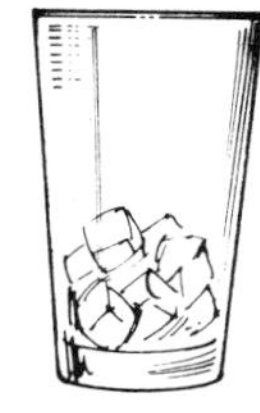

step 2: With the scoop, fill the mixing glass ⅓ full of cube ice.

step 3: Measure liquor and vermouth and add to the mixing glass.

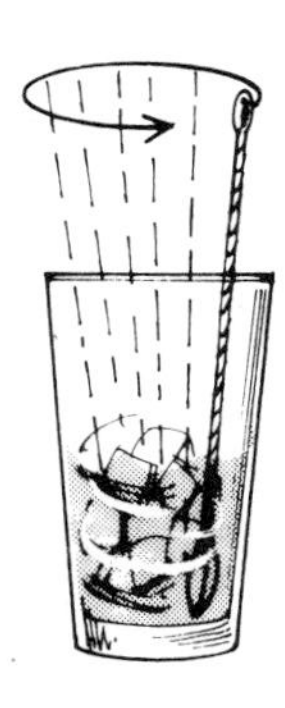

step 4: Stir briskly in one direction 8 to 12 times.

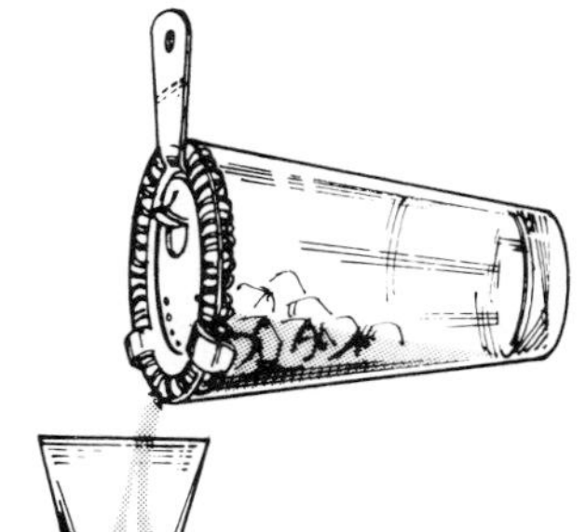

step 5: Strain the liquid into the cocktail glass.

step 6: Add the garnish, using tongs, pick, or condiment fork. Serve on a cocktail napkin.

figure 10.1

—one served in a chilled stemmed cocktail glass with no ice in the drink itself. Nowadays Martinis and Manhattans and all their relatives are served nearly as often on the rocks as they are straight up. We'll talk about that shortly.

There are several things to consider carefully here. The first is the chilled glass. The cold glass is absolutely essential to the quality of the drink, since there is no ice in the drink itself. You handle the chilled glass by the stem so that the heat of your fingers does not warm it or leave fingerprints on the frosty bowl. If you do not have a chilled glass to start with you must chill one. You do this by filling it with ice before you begin Step 1. The cocktail glass chills while you are completing Steps 2 through 4. Then you pick up the glass by the stem, dump the ice into your waste dump, and proceed with Step 5.

The purpose of the stirring in Step 4 is twofold: to mix the wine and liquor without producing a cloudy drink, and to chill them quickly without unduly diluting the mixture. If you stir a drink containing wine too enthusiastically or if you shake it, the clarity of the drink will be lost. If you stir too long, melting ice will weaken the drink's flavor. So the stirring is just enough to blend and chill these two easily blended ingredients and to add about an ounce of water—no more. Notice that the ice is cube ice; crushed ice would dilute the drink unduly.

In Step 5 you use the strainer to keep the ice out of the glass. Figure 10.2 shows you how to hold it.

Let's put these two parent drinks into a recipe format:

Martini Stir

Chilled 4-oz cocktail glass
6 parts gin
1 part dry (French) vermouth
Olive or lemon twist

Manhattan Stir

Chilled 4-oz cocktail glass
6 parts whisky
1 part sweet (Italian) vermouth
Maraschino cherry

Here you see a 6:1 ratio for both drinks. In terms of amounts it is commonly 1½ ounces of liquor and ¼ ounce of vermouth. Allowing for a small amount of melted ice and the space taken up by the garnish, you will need a 4- to 4½-ounce glass.

A 6:1 drink is fairly dry. Not long ago the accepted standard was 4:1, but among straight-up cocktail drinkers the recent trend has been toward drier drinks. If today's customer asks for a *dry* Martini you can decrease the vermouth in the recipe or increase the gin, depending on house policy. For a *very dry* Martini you use only a dash of vermouth or none at all. Bartenders develop their own forms of showmanship about this, such as using an eyedropper or atomizer, or passing the glass over the vermouth bottle, or facing toward France and saluting.

On the other hand, if a customer orders a Dry Manhattan you keep the

figure 10.2 How to use the strainer. (Photo by Pat Kovach Roberts)

same proportions but use dry vermouth in place of sweet and change the garnish to a lemon twist.

If you use equal parts of dry and sweet vermouth in either a Martini or a Manhattan, it becomes a Perfect Martini or a Perfect Manhattan. The garnish usually becomes a lemon twist in each case.

If you change the olive to a cocktail onion in the original Martini you have made a Gibson.

A Martini is always made with gin unless the customer specifies another liquor. Vodka is the most common alternative; this is sometimes called a Vodkatini. Then there is the Rum Martini; and if you add a dash of bitters or a little lime juice and sweeten it with grenadine and curaçao it becomes El Presidente. There's a Tequila Martini, or Tequini, and there's a Silver Bullet, which keeps the gin but substitutes scotch for the vermouth or uses both scotch and vermouth, floating the scotch. There are at least 20 or 30 other variations of the Martini; they substitute ingredients or change proportions or add flavor accents to the original recipe. You should be familiar with any that are regional favorites. They are also material for house specialties.

There are many similar variations for the basic Manhattan recipe. Even the original recipe has regional variations—bourbon in the South, blended whisky, or "rye," in the East. Brandy is likely to be the liquor for a Manhattan in Wisconsin unless the customer specifies whisky. Then there is the Rob Roy, which is a Perfect Manhattan made with scotch. And there's the Dry Rob Roy. Is this a Dry Manhattan made with scotch or a Martini made with scotch? Think of it either way, use a lemon twist as a garnish, and you'll be making the same drink.

A Manhattan can be made with other liquors too. Irish whiskey makes a Paddy. If you use rum you may call it a Little Princess, or if you use equal parts of rum and sweet vermouth you'll be making a Poker. If you make a Manhattan with Southern Comfort as the whisky, use dry vermouth to cut the sweetness of the liquor.

All these versions of both Martini and Manhattan are made in the same way, stirring in the mixing glass and straining into the chilled cocktail glass. Straight-up cocktails made with other fortified wines such as sherry or Dubonnet are made the same way.

As we noted earlier, all these drinks we have been describing are also often served on the rocks. In this case you have a choice of mixing methods. You can make a drink as you do the straight-up cocktail, simply straining the contents of the mixing glass into a rocks glass three-quarters full of cube ice. Or you can build the drink in the rocks glass as you do the two-liquor drinks on ice. The latter method, described in Chapter 9, is the easiest and fastest and by far the most common. If you build in the glass it is wise to pour the vermouth first. Then if the mingling of the ingredients is less than perfect, the customer tastes the liquor first.

If volume warrants, Martinis and Manhattans can be premixed in quantity. To do this, follow these steps:

step 1: Fill a large small-necked funnel with ice cubes and put it into the neck of a quart container.

step 2: Pour 4 ounces of the appropriate vermouth and a fifth or 750-milliliter bottle of the appropriate liquor over the ice into the quart container.

step 3: Stir with a long-handled barspoon.

step 4: Keep chilled in the refrigerator until used.

step 5: To serve, measure out 3½ ounces per drink into a chilled cocktail glass.

Sours and other sweet-sour cocktails . . .

The idea of combining sweet and sour flavors with liquors has been around a long time, as demonstrated in the West Indies rum punch described in Chapter 9. It is no accident that several of the drinks in the sweet-sour cocktail family originated in tropical climates where lemons or limes grow in profusion.

Family Characteristics

- ***Ingredients.*** Liquor, lemon or lime juice, and a sweetener—"sweet, sour, and strong."
- ***Glass.*** Sour or cocktail, chilled.
- ***Mixing method.*** Shake (or Blend or Shake-Mix).

The subgroup of drinks known as sours use lemon rather than lime, have a standard garnish of cherry and orange, and are traditionally served in a sour glass of 5 or so ounces, whose size and shape accommodate the garnish attractively. Sometimes a sour is made with egg white or a mix containing frothee, giving the drink an appetizing fizz topping.

The other cocktails in this family are served in a standard cocktail glass of 4 to 4½ ounces. Some use lime in preference to lemon; some use a sweet liqueur or syrup in place of sugar. Most have no standard garnish.

Any of these drinks may also be served over ice in a rocks glass if so ordered. Some of them are also made in a frozen version or a fruit version; more of that later.

The citrus juices and the sugar demand that these drinks be shaken or blended or mechanically mixed, whether you make them from scratch or use a sweet-sour mix. Neither the sugar nor the fruit juices can be smoothly combined with the liquor by stirring.

The Shake method. The cocktail shaker was a symbol of the good life in the Gay Nineties, as it was for the joyous return to legal drinking after Prohibition. The shaking of the drink was a ceremony of skill that whetted the customer's appetite while commanding the admiration of the audience. When mechanical mixers were invented it was quickly discovered that they made a smooth drink a great deal faster than the hand shaker. Today most bars use shake mixers and blenders, and those that use hand shakers do so for reasons of tradition or showmanship. There are bartenders who do not even know how to shake a drink by hand.

Figure 10.3 shows the making of a sour using the hand shaker. You will notice that the first three steps are essentially the same as the Stir method. It is Step 4 that is the heart of the matter. Let us look at this technique more closely.

The cup of the shaker fits tightly over the glass, since a certain amount of flex in the metal makes for a good fit. It should be put on at an angle with one side of the cup running along the side of the glass. This makes it easier to separate again. (Sometimes shaking creates a vacuum, and the cup adheres to the glass.) Shake vigorously, using long strokes that send the contents from one end to the other. Some people shake up and down; others shake back and forth over the shoulder.

If you have trouble separating the glass from the cup, don't yield to the temptation of banging the cup on the rail. You can easily break the glass

THE SHAKE METHOD: HOW TO MAKE A SOUR IN A HAND SHAKER

Ingredients

Liquor
Lemon juice, Sugar or simple syrup, Egg white (optional) } Or sweet-sour mix
Cherry/orange garnish

Glass

Sour glass (5 to 5½ ounces)

Mixing Method

Shake

Equipment and Accessories

Shaker: mixing glass with stainless-steel cup
Strainer
Jiggers
Barspoon
Ice scoop
Tongs or pick
Straws
Cocktail napkin

step 1: Place a chilled sour glass on the rail, handling it by the stem.

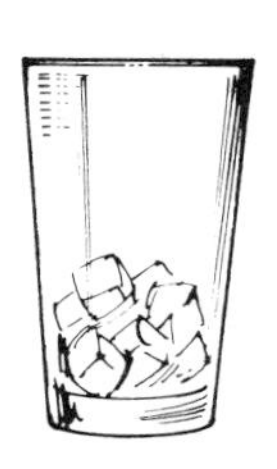

step 2: Fill the mixing glass ⅓ to ½ full of cube ice.

step 3: Measure liquor, lemon juice, and sugar (or mix) and add to the mixing glass.

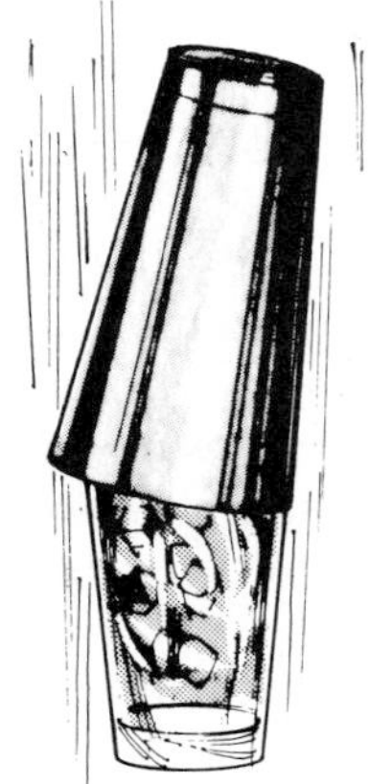

step 4: Place the cup over the glass and shake 10 times.

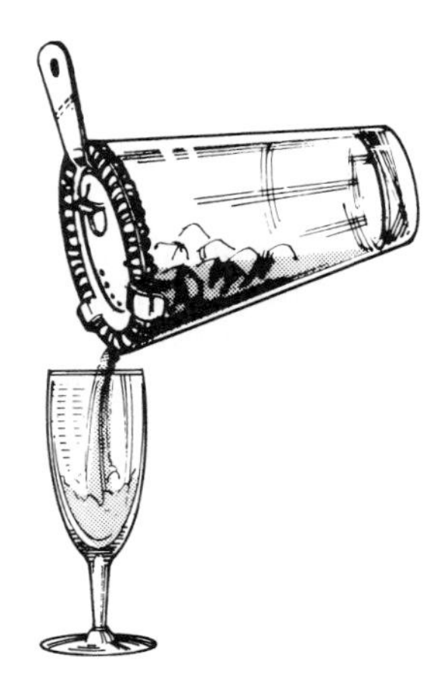

step 5: Remove the cup and strain the drink into the chilled glass.

step 6: Add the garnish, using tongs or a pick. Serve on a cocktail napkin.

step 7: Wash both shaker parts and invert them on the drainboard.

figure 10.3

this way. You may also dent the cup. Instead, hit the cup with the heel of your hand halfway between the point where the cup touches the glass and the point where it is farthest away from the glass.

In Step 7, the washing is necessary because sugar and fruit juices may cling to the sides of the containers after shaking.

The Shake-Mix method. If you use the shake mixer to "shake" your cocktail, you substitute the mixer can for the mixing glass and proceed as shown in Figure 10.4. As you can see, the procedures are very similar, but there are some noteworthy differences.

Notice that in Step 2 you use only ¼ can of ice. That is because the mixer can is bigger than the mixing glass. You need only enough ice to chill the drink.

In Step 4 you substitute the mixer can for the hand shaker. To estimate 10 seconds, count "one-hundred-one, one-hundred-two," and so on up to "one-hundred-ten."

The Blend method. To make the same drink using a blender, you substitute the blender cup for the shaker glass or mixer can. The mixing method is like that for the shake mixer, as you can see in Figure 10.5. Set the blender speed on high.

In carrying out Step 4, do not blend longer than the specified time. You do not want to incorporate bits of ice into the drink; you only want the ice to chill it. Blending too long will turn it into frozen slush.

Some sweet-sour drinks. Here is a pair of well-known sweet-sour cocktails.

Whisky Sour	Shake, Shake-Mix, or Blend
5-oz sour glass, chilled	
1 jigger whisky	
1 jigger lemon juice	
1 tsp sugar or 1½ tsp simple syrup	
Lemon or orange slice	
Cherry, orange slice, straws	

Daiquiri	Shake, Shake-Mix, or Blend
4½-oz cocktail glass, chilled	
1 jigger light rum	
1 jigger lime juice	
1 tsp sugar or 1½ tsp simple syrup	

A Whisky Sour is usually made with bourbon or a blended whisky. A sour can be made with any other liquor, such as gin, brandy, scotch, rum, tequila, vodka. For speed production you can substitute a jigger of sweet-sour

THE SHAKE-MIX METHOD: HOW TO MAKE A SOUR IN A SHAKE MIXER

Ingredients

Liquor
Lemon juice } or sweet-sour mix
Sugar or simple syrup }
Egg white (optional) }
Cherry/orange garnish

Glass

Sour glass (5 to 5½ ounces)

Mixing Method

Shake-Mix

Equipment and Accessories

Shake mixer
Strainer
Jiggers
Barspoon
Ice scoop
Tongs or picks
Straws
Cocktail napkin

step 1: Place a chilled sour glass on the rail, handling it by the stem.

step 2: Fill the mixer can ¼ full of cube ice.

step 3: Measure liquor, lemon juice, and sugar (or mix) and add to the mixer can.

step 4: Place the mixer can on the mixer and mix for 10 seconds.

step 5: Remove the can and strain the drink into the chilled glass.

step 6: Add the garnish, using tongs or a pick. Serve on a cocktail napkin.

step 7: Wash the mixer can and invert it on the drainboard.

figure 10.4

THE BLEND METHOD: HOW TO MAKE A SOUR IN A BLENDER

Ingredients

Liquor

Lemon juice ⎫ or
Sugar or simple syrup ⎬ sweet-sour
Egg white (optional) ⎭ mix

Cherry/orange garnish

Glass

Sour glass (5 to 5½ ounces)

Mixing Method

Blend

Equipment and Accessories

Blender
Strainer
Jiggers
Barspoon
Ice scoop
Tongs or pick
Straws
Cocktail napkin

step 1: Place a chilled sour glass on the rail, handling it by the stem.

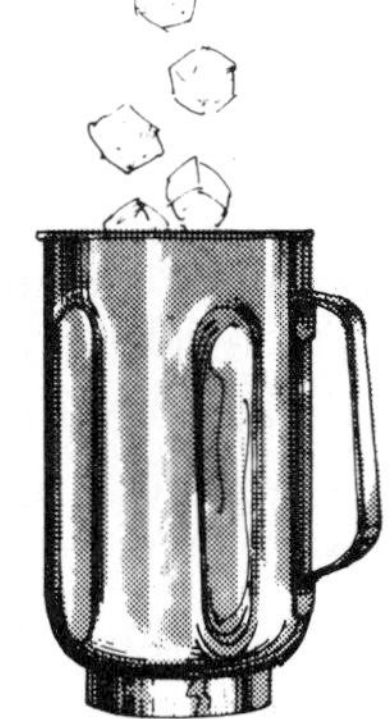

step 2: Fill the blender cup ¼ full of cube ice.

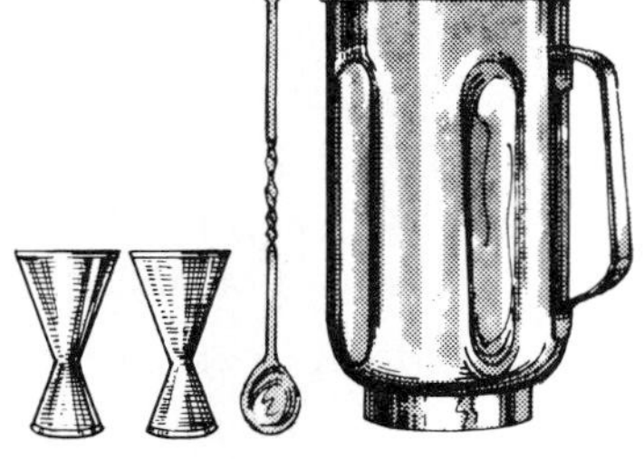

step 3: Measure liquor, lemon juice, and sugar (or mix) and add to the blender cup.

step 4: Place the blender cup on the electric blender and blend for 10 seconds.

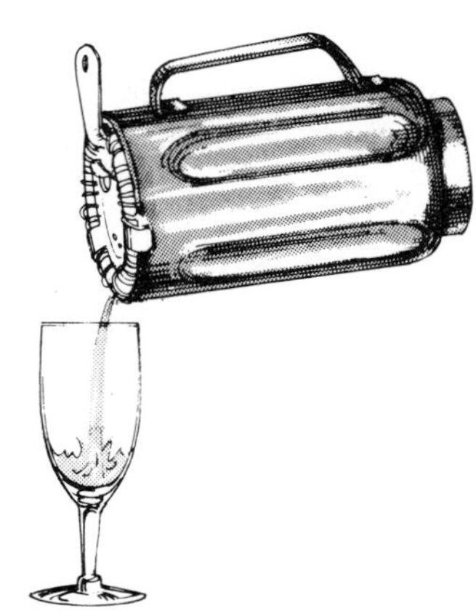

step 5: Remove the cup and strain the drink into the chilled glass.

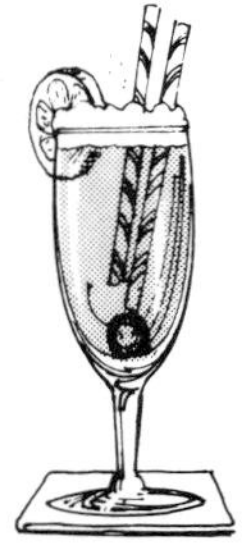

step 6: Add the garnish, using tongs or a pick. Serve on a cocktail napkin.

step 7: Wash the blender cup and invert it on the drainboard.

figure 10.5

mix for the lemon juice and sugar, at considerable sacrifice in quality. In the other direction, a special quality touch would be the addition of a teaspoon of egg white before the ingredients are blended.

The Daiquiri dates from the Spanish-American War and was named for the Daiquiri iron mines in Cuba. It is the prototype for a number of other drinks made with different spirits. Most similar is the Bacardi; it is essentially a Daiquiri made with Bacardi rum, with a dash of grenadine replacing half the sugar.

Change the liquor, substitute a liqueur or a syrup for the sugar, and you find these family members, some familiar, some passé but still occasionally called for:

- Ward 8, a Bourbon Sour with grenadine added.
- Margarita, tequila, lime juice, triple sec,with a salted rim—one of the most popular cocktails in the West and Southwest.
- Side Car, brandy, lemon juice, Cointreau, with a sugared rim. This drink was invented by a World War I captain who rode to his favorite Paris bistro on a motorcycle with a sidecar.
- Between the Sheets, half brandy/half rum, lemon or lime juice, triple sec—a variation of the Side Car.
- Jack Rose, apple brandy, lemon or lime juice, grenadine.
- Clover Club, gin, lemon or lime juice, grenadine, egg white.
- Tequila Rose, tequila, lime juice, grenadine.
- Pink Lady (*yesterday's version*), gin, apple brandy, lemon or lime juice, grenadine, egg white. (Today's Pink Lady adds cream and often omits the juice and the brandy.)
- Gimlet, gin, Rose's lime juice (sweet) or fresh lime juice and sugar. The Gimlet is said to have been the creation of the British in colonial India. If you use Rose's lime juice, as nearly everyone does, you can stir it instead of blending or shaking.
- Scarlett O'Hara, Southern Comfort, cranberry juice, lime juice. In this dry version the liqueur provides the sweet. A sweeter version uses Southern Comfort, grenadine, and lime juice—a totally different drink, no doubt invented by a bartender who had no cranberry on hand.

Another version of the Daiquiri is made by adding fresh fruit. You can blend in half a crushed banana to make a Banana Daiquiri, garnishing it with a banana slice. Or blend in crushed fresh or frozen strawberries for a Strawberry Daiquiri, using a whole fresh strawberry as a garnish. Fruit Daiquiris are often made as frozen drinks—a type of drink we will examine shortly.

You can serve any sour on the rocks in a rocks glass. You shake or blend or shake-mix it in the same way you make a straight-up sour, and then pour it over ice in a rocks glass.

Sour-related drinks . . .

You can start with the ingredients of the sour and make several other drink types by adding another basic ingredient. This gives us another set of drink families—collins, fizz, sling, daisy. Like the cocktail, these drinks originated in the Victorian era and have changed to keep pace with the times.

The collins is simply a sour with soda added, served over ice in a tall glass. We noted in Chapter 9 that today's collins is usually just a blend of liquor and mix, built in the glass. But if you break the drink down to its components, you can see there are other ways of making it: you can make it from scratch with fresh ingredients, or you can make it with a sweet-sour mix and soda. You can also see by looking at its structure that you can make other new drinks by changing or adding an ingredient or two.

Figure 10.6 shows you a collins-from-scratch using a hand shaker and freshly squeezed lemon juice. The essential point to grasp here is that the drink is made by combining two methods: first Shake, to mix together liquor, sugar, and fruit juice, then Build, to incorporate the soda without losing its bubbles.

You can also make a collins from scratch using a blender or a shake mixer as in Figures 10.4 and 10.5 and adjusting the ice measurement in Step 2. You can make this substitution in any drink you can shake by hand. Thus, when a recipe says Shake, you have your choice of three methods.

If you substitute sweet-sour mix for the lemon and sugar in Figure 10.6, you will still make the drink the same way, though you will no longer be making a collins from scratch.

A number of drinks take off from the collins-with-soda by substituting or adding other ingredients:

- French 75, a Tom Collins with champagne in place of soda.
- French 95, a John Collins with champagne in place of soda.
- French 125, a Brandy Collins with champagne in place of soda.
- Skip and Go Naked, a Vodka Collins or Tom Collins with beer instead of soda or in addition to it.

A fizz, in its bare essentials, is like a collins except that it is a shorter drink, served in a highball glass or a stem glass of highball size. At one point in the history of drinking, a fizz was designed to be gulped down like an Alka-Seltzer, and for the same reasons. To make it as bubbly as possible it was shaken long and hard with ice, the soda was added under pressure from a seltzer bottle, and the drink was served foaming in a small glass without ice.

Today, because of our modern ways with ice, a simple fizz is more like a short collins or a cross between a sour and a highball. However, some of the elaborations on the basic fizz make it a good deal more than a simple drink. Here is today's basic fizz:

HOW TO MAKE A COLLINS FROM SCRATCH

Ingredients

Liquor
Lemon
Sugar
Soda
Cube ice
Cherry, optional orange slice

Glass

Collins (10 to 12 ounces)

Mixing Method

Shake/Build

Equipment and Accessories

Shaker (or blender or shake mixer)
Strainer
Jigger
Barspoon
Ice scoop
Fruit squeezer
Long straws
Pick
Cocktail napkin

step 1: Fill collins glass ¾ full of cube ice and place on rail.

step 2: Fill mixing glass ⅓ to ½ full of cube ice. Measure and add liquor, sugar, and lemon.

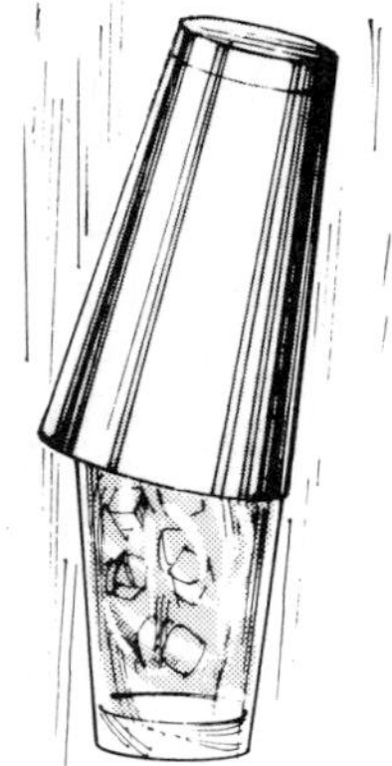

step 3: Shake the contents 10 times.

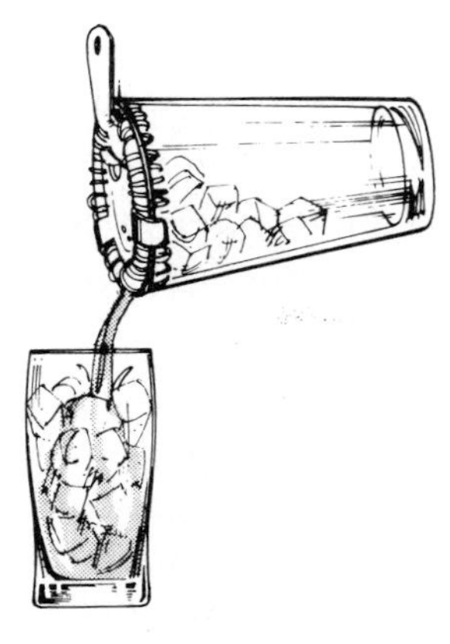

step 4: Strain shaker contents into collins glass.

step 5: Add soda to within ½ to 1 inch of rim. Stir gently (2 or 3 strokes).

step 6: Garnish, add long straws, and serve on a cocktail napkin.

figure 10.6

Family Characteristics

- ***Ingredients.*** Liquor, lemon, sugar, soda, cube ice.
- ***Glass.*** Highball or 8-ounce stem glass.
- ***Mixing method.*** Shake/Build.

To make a fizz, you follow the collins-from-scratch method. The liquor can be any type. Gin is the most common, but rum, scotch, brandy, or sloe gin are called for now and then.

The Gin Fizz is the one most frequently elaborated upon:

- Silver Fizz, a Gin Fizz with an egg white. If you add grenadine it becomes a Bird of Paradise Fizz.
- Golden Fizz, a Gin Fizz with an egg yolk.
- Royal Fizz, a Gin Fizz with a whole egg and sometimes cream.
- New Orleans Fizz or Ramos Fizz, a Gin Fizz made with both lemon and lime, with added egg white, cream, and a few dashes of orange flower water, served in a tall glass.

Most recently these more elaborate gin fizzes are undergoing further transformation by being made with ice cream—which introduces new methods we'll examine shortly.

Other fizzes include the Morning Glory Fizz, made with scotch and a little pernod, and the Sloe Gin Fizz. If you add cream to the Sloe Gin Fizz you have a Sky Ride Fizz.

A sling is like a collins to which something more is added—a liqueur, a special flavor, a special garnish. It is usually made by the collins-from-scratch method and is served in a collins glass with fruit garnishes. The most famous sling—and probably the one made most often today—is the Singapore Sling. We give you a modern recipe, adapted from the original made in the Raffles Hotel in Singapore at the turn of the century:

Singapore Sling Shake/Build

12-oz glass
¾ glass cube ice
1 jigger gin
½ jigger cherry-flavored brandy
½ jigger lemon juice
Soda to fill
Lemon or lime slice

Notice that this drink is just like a Tom Collins except that it uses a cherry liqueur (sweet) in place of sugar and has a different garnish. The sling as a species has great potential for creating your own specialty drinks.

A daisy is nothing more than a sour made with grenadine as the sweet, served in a larger glass over crushed ice and garnished lavishly with fruit. Sometimes it is served in a silver mug and stirred until the mug frosts, like a julep.

Tropical drinks . . .

The term tropical drinks as used in the trade comprises a loose collection of drinks originating in the resorts of the tropics or in restaurants having a tropical ambience such as Trader Vic's and Don the Beachcomber. The family characteristics are diffuse; there are no indispensable ingredients that tie them all together. Generally they have various kinds of rum as their base and make lavish use of fruit juices, liqueurs, syrups, and flower and fruit garnishes. They are showy, often expensive to make, and command a high price tag. Cheaper and easier versions of some can be made using bottled mixes.

Family Characteristics

- ***Ingredients.*** Rum (occasionally brandy, and once in a while gin), fruit juices, liqueurs, syrups, coconut milk, fruit garnishes, flowers, fresh mint.
- ***Glass.*** Anything from a cocktail glass to a whole coconut or pineapple.
- ***Mixing method.*** Shake (or Blend or Shake-Mix).

Among the fruit juices you have pineapple, papaya, coconut milk, and such other exotics as kiwi and mango, in addition to the usual lemon, lime, and orange. Among the syrups, grenadine, orgeat, falernum, and passion fruit syrup are popular. Among the liqueurs frequently called for are fruit-flavored brandies, cherry liqueur, curaçao, and pernod or some other absinthe substitute. Among the garnishes are pineapple cubes, coconut, mint leaves, the usual oranges, limes, and cherries, and orchids if available.

Here are two of the best-known tropicals:

Mai Tai *(from scratch)* Shake, Blend, Shake-Mix

12-oz glass
¾ glass cube or crushed ice
1 jigger light rum
1 jigger dark rum
1 lime (juice and peel)
½ oz orange curaçao
½ oz simple syrup
½ oz orgeat
Pineapple stick, cherry, mint sprig

Pinã Colada *(from scratch)* Blend

12-oz glass
¾ glass ice, cube or crushed
1 jigger light rum
1 jigger cream of coconut or coconut milk
1–2 jiggers pineapple juice or crushed pineapple
Cherry, pineapple, lime

Both these drinks can also be made from prepared mixes, and usually are.

The Mai Tai, like many other tropical drinks, was created by Trader Vic in the '40s and is still going strong. In Hawaii, where orchids grow on trees, your bountiful glass of Mai Tai will be topped with several of them.

In the Piña Colada the pineapple can be fresh or canned. If you use the crushed fruit be sure to blend at high speed until smooth. A variation of the Piña Colada is the Chi Chi, which substitutes vodka for rum.

Other classic tropical drinks are the Planter's Punch, the Scorpion, and the Zombie. It is said that buckets of Planter's Punch were carried to workers in the sugarcane field; another story says the drink was a specialty of the famous Planter's Hotel in St. Louis (both stories could be true). These three drinks are typically finished off with a float of 151° rum, so that the customer's first sip is the sting of the scorpion, the punch of the planter, or the kick of the zombie. The Zombie made its fame with the kick, the name, the challenge "only one to a customer," and in some cases the recipe (some recipes have four or five kinds of rum).

Recipes for these drinks are included in the Dictionary of Drinks at the end of the book (the Zombie recipe is a mild version). Look up the Pineapple Rum Royal too, for one of the ultimates in tropical merchandising.

Cream drinks . . .

Cream drinks are smooth, sweet after-dinner drinks made with cream and usually served straight up in a cocktail or champagne glass.

Family Characteristics

- ***Ingredients.*** Cream, one or more liqueurs or a liquor-liqueur combination.
- ***Glass.*** Cocktail or champagne, chilled.
- ***Mixing method.*** Shake (or Blend or Shake-Mix).

The proportions of the ingredients vary from one house to another. Some use equal parts (from ½ ounce to 1 ounce of each), some use up to 2 ounces of cream and smaller amounts of the other ingredients; some use more of the predominant flavor or the major liquor if there is one. The total ingredients should add up to about 3 ounces; any more and you may have to use a larger glass. Light cream or half and half is usually used, but heavy cream makes a better drink. The cream must be very fresh.

Whether you blend, shake, or shake-mix a cream drink, you follow the steps given for a sweet-sour cocktail. You may want to serve it with a pair of short straws. After mixing you must wash and rinse both your jigger and your glass or cup, because the cream and liqueurs cling to the sides.

Here are two familiar cream drinks:

Brandy Alexander Shake, Blend, Shake-Mix

Cocktail or champagne glass
¾ oz brandy
¾ oz dark crème de cacao
1 oz cream

Grasshopper Shake, Blend, Shake-Mix

Cocktail or champagne glass
¾ oz green crème de menthe
¾ oz light crème de cacao
1 oz cream

You can make an Alexander with any base liquor, substituting it for the brandy in this recipe. Apparently the earliest was the Gin Alexander, invented to disguise the bathtub gin of Prohibition days. Light crème de cacao is used when the base liquor is a light color, such as vodka, rum, or tequila, while dark crème de cacao is used with brandy and whiskies. A Vodka Alexander is sometimes called a Russian Bear or a White Elephant. A Rum Alexander is a Panama.

The original cream concoctions spawned a whole menagerie of animal drinks:

- Pink Squirrel, light crème de cacao, crème de noyaux, cream.
- Brown Squirrel, dark crème de cacao, amaretto, cream.
- Blue-Tailed Fly, light crème de cacao, blue curaçao, cream.
- White Monkey, light crème de cacao, crème de banana, cream. Also known as a Banshee.
- Purple Bunny, light crème de cacao, cherry-flavored brandy, cream.

In addition to the animals we have:

- Golden Cadillac, light crème de cacao, Galliano, cream.
- White Cadillac, light crème de cacao, Cointreau, cream.
- Velvet Hammer, vodka or Cointreau, light crème de cacao, cream.
- Cucumber, green crème de menthe, cream. Another version adds brandy or gin as the base ingredient.
- White Russian, vodka, Kahlúa, cream (a Black Russian with cream).
- Golden Dream, Galliano, triple sec, orange juice, cream.
- Pink Lady, gin, grenadine, cream.

Any of the cream drinks may be served on the rocks if the customer requests it. A cream drink on the rocks must be blended or shaken as for a straight-up drink, then strained over cube ice in a rocks glass. Sometimes it is built in the glass without stirring, with the cream as a float. This makes a very different drink.

Another spinoff from the after-dinner cream drinks is to add a mixer and make it into a highball-size drink. Thus the Colorado Bulldog starts off as a vodka-Kahlúa-cream drink shaken and poured over ice in a highball glass; then the glass is filled with Coke. It's a Shake/Build method.

Other dairy drinks . . .

In addition to cream, other dairy products are sometimes used in mixed drinks. These are usually long drinks rather than cocktails—pick-me-ups or nightcaps rather than appetizers or desserts. They are too filling to precede or follow a meal.

It is hard to say whether the current crop of milk drinks is a logical extension of today's popular cream drinks or a modern version of old colonial libations. There seem to be some of each type.

Today's milk punches are clearly descendants of the older punch drinks, using milk in place of water and served iced instead of hot. (Sometimes they are served hot too.)

Family Characteristics

- ***Ingredients.*** Liquor, sugar, milk, cube ice, nutmeg.
- ***Glass.*** Collins.
- ***Mixing method.*** Shake (or Blend, or Shake-Mix).

Liquors most commonly called for are brandy, whisky, rum, and gin. Make your basic milk punch with brandy, add an egg, and you have an eggnog.

Rum Milk Punch Shake, Blend, Mix

12-oz glass
¾ glass cube ice
1 jigger rum
1 tsp sugar
4 oz milk
Sprinkle of nutmeg

Eggnog Shake, Blend, Mix

12-oz glass
1 egg
1 jigger brandy
1 tsp sugar
4 oz milk
Sprinkle of nutmeg

Notice that you don't use ice in the eggnog glass. The egg will give you extra volume, and the drink does not demand the ice-cold temperatures of most. Do shake it with ice, however. You can add half a jigger of brandy to the Rum Punch if you wish, to pep it up.

Among the newer drinks are some that use milk as a mixer in a highball-type drink. Here are two examples:

Aggravation Build

8-oz glass
¾ glass cube ice
1 jigger scotch
½ jigger Kahlúa
Milk to fill

Smith & Kerns Build

8-oz glass
¾ glass cube ice
1 jigger dark crème de cacao
Milk to ¾ full
Soda to fill

Another milk drink is Scotch and Milk. It is served over ice in a highball glass like any scotch highball. You can also substitute milk for cream in cream drinks if a calorie-conscious customer requests it.

Another egg drink is the flip, a cold, straight-up drink of sweetened liquor or fortified wine that is shaken with egg and topped with nutmeg. This is the descendant of the colonial flip, which was drunk piping hot, either simmered over the fire or heated with a hot poker or loggerhead.

Ice cream drinks . . .

An ice cream drink is any drink made with ice cream. Many of the ice cream drinks are variants of cream drinks, with ice cream replacing the cream. Others are made by adding ice cream to another drink such as a fizz. Figure 10.7 tells the whole story, including the family characteristics and the step-by-step mixing method.

In addition to the equipment listed you will need a special freezer chest at the serving station for storing the ice cream, and your health department will probably require you to have a special well with running water and an overflow drain into which to put your scoops between uses. An alternative to all this is a soft-serve machine.

Here are two ice cream drinks, a short one and a long one:

Grasshopper	Blend or Shake-Mix
8-oz stem glass, chilled	
1 scoop vanilla ice cream	
½ jigger green crème de menthe	
½ jigger white crème de cacao	
Straws	

Ramos Fizz	Blend/Build or Shake-Mix/Build
12-oz glass, chilled	
3 scoops vanilla ice cream	
1 jigger gin	
1 oz lemon juice	or sweet-sour-frothee mix
½ oz lime juice	
1 egg white	
1 tsp sugar	
3–4 dashes orange flower water	
Soda to fill	
Straws	

In both drinks ice cream replaces the cream of the original recipe.

Other popular ice cream drinks include the following, in which vanilla ice cream replaces cream: Brandy Alexander, Velvet Hammer, White Russian, Pink Lady, Golden Dream. Other fizzes, such as the Royal Fizz and Silver Fizz, are sometimes made with ice cream.

Ice cream drinks make good house specialties. You can invent your own: not only do you have many drinks to start from but you do not have to stick to vanilla ice cream. Consider special glasses for your creations.

HOW TO MAKE AN ICE CREAM DRINK

Ingredients
Liquor
Ice cream
Optional ingredients
Optional garnish

Glass
8- or 12-oz, chilled

Mixing Method
Blend or Shake-Mix

Equipment and Accessories
Blender or shake mixer
Ice cream scoop, #20 or #24
Jigger
Barspoon
Straws
Cocktail napkin

step 1: Place prechilled glass on the rail.

step 2: Scoop ice cream into blender or mixer cup.

step 3: Add the liquor and other ingredients.

step 4: Blend or mix until ice cream has liquefied.

step 5: Pour the entire contents into the glass.

step 6: Add garnish and straws and serve on a cocktail napkin.

figure 10.7

Frozen drinks . . .

Frozen drinks are moneymakers, especially in warm climates. The method is simple and straightforward: you simply blend crushed ice along with the ingredients of the drink until everything is homogeneous and the ice has refrozen to the consistency of slush. See Figure 10.8 for details.

Several points are critical in making a successful drink:

- In Step 3, the right amount of ice is important. Too little, and you have a drink without body. Too much, and you have a drink without taste.
- In Step 4, it takes quite a long time to reach the right consistency, several times as long as blending any other drink. Listen to it! When you no longer hear the bits of ice hitting the blender cup it is ready to serve.
- In Step 5, use the barspoon to scrape everything out of the cup.

The various sours make the best frozen drinks, because of their tangy flavor. A bland drink will be even blander when frozen. Many frozen drinks are sweet-and-sour drinks with fruit added, such as the two below:

Frozen Strawberry Daiquiri Blend

8- to 10-oz stem glass, chilled
1 jigger rum
1 tsp lemon juice
2 tsp simple syrup
1 jigger pureed strawberries or six fresh berries cut up
Crushed ice to submerge liquids
Whole berry garnish

Frozen Peach Margarita Blend

8- to 10-oz stem glass, chilled
1 jigger tequila
½ jigger peach liqueur
1½ jiggers pureed peaches
Crushed ice to submerge liquids
Fresh peach wedge in season

These recipes are for use with fresh fruit. You can substitute frozen strawberries in the Daiquiri, omitting the simple syrup. The Peach Margarita is a specialty at Crackers Restaurant in Dallas. They use fresh peaches in season, blending the puree in advance with simple syrup and a dash of lemon juice. In winter they puree canned peaches without added sugar, and yes, they do sell them all winter.

Filling drink orders . . .

Most drinks are made to order drink by drink, but the orders seldom come in one drink at a time. Here are a few tips on handling orders. They are aimed at the following objectives:

- Speed—keeping up with the orders.
- Quality—getting the drink to the customer at its peak of perfection.

HOW TO MAKE A FROZEN DRINK

Ingredients
Liquor
Optional ingredients
Optional garnish
Crushed ice

Glass
8- to 12-oz stem glass

Mixing Method
Blend

Equipment and Accessories
Heavy-duty commercial blender
Jigger
Barspoon
Ice scoop
Ice crusher
Short straws
Cocktail napkin

step 1: Place prechilled glass on the rail, holding it by the stem.

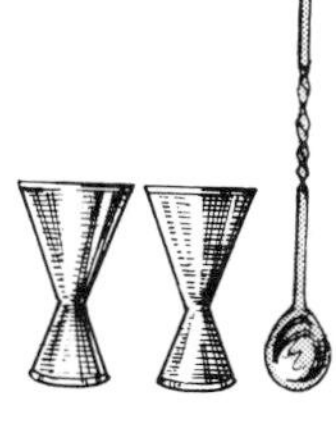

step 2: Pour cocktail ingredients into blender cup.

step 3: Using the scoop, add crushed ice to come just above liquor level.

step 4: Blend on high speed until mixture blends and refreezes to a slush.

step 5: Heap contents of the cup into the glass.

step 6: Garnish, add straws, and serve on a cocktail napkin.

figure 10.8

- Accuracy—the right drink to the right customer.

First off, deal with one set of orders—one server's guest check or one party of bar customers. Set up all the glasses at once; it will help you remember what was ordered. Group them according to the base liquor, setting them up in the same sequence as the liquor bottles in the well. (Have your servers call drinks in this order too.) In this way a good bartender can handle a fairly long list without taking time to refer back to the written ticket.

If there is more than one drink of the same kind that is not built in the glass, make them together. Put extra ice in the mixing glass or the blender or mixer cup, multiply each ingredient by the number of drinks, and proceed as for a single drink. Divide the finished product among the glasses you have set out for these drinks, but not all at once. Fill each glass half full

table 10.1 GUEST-CHECK ABBREVIATIONS

liquors, etc.		*drinks*		*call brands*	
liquors					
Bourbon	B	Bourbon and Water	B/W	Beefeaters	BEEF
Brandy	Br	Black Russian	BRUS	Canadian Club	CC
Gin	G	Brandy Alexander	BR ALEX	Chivas Regal	CHIVAS
Rum	R	Bacardi	BAC	Cutty Sark	CUTTY
Scotch	S or SC	Banana Daiquiri	BAN DAQ	Dewars White Label	WHITE or WL
Tequila	TEQ	Bloody Mary	MARY	Drambuie	DRAM
Vodka	V	Daiquiri	DAQ	Early Times	ET
		Godfather	GOD	Four Roses	ROSES
mixes		Gibson	GIB	Grand Marnier	MARNIER
Coke	C	Gimlet	GIM	Hennessy	HENN
Ginger ale	G	Grasshopper	GRASS	J & B	JB
7-Up	7	Irish Coffee	IRISH C	Jack Daniels	JD
Soda	S	Harvey Wallbanger	BANGER	Jim Beam	BEAM
Sprite	SP	John Collins	JOHN	Johnny Walker Black	BLACK S
Tonic	T	Manhattan	MAN	Johnny Walker Red	RED S
Water	W	Margarita	MARG	Old Fitzgerald	FITZ
		Martini	MT	Old Grand-Dad	DAD
garnishes		Old-Fashioned	OF	Seagrams 7 Crown	7
Lime	LI	Ramos Fizz	RAMOS	Seagrams V.O.	VO
Olive	OL	Rob Roy	R ROY	Sloe gin	SLG
Onion	ON	Rusty Nail	R NAIL	Smirnoff	SMIRN
Twist	TW or ~	Screwdriver	DRIVER	Southern Comfort	SO C
		Scotch and Water	SC/W	Tanqueray	TANQ
special instructions		Tequila Sunrise	SUNRISE		
Double	DBL	Tom Collins	TOM		
Dry	X	Vodka Martini	V MT		
Extra dry	XX	Whisky Sour	WS		
On the rocks	R				
Straight up	Up or ↑				
Frozen	Z				

the first time around; then add a little more to each glass in another round or two until you complete all these drinks evenly.

Make drinks in the following sequence:

1. Start frozen drinks and ice cream drinks (they make on their machines while the rest are being poured).
2. Straight liquor drinks (straight shots, liquor on rocks).
3. Juice drinks and sours.
4. Cream drinks and hot drinks.
5. Highballs with carbonated mixers.
6. Draft beer.

The reason for the sequence is to make first those that keep best and to make last those that don't hold well. Some places have the server call drinks in this sequence instead of in the well order.

When writing an order on a guest check, use a standard set of abbreviations for drinks, liquors, brand names, mixes, and special garnishes. Table 10.1 gives you some common ones. You can adapt them to your needs; just be sure everyone uses the same ones.

In writing a drink order, a slash is used to separate the items in the instructions. For example, a very dry Vodka Martini on the rocks with a twist is written V MT/XX/R/TW.

When the server takes a table order, the best way to get the right drink to the right person is to pick out one seat as Number 1—say the one closest to the bar. Then each seat is numbered in order around the table. Each drink is written on the check following the number of the customer's seat.

Figure 10.9 is an example of a guest check for a party of six.

Developing drink menus and specialty drinks . . .

A thorough knowledge of drinks and the ways they are made makes it clear why planning the drink menu—that is, the range and types of drinks you will serve—is one of the most important things you do. The drinks you serve will determine the sizes of glassware, the number and type of ice machines, the refrigerator and freezer space, the small equipment and utensils, the space on the backbar. They will also determine the skill levels you require of your bartenders and servers. And of course they will determine the kinds of liquor and supplies you buy and the number of items you must keep in inventory.

The unlimited bar concept. If your menu concept is an unlimited bar serving the full spectrum of drinks, you must be able to produce those

CRACKERS
RESTAURANT & BAR
2621 McKINNEY
DALLAS, TEXAS 75204

Date	Table No.	Persons	Server	Check No.
8-2-81	10	6	L. S. D.	113022

Seat	Item	Amount
1	V/MT/R/Tw	2.25
	XX/TANQ/MT/R/ON	2.50
	7/7	2.25
	Chivas/W	2.25
	Br/MAN/R	2.25
6	MARG/Z/Cuervo	2.75
		14.25
	Ⓛ 14.25 FOOD TOTAL	——
	B BAR TOTAL	14.25
	W TAX	——
	SALES TOTAL	14.25
	TOTAL AMOUNT	

figure 10.9 Guest check using typical abbreviations. Seat numbers are in left column. (Courtesy Crackers, Dallas)

drinks. This means equipment producing both cube and crushed ice, glassware that will accommodate everything from the after-dinner liqueur to the Zombie, a freezer for ice cream, a means of chilling cocktail glasses, all the necessary small equipment, an ample draft-beer setup, and 100 or more different beverages in your inventory. In addition you must have skilled and knowledgeable bartenders as well as servers who know how to take and transmit orders.

There are many types of enterprise where the versatility of the unlimited bar is part of the bar's image. Even though the customer may order the same drink time after time (often without knowing what is in it or how much it costs), the assurance that that drink is available at that bar is important. Furthermore, it is possible for a whole party of people to order widely different types of drinks to suit each individual taste. Usually it is the bars where little if any food is served for which the unlimited drink menu is important. It is also appropriate to the expensive restaurant where excellence in everything is the image.

The limited drink menu. A recent development has been the appearance of printed drink menus. You may find them in restaurants where drinks are secondary to food or in the trendy neighborhood bar/grills. They usually feature specialty drinks, along with old favorites, with descriptions that raise the thirst level as they list the ingredients. Or there may be a list of special drinks chalked on a blackboard at the bar. These drink menus are proving to be good sales stimulators. At the same time, by focusing attention on a limited selection, they can avoid some of the costs of a full-spectrum bar.

For example, a well-designed specialty menu can reduce the extensive liquor inventory required in an unlimited bar. If you offer an attractive selection of 15 or 20 drinks all based on a few liquors, liqueurs, and mixes, you can cut the number of items you must have in your inventory by 50% at least. You should still be prepared to serve the standard highballs and Martinis and Bloody Marys, and you will still carry a small selection of the popular call brands, but your customers will order up to 90% of their drinks from your printed menu.

The limited drink menu applies the philosophy of the limited food menu: instead of offering everything anyone might want, you specialize, in the same way that you develop a successful food menu. You combine a few basic ingredients using a skillful mix-and-match technique, in the same way that an Italian restaurant offers a long list of entrees by mixing and matching pastas and sauces.

The limited drink menu also shares some other advantages of the limited food menu. Properly developed, it can mean that less equipment is needed at the bar as well as less space for the smaller inventory, and thus less investment overall. It can mean that fewer skills and less experience are required of the bartender, so you do not need to pay high-skill wages. Your own training of your own personnel to prepare your own selection of drinks can produce that sought-after consistency of product. Also you can choose your base ingredients with an eye to keeping down costs: vodka and rum are cheaper than the whiskies, and they mix well with a variety of flavor additions. And buying larger quantities of fewer items may give you better quantity discounts.

For a limited menu to be successful, the first requirement is that it must reflect the tastes of your customers. If you are already in business you have data on your most popular drink types. Include the favorites and go on from there to make new drinks by changing or adding flavors and flavor accents to the popular drink types. If yours is a new enterprise, find out what your target population is drinking in other places. Be sure to include house wines and a selection of popular beers—something for everybody.

The printed menu must catch the customer's eye and whet the appetite and the thirst. It must spell out the ingredients in each drink, since your specialties will be new to the customer and the names you give them won't mean anything. (They may also be interested to read what their old favorites are made of.) Depending on your clientele and your budget, you may even want to illustrate your menu with inviting photos or sketches of your drinks. Make it interesting to browse through; like your food menu, it is a promotion piece.

Several years ago a chain of Mexican restaurants, by introducing such a menu, cut its inventory down to 35 liquors and 9 mixes plus 4 wines and 4 beers—about a third of the average unlimited bar. The four-color printed menu offered 20 mixed drinks—frozen drinks, cocktails, and slings, some new, some old, all aimed at the unsophisticated, sweet-tooth tastes of their youngish, blue-collar clientele. All the drinks were made from the same few base liquors, liqueurs, and made-to-order bottled mixes. This made it possible to hire persons of no previous experience, train them thoroughly to mix each drink on the menu, and serve the customer the same drink in Denver as in Tallahassee at an attractive menu price.

Developing specialty drinks. You do not need to have a special printed drink menu to take advantage of some of its best features. A few specialty drinks on a table tent or menu flier can perk up sales, enhance your image, and often reduce your inventory by focusing customer orders. Here are a few pointers about developing specialty drinks:

- Cater to your clientele and their preferences.
- Observe the basic drink structure discussed in Chapter 9. A successful drink has a base liquor plus one or more flavor modifiers or flavor accents. The base liquor should be at least 50% of the liquor in the drink.
- Choose flavor combinations that are compatible; don't combine at random; don't mix orange juice and chocolate. Work with popular flavors. Try adding a trendy flavor as a float atop a familiar drink, or be the first to make an old drink with a new product, *if* you have a clientele this would appeal to.
- Consider your equipment, glassware, and space. If you want to feature frozen drinks you must have an ample supply of crushed or flake ice and plenty of blenders, or enough demand for a single specialty to invest in a frozen-drink machine. If you want ice cream drinks you must have ice cream equipment at the bar.
- Consider your bartenders' skill level. If you want to serve pousse-cafés be sure your personnel can make them.
- Keep the drinks simple so they can be made quickly.
- Plan attractive visual effects, dream up a catchy name, and blend it all into your image.

Summing up . . .

The systematic development of your drinks and your drink menu provides you with performance standards and products of consistent quality. It also makes it easy to train your bartenders and cocktail servers. All this makes for a smooth operation and satisfied customers. Standardizing also facilitates pricing drinks, controlling costs, and cutting losses, all of which enhance the profit picture. These things are discussed in chapters to come.

Mixology and drink development are fascinating subjects. A thorough knowledge of drinks opens the door to invention, not only by you but by your bartenders and servers. Intriguing menus and specialty drinks are excellent merchandising devices, and getting your employees involved is a good way to increase sales.

chapter 11 . . .

PURCHASING, RECEIVING, AND STORAGE

Our focus shifts now from the front of the house to what goes on behind scenes. The next few chapters will deal with managing the business side of the bar. This begins with providing the alcoholic beverages to be sold to the customers. Maintaining a steady supply is one side of this function. Keeping track of the supply to ensure that what is bought produces sales is another aspect. Managing the storeroom is a third; this includes physical care to maintain quality and watchdogging to maintain quantity.

This chapter examines purchasing policies and decisions; the routines of purchasing, receiving, and issuing; the inventory records and procedures commonly used; and the use of inventory figures to measure bar cost and purchasing efficiency.

You may wonder how a straightforward subject like buying liquor and supplying it to the bar could take up so many book pages. Part of the answer is that alcoholic beverages are an investment in income-producing stock, and there are many facets to making the best investment for the least money. Part of it is that purchasing and the recordkeeping it generates are of critical importance to other aspects of the business. Part of it is that alcohol has an irresistible attraction for many people, and if you don't keep track of it, it will evaporate, so to speak.

The goal of beverage purchasing is to provide a steady supply of ingredients for the drinks you sell, at costs that will maximize profits. The purchasing function moves in a continuous cycle having several distinct phases:

- ***Planning and ordering.*** Selecting what you need at the most advantageous prices.
- ***Receiving.*** Taking delivery of exactly what you have ordered—brands, sizes, and quantities, at specified prices, in good condition.
- ***Storing.*** Keeping your beverage supplies until needed, in a place secure against theft and deterioration.
- ***Issuing.*** Transferring your beverages from storeroom to bar, where they will be used to make drinks for your customers.

The sale of these drinks keeps the cycle revolving, since you must constantly replenish the supplies consumed. But it doesn't revolve by itself; it has to be managed. It must be responsive to needs (sales volume and customer tastes), to the market (supply and price), to cash flow (money available for investment), and to indicators of change in any of these. The beverage manager must know what is going on all the time.

Planning the purchasing . . .

Many factors go into the making of purchasing decisions. Some are management policies that, once established, should be followed to the letter. Others are day-to-day decisions based on current situations. The most important decisions have to do with what to buy, where to buy it, how much to buy, when to buy, and what to pay.

What to buy. Deciding what to buy involves two basic policy decisions: the quality of the beverages you will pour and the variety of items you will have available. Take quality first. To begin with, it will depend on your clientele and the quality they expect and are willing to pay for. It would be foolish to buy premium brands or fine French wines for a low-budget clientele, and equally foolish not to offer such items in a luxury restaurant. As in everything else, you must *know your clientele*.

Beyond this, there is the quality of your ***well brands*** to consider—the liquors you pour in mixed drinks when the customer does not call for a specific brand. Choose a set of well brands and stick to them, for the sake of consistency in your drinks. Many bars use inexpensive brands in the well on the theory that customers can't tell the difference in a mixed drink. Others use familiar advertised brands in the middle price range. Still others use premium brands—this is sometimes called a ***super well***—which they feature with pride in their merchandising. In addition to making an impression it means one less item to carry in inventory, since they would be stocking these as call brands anyway.

The one thing to avoid in choosing your well brands is using cheap liquors while charging quality prices. If you think you are fooling your customers you are fooling yourself. Taste your own drinks and you will see the difference just as your customers will. You may save a few pennies per drink but you will never know how many customers never came back, and how many others they told about your cheap high-priced drinks. Besides, you don't save all that much. Take, for example, a liter of scotch costing $10.00 and one costing $6.00. If the portion is 1 ounce, the $10.00 scotch costs 29¢ per drink and the $6.00 scotch costs 18¢, a difference of 11¢ per drink. Is the savings worth it when you may lose customers?

The second major decision is the variety of items you will stock. Some bars take pride in never having to tell a customer, "I'm sorry but we don't have that." Such a policy, while it can be good merchandising for some types of clientele, has the potential of expanding inventory indefinitely with items that do not move. Liquor that does not sell does not earn a penny of profit, except as it contributes to atmosphere and image in a backbar display.

Many bars prefer to limit their offerings to popular and well-advertised brands of each item, the number of brands varying with the type and size of the operation. If a customer calls for a brand you don't carry, you probably won't lose either the sale or the customer if you can offer a well-known brand of comparable quality. But *offer it*—don't try to substitute another brand without the customer's knowledge. That is a very bad practice and will only create an image of your place as one that can't be trusted.

Still another approach is to deliberately limit the number of brands and items stocked by developing a printed drink menu based on a small number of beverages, as discussed in Chapter 10. Most customers will respond to such a menu by ordering from the drinks listed, and the questions of call brands and unusual drinks seldom come up.

Where you draw the line on brands and items to be stocked will depend on your clientele, your type of enterprise, your volume of business, and the money available for such investment. But it is wise to draw the line somewhere and then hold to it. One way to avoid proliferation of brands and bottles is never to add a new item of unpredictable demand without eliminating a slow-moving item from your list. You really have to watch the sales of individual items closely in order to keep your supplies honed down to what is current and what moves.

You also need to keep up with new products and to anticipate changes in customer tastes. One way of doing this is to consult frequently with sales personnel and with the buyers for the wholesalers you deal with. They know who is launching a huge advertising campaign or coming out with a new light beer or soft wine, and what new liqueur or imported beer is big in California or New York and coming your way. Many buyers have a regular time for sales personnel to call. In this way the buyer is not interrupted while immersed in some other task, and the salesperson has a receptive audience.

Do not, however, let anyone tell you what you should buy. Consider suggestions, but measure them against your own ground rules and current needs. Sales representatives will push their own products; that is their job. Compare what they say with what their competitors say, and make up your own mind.

As for choosing the individual items, buying spirits and beers is mostly a matter of brand selection. Where the brand concerns the customer, you buy the brands your customers will buy. For your well, select the brands that make the quality of drinks you want to pour. This goes for your vermouths and liqueurs as well. It is a good idea to taste your own mixed drinks with different brands of spirits and liqueurs. Generic liqueurs in particular can taste quite different from one brand to another, and the expensive imported brands are not necessarily the best.

Buying wines is somewhat more complicated. For house wines you can buy jug wines by brand, and the wine will be the same from one bottle to the next. Wines you will serve by the bottle are another matter. In many instances the same wine will vary from one vintage to another, and will change in the bottle too. Customer demand is also less clear-cut than in beers and spirits. Taste before you buy, then choose according to what you know of your customers' tastes, and get as much expert advice as you can. You may find it helpful to reread Chapter 6 on developing a wine list.

Where to buy. The beverage buyer does not always have a great deal of choice about where to buy. State laws govern the purchase and sale of alcoholic beverages, and these laws vary from one state to another. Local laws also come into play. The first thing a buyer should do is to study the laws of the state, county, city, and even precinct as they apply to liquor purchase.

In 18 states the retailer (the seller of alcoholic beverages to the consumer) must buy from state stores. These states are known as ***control states*** or ***monopoly states***. In the remaining states—known as ***license states***—and

table 11.1 **CONTROL STATES AND LICENSE STATES**

control states	*license states*	
Alabama	Alaska	Mississippi (retail only)
Idaho	Arizona	Missouri
Iowa	Arkansas	Nebraska
Maine	California	Nevada
Michigan	Colorado	New Jersey
Mississippi (wholesale only)	Connecticut	New Mexico
Montana	Delaware	New York
New Hampshire	Florida	North Dakota
North Carolina	Georgia	Oklahoma
Ohio	Hawaii	Rhode Island
Oregon	Illinois	South Carolina
Pennsylvania	Indiana	South Dakota
Utah	Kansas	Tennessee
Vermont	Kentucky	Texas
Virginia	Louisiana	Wisconsin
Washington	Maryland	Wyoming (retail only)
West Virginia	Massachusetts	District of Columbia
Wyoming (wholesale only)	Minnesota	

the District of Columbia, the buyer is typically allowed to purchase from any wholesaler licensed by the state and in some states from licensed distributors and manufacturers as well. (Table 11.1 shows which states are which.) There may be local restrictions, however. For example, a county may have a law against buying in another county. To further complicate matters, the requirements and limitations are usually not the same for beers and wines as they are for spirits.

In license states the buyer's choice is further limited by the structure of the market, which works to give some items to some sellers but not to everybody. The market has several tiers or levels. There is the manufacturer —the distiller, brewer, or vintner. The manufacturer may have its own distribution network, as breweries typically do, or it may sell through a distributor who handles the products of many manufacturers, such as an importer who handles wines and spirits from all over the world. A distributor typically has the exclusive right to sell a given brand or product in a given territory. It may sell directly to the retailer or it may sell to wholesalers, or both, depending on the requirements of state law. If it sells to wholesalers it may sell to all or to only one or two. As a result you will have to deal with several suppliers, and you will have to find out who carries what items and what brands at what prices.

In most license states a master list is published monthly containing the names of all the wholesalers in the state, the lines they carry, and the prices they charge. Individual wholesalers also have lists of their products and prices, quantity discounts, and special sales and promotional materials. Control states publish lists of all brands available and their prices, along with the addresses of state stores.

Easier than working with lists is dealing with sales representatives. Though you may not be able to buy from them directly, they can tell you where you can buy their products. They can also keep you posted on special sales and promotions.

Where you have a choice, price is certainly one reason for buying from one supplier rather than another. But consider also what services the supplier offers—or doesn't offer. For instance:

- How often does a given supplier deliver? The oftener the better, and daily is best, even though you do not order daily. You do not have to stock so much, and you can get something quickly in an emergency.
- Where is the supplier located? Suppose you are in a small country town and your suppliers must travel 50 to 100 miles to reach you. Does distance affect the delivery schedule? What about the weather—will your wine or beer travel for hours in the hot sun? Will snow and ice interrupt service?
- What resources does the supplier have—a large and varied inventory kept well stocked, or a small stock that is constantly being depleted? Temperature-controlled warehouse facilities? Refrigerated trucks?
- Does the supplier give proper and systematic care to goods in storage or in transit? Are wines kept at proper temperatures, bottles on their sides or upside down? Are draft beer and unpasteurized package beer kept refrigerated? It is a good idea to pay a visit to the warehouse to see how things are cared for.
- Must you buy a certain minimum per order? Can you adjust your orders to meet these requirements? Is it worth it?
- Does the supplier extend credit and what are the terms?
- Can you buy mixes and accessories from the supplier at advantageous prices and quantities?

How much to buy. How much to buy is a central question for the beverage buyer. The answer is *enough but not too much*—enough to serve your customers what they want but not so much that numbers of bottles stand idle on shelves for long periods tying up money you could put to better use. You should never run out of your well brands, your house wines, your draft beers, the popular bottle beers, and the popular call brands of spirits. But you might not reorder a slow-moving item until your last bottle is half gone.

The place to start in setting a standard for an adequate supply is to establish your par stock for each bar in your facility, using a form similar to Figure 8.2 (page 201). You can determine par stock needs from your detailed sales records. (If you are still in the planning stage you can guesstimate your rate of sale.) A general rule is to have enough of each type and brand to meet 1½ times the needs of your busiest day of the week. For a small restaurant bar averaging $500 in daily sales this might work out to be the open bottle plus two full ones for each fast-moving brand and the open bottle plus one for each slow-moving brand.

From par stock needs for each bar you can figure what you should have in the storeroom to back them up. This then becomes your par stock for the storeroom, your normal storeroom inventory.

You can also use par stock to measure your daily consumption. The bottles it takes to bring the bar stock to par represent roughly the consumption of the day before. Over a week's time these bottles will yield an accurate figure of average daily consumption. This can guide your rate of purchase. Par stock is also one way of keeping up with customer tastes. You know what brands are moving quickly because you have par stock as a measure of their popularity. Thus it tells you what to buy as well as how much you are using.

Whatever your buying interval, it is a good idea to set ***minimum*** and ***maximum stock levels*** for each item to maintain your storeroom inventory. The minimum level may be supplemented by a ***reorder point*** that gives you lead time, so that the stock does not drop below the minimum level before you receive delivery. Your maximum level represents the dividing line between enough and too much. Like the par stock level at the bar, it should represent 1½ times what you expect to need before you replenish your supply.

Why be concerned about having too much? There are several arguments for keeping a small inventory geared closely to your sales volume. The major point is that beverages are expensive, even at discount rates, and they tie up money that is not earning anything and may be needed elsewhere. When finally you do use a bottle of liquor you bought months ago at a discount, it may have cost you far more in lost use of the money you paid than the money you saved in buying it.

There are other problems with large inventories. One is security: the more liquor you have, the more tempting it is, the harder it is to keep track of, and the easier it is to steal. The larger the inventory, the more space and staff are needed and the greater the burden of recordkeeping and taking physical inventory. Perishable items such as beers should never be overstocked; some wines also deteriorate quickly. Customer tastes may change suddenly, leaving you with items you will never use.

On the other hand, some wines should be bought in quantity because they are scarce and will quickly disappear from the market. You will want to buy enough of them to last as long as your printed wine list does.

Beverage wholesalers sell mostly in case lots, sometimes offering discounts for a certain number of cases. You can buy cases of spirits and wines in the bottle sizes shown in Table 11.2. (Some imported wines have

table 11.2 **BOTTLE AND CASE SIZES**

bottle size[a]	*fluidounces*	*units per case*	*replaces*
Distilled spirits			
50 milliliters	1.7	120	1, 1.6, and 2 ounces
200 milliliters	6.8	48	1/2 pint
500 milliliters	16.9	24	1 pint
750 milliliters	25.4	12	4/5 quart (fifth)
1 liter	33.8	12	1 quart
1.75 liters	59.2	6	1/2 gallon
Wine			
100 milliliters	3.4	60	2, 3, and 4 ounces
187 milliliters	6.3	48	2/5 pint
375 milliliters	12.7	24	4/5 pint
750 milliliters	25.4	12	4/5 quart (fifth)
1 liter	33.8	12	1 quart
1.5 liters	50.7	6	2/5 gallon
3 liters	101	4	4/5 gallon
4 liters	134.8	4	1 gallon
Beer[b]			
6 ounces	6	24 or 32	
7 ounces	7		
8 ounces	8		
10 ounces	10		
12 ounces	12		

[a] Not all sizes are legal in all states.
[b] For draft beer container sizes see Table 11.3.

variant sizes, such as 700 ml.) The size bottles you buy will depend on the type and size of your establishment, and even more on the way you serve your beverages.

Since every wine bottled since 1978 and every spirit since 1979 is in metric sizes, you will find fewer and fewer gallons, quarts, and fifths available. But it is still worth pointing out that the half-gallon of 64 ounces contains almost 5 ounces more than the 1.75-milliliter size. If you find half-gallons this difference may possibly give you a lower price per ounce.

Some suppliers will sell a ***broken case*** (a case of 12 bottles made up of several brands or items of your specification), and some will sell some items by the single bottle or, for wines, a minimum of three bottles. The cost per bottle is higher for items in a broken case or by the bottle; the cost per bottle is least when purchased by the case or in multiple-case lots. Certainly you should buy by the case anything you use a lot of—your well brands, house wines, domestic beers, and popular brands—to take advantage of cost savings and because you will probably need these in cases. From there on it becomes a matter of calculating your rate of use for each item and matching it with your purchasing intervals.

When to buy. When and how often you place an order will depend on the volume of your business, the size inventory you are willing to stock, the requirements and schedules of the suppliers, the scheduling of your receiving people, the specials you want to take advantage of, and such variables as holidays and conventions and special events or a run of bad weather. Some enterprises order daily, some once per accounting period, others somewhere in between. Some might buy wine once a year, spirits weekly, and beer every day.

The more frequently you buy, the less inventory you have to cope with. On the other hand, every order initiates the whole purchasing-receiving-storing routine, which may not always be labor-efficient.

Other factors affecting the timing of purchase are your cash position and the payment or credit requirements imposed by state regulations or by the purveyor. Some states allow no credit at all; all sales must be cash. Other states require payment monthly or semimonthly or within some other interval such as 10 days after the week of sale or the second Monday after delivery. Fines and publicity may accompany late payments or nonpayment. These requirements make it very important to integrate your purchasing times with your cash flow so that you are not caught short on the due date.

What to pay. In control states price markups are typically fixed by law and prices will be the same in all state stores. The only price decision the buyer must make is whether to take advantage of an occasional special or quantity discount.

In license states it is often worth shopping around to find the best deal on the brands you want to buy. There are seldom large price differences because state laws are typically designed to avoid price wars. Manufacturers and distributors must give the same deals to everyone they sell to, so the price structure is fairly homogeneous. Nevertheless in some areas suppliers are free to set their own markups, grant their own discounts, and run their own specials, and they do. So it is worthwhile to study price lists and talk to sales contacts and look at the ads in the trade journals.

When you do find good buys you must weigh the money you save against your inventory size, rate of use, and what else you could do with the money. Sometimes the best price is not the best buy. Be wary, too, of the motives behind supplier specials. Sometimes these are intended to get rid of wines or beers approaching their limits. Fortunately you do not have to worry about this in spirits, which do not deteriorate.

You will also want to look at the cost per ounce of the products you purchase, particularly the spirits, since you are going to be pouring and pricing them by the ounce. The cost per ounce is highest when the bottle size is smallest, and decreases as the bottle size increases. Thus fifths and 750 mls cost most per ounce, quarts and liters cost less, half-gallons and 1.75 liters still less, and gallons and 3- and 4-liter bottles least. The cost per ounce is higher for items purchased by the bottle or in a broken case than when purchased by the case.

Cost per ounce is not the only consideration, however; the size bottle that is most efficient in action is the decisive factor. In your well you will probably use quarts and liters, for speed and ease of handling. For a remote pouring system you will use the half-gallon or 1.75-liter size. You will use these or the 3- or 4-liter bottles for pouring wines by the glass or carafe, to take advantage of low cost per ounce. But if the lowest cost per ounce does not give you what you want, it is not a good buy.

Placing the liquor order . . .

Once you have established your purchasing plans, you need to develop a standard procedure for placing your orders. These procedures will be influenced by size and sales volume. At one extreme is the multicopy purchase order typically used by the large organization. At the other extreme is the informal verbal order, given by phone or to a visiting salesperson.

But even the verbal order should have a complete paper record to back it up. Every record, formal or informal, should contain the following information:

- Date of the order.
- Name of the purveyor (seller).
- Name of the salesperson.
- Purveyor's phone number.
- Anticipated date and time of delivery.
- Items, brands, vintages ordered.
- Sizes of containers (bottle and case).
- Numbers of bottles or cases.
- Unit prices (price per bottle for each item).
- Name of person placing order.

There are four good reasons for keeping a written record of each order:

- It gives the person receiving the delivery the data needed for checking it.
- It gives the person paying the bills data needed for checking the bill.

- It gives everyone concerned with the order—whether buyer, receiving agent, storeroom staff, accountant, bar manager, banquet manager, or bartender—access to exact data.
- It minimizes uncertainty, misunderstanding, and argument.

In a small operation the record may simply be a memo of a phone order written on a form devised by the house, or it may be an order written out by a visiting salesperson. The only one concerned may be the owner/manager who buys, receives, and stores the merchandise, stocks and tends the bar, and pays the bills. In a large organization, where responsibilities are divided among many departments, a formal multicopy ***purchase order*** may be used (Figure 11.1), with the original going to the purveyor and copies sent to all concerned.

Every purchase order has an order number ***(P.O. number)*** which is a key element in a network of paper records. It will be referenced on the purveyor's invoice, thus becoming the link between the two.

The ***invoice*** (Figure 11.2) is the purveyor's response to the buyer's order. It reflects the information on the buyer's order sheet from the seller's point of view. It accompanies the delivery and must be signed by the buyer or the buyer's agent when delivery is received—which brings us to the second phase of the purchasing cycle.

Receiving . . .

The signing of the invoice by the purchaser has legal significance: it is the point at which the buyer becomes the owner of the merchandise. The delivery must be carefully checked before the invoice is signed. The person you give this responsibility to must be someone you trust who has a good head for detail and has been trained well for this assignment.

The first step is to check the invoice against the purchase order or memo. This must be an item-for-item check to see that the quantities, unit and case sizes, brands, vintages, and so on are listed as ordered and that the unit and case prices are quoted correctly. Then the math must be checked —the total costs per item (called ***extensions***—number of units × unit cost) and the invoice total.

The second step is to check the delivery itself, to see that it matches the invoice. Each item must be checked as to quantity, unit, brand, vintage, and any other specification. Open cases should be verified in this way bottle by bottle, and examined also for breakage, missing or broken stamps, loose corks. Sealed cases should be examined for evidence of leaking bottles or weighed. The weight should agree with the weight printed on the case; a broken bottle will give a short weight. Beer should be checked for freshness by reading the pull dates on the containers. Kegs should be examined for

PURCHASE ORDER
Name of Bar
Address

Date ______________ Purchase Order No. ______________

Purveyor ______________ Salesperson ______________

Phone ______________ Date Needed ______________

Address ______________ Hour of Delivery ______________

Item #	*Quantity*	*Unit*	*Brand*	*Unit Price*	*Total*

P.O. Order Total ______________
Ordered by ______________

figure 11.1

WHOLESALE DIVISION
10301 HARRY HINES BLVD.
DALLAS, TEXAS 75220
(214) 350-5806

Goody-Goody Liquors, Inc.
INVOICE

Crackers
2621 McKinney Ave.
Dallas TX 75204

ORDER NO.	ORDER DATE	ACCOUNT	LICENSE NO.	SALES REPRESENTATIVE	SHIP VIA	TERMS	INVOICE NO.	INVOICE DATE	PAGE
4566	10/01/80	001638	MB 98376		Delivery	COD - beer only	002984	10/01/80	1

ITEM NO.	QTY. ORDERED	QTY. SHIPPED	PACKAGE	UNIT	DESCRIPTION	PRICE	*	TOTAL
3256	1	1	unit	litre	W. L. Weller 90 ✓	7.67		7.67
17346	1	1	unit	1/5	Courvoisier V. S. Cognac 80 ✓	12.25		12.25
4418	1	1	unit	quart	Tanqueray 94 ✓	9.04		9.04
11885	1	1	unit	litre	John Jameson 80 ✓	9.43		9.43
17838	4	4	unit	litre	Amaretto Dicupera 56 ✓	6.10		24.40
17905	1	1	unit	750 ml	B & B 86 ✓	13.56		13.56
20540	2	2	unit	750 ml	Tia Maria 63 ✓	9.89		19.78
					Total liquor			96.13
9635	1	1	unit		Picks, 3" sword 1000 per box ✓	3.54		3.54
9520	1	1	unit	1/2 gal	Cherries ✓	5.55		5.55
					Total supplies			9.09

PERMIT NO. 85512

GALLONAGE

Crackers
2621 McKinney Ave.
Dallas TX 75204

SHORTAGE MUST BE REPORTED ON DELIVERY

QTY.	UNIT	(1) OVERS, (2) SHORTS, (3) BREAKAGES	AMOUNT
Deliver	in the	afternoon	
Mary	Delivery	10/01/80	

C.O.D. AMOUNT	
AMT. CHARGED	105.22
TOTAL INVOICE	105.22
ON OR BEFORE	10/25/80

CUSTOMER COPY

figure 11.2 Liquor invoice. (Courtesy Goody-Goody Liquors and Crackers, Dallas)

signs of leakage. Their contents can be checked by weighing the keg and subtracting the empty keg weight later. Table 11.3 gives the correct net weights for draft beer.

The third step is to request a ***credit memo*** for any discrepancies between the order memo and the invoice or the delivery itself. This will include items invoiced but not delivered, items invoiced and delivered that were not ordered, wrong merchandise (sizes, brands, etc.), items refused (overage beer, broken bottles, missing stamps, swollen beer cans), wrong prices, math errors. The credit memo at this point is usually a notation right on the invoice showing the item, problem, and amount of credit (see the bottom of the printed invoice in Figure 11.2), or it may be a separate credit slip on which the invoice date and number are written. In either case it must be initialed or signed by the delivery person and it too must be checked for accuracy.

table 11.3 **BARREL AND KEG SIZES**

container	*gallons*	*liters*	*net wt/lb*	*net wt/kg*	*fluidounces*
Barrel	31	111.33	248	112.48	3968
½ barrel (keg)	15.5	58.67	124	56.24	1984
¼ barrel (½ keg)	7.75	29.33	62	28.12	992
⅛ barrel (¼ keg)	3.88	14.67	31	14.06	496

Only when everything has been checked, settled, and initialed does the receiving agent accept delivery by signing two copies of the invoice, one for you and one for the purveyor.

The purveyor will follow up the credit notation or memo with a confirming memo carrying the invoice date and number (Figure 11.3). This is in effect an amendment to the invoice and must be coupled with the invoice when the bill is paid. If not, or if you pay an unsigned invoice, you may end up paying for merchandise you never received.

The receiving routine must proceed as quickly as possible and with undivided attention. Then just as quickly the delivery must be taken to the storeroom and signed in by the person in charge there. Liquor being received and transported is very vulnerable to pilferage and there is no point in increasing temptation.

A large organization may require the storeroom manager to repeat the check of merchandise using another invoice copy. When the storeroom manager signs for the beverages he or she accepts responsibility for them from the receiving agent. There it rests until they are requisitioned for bar use.

The careful checking of deliveries has a double purpose: it is one part protection against the purveyor's errors and one part security against pilferage. Receiving a verified delivery is the first checkpoint along the system of controls that should follow your beverages from the time they enter your doors until they are poured at your bar and paid for by your customers.

Storage . . .

The storeroom is the setting for the third phase of the purchasing cycle. It performs three functions: security from theft, physical care to maintain quality, and inventory maintenance and recordkeeping.

The first essential for storeroom security is to limit access. This room is off limits for all but authorized personnel. Anyone withdrawing beverages must not enter the room but must request what is needed from the store-

CREDIT MEMO

Customer:

Credit Memo No. ________ Invoice No. ________ Date ________

No	*Item Description*	*Quantity*	*Unit Price*	*Total*
1				
2				
3				
4				
5				

Amount of credit ________

Reason for credit ________

Purveyor Signature ________

Receiving Bar Signature ________

figure 11.3

room staff or whoever has responsibility in a small operation. When open, the room must never be left unattended. If it must be left empty even briefly, the door must be locked. This should be a substantial door with a deadbolt lock and only two sets of keys, one for the storeroom manager and one for emergencies, to be kept in the safe. Or it might have a combination lock that can be reset frequently, with only two people knowing the combination. If keys are used, locks should be changed often, in case someone makes duplicate keys, and always when someone who has had keys leaves your employment.

Windows should be barred or covered with barbed wire. Alarm systems are widely used to protect against off-hour break-ins. Some of these depend on light or noise to scare away intruders or summon help; some alert police or a private security system directly.

An orderly storeroom is a security measure as well as a necessity for efficient operation. When everything is systematically in place, something amiss is soon noticed. Each beverage should have its designated place in a logical arrangement grouping similar items on adjacent shelves. Opened cases should be emptied immediately and their contents shelved, rotating the stock to put the older stock in front. Never leave a case half empty; flatten empty cases and remove them promptly. It could be easy to spirit away hidden bottles along with the trash.

Shelving should be heavy and well braced, because liquor is heavy. Sealed cases can be stacked on low platforms until you need their contents.

Store wines on their sides or upside down in the sealed case. They are perishable, subject to deterioration by light, warmth, agitation, and old age. They need a cool dark environment between 50° and 70°F (7° to 21°C), the ideal temperature being a constant 55° to 60°, called ***cellar temperature***. Move them as little as possible and handle them gently. Agitating a wine may upset both its chemistry and its sediment, making it unservable until the sediment settles again. However, since wines have limited life spans, rotating the stock becomes particularly important, lest a wine go over the hill before you use it.

Beer has the most limited shelf life of all. Canned and bottled beers should be stored below 70°F in a dark place and their pull dates should be checked periodically. Draft beers and unpasteurized canned or bottled beers (Coors, for example) should be kept refrigerated. Draft beer must be kept at an even 36° to 38°F (2.2° to 3.3°C) and should be used within 30 to 45 days.

As custodian of the storeroom contents, the storeroom manager has responsibility for keeping track of the stock of each item at all times. This is an important part of any system for minimizing pilferage. Some storerooms use a system of individual ***bin cards*** to log stock in and out. A card for each item, such as those shown in Figure 11.4, is attached to the shelf where the item is stored. A typical bin card shows the brand name, bottle size, quantity on hand, and sometimes a bin number or code number. The amount of stock delivered is immediately added to the quantity on hand, and the amount of stock issued for use is immediately subtracted. The number of bottles shown on the bin card should always agree with the actual number of bottles of that item. Spot-checking stock against bin cards helps to keep track of inventory.

The minimum and maximum stock levels may also be recorded on the bin cards, making it easy to be aware of purchasing needs. The cards are also a quick index to rate of use.

Issuing . . .

The issuing of beverage supplies from the storeroom to the bar fulfills the ultimate purpose of the purchasing function by supplying the beverages to be sold. Like the other key points in the purchasing cycle, the issue of stock

figure 11.4 Storeroom at Jack Tar Hotel. Bin cards identify and record each liquor item. Bin numbers (top shelf) coordinate with wine list. (Photo by Charlotte F. Speight)

must be duly recorded and responsibility formally passed from the storeroom to the bar.

The written record in this case is a ***requisition (issue slip)*** such as the one in Figure 11.5. You might think of it as a sort of in-house purchase order, with the bar as the "buyer" and the storeroom as the supplier. The bar lists the brand, size, and number of bottles for each item required, with the date and the signature of the person requesting the issue. The storeroom adds the cost and value information to complete the record, and the person issuing also signs. The person receiving the stock at the bar adds his or her signature to complete the transfer of responsibility.

Many bars require the bottles emptied the previous day to be turned in with the requisition. This is known as the ***one-empty-for-one-full*** system. The empties are collected and the requisition made out at closing time, as noted in Chapter 8. The manager double-checks the requisition and the bartender's bottle count to be sure they agree. Both are turned in to the storeroom the following day by the opening bartender, and replacement stock is issued. In this way the supply is automatically maintained at par-stock level. As an additional benefit the empty-bottle system enables management to check the par stock at any time against the par-stock list to see

FORM NO 159 REV. 6-71

SKY CHEFS, INC., & CONSOLIDATED SUBSIDIARY BEVERAGE REQUISITION

TO ______ REQ. NR. ______

FROM ______ DATE ______

ITEM (BRAND DESCRIPTION)	UNIT	QUANTITY REQUESTED	QUANTITY ISSUED	UNIT COST	TOTAL COST	UNIT RETAIL VALUE	TOTAL RETAIL VALUE

REQUISITIONED BY ______ DATE ______

FILLED BY ______ DATE ______

DELIVERED BY ______ DATE ______

RECEIVED BY ______ DATE ______

figure 11.5 Typical beverage requisition form. Bartender completes first four columns; storeroom completes last four. (Courtesy Sky Chefs)

if anything has disappeared. There should be a bottle for every bottle on the list, whether that bottle is empty or full.

Some enterprises stamp each bottle with an identifying stamp when it is issued. Then if an empty comes back without a stamp you know somebody is up to something.

In multiple-bar operations each bar has its own par stock, which is requisitioned and issued separately. For public bars and service bars the one-empty-for-one-full system works well, but a different system must be used in supplying liquor for one-time events such as banquets and conventions, a common situation in hotels. In this case there are no empty bottles to replace, and there is no one par stock. The brands and amounts are estimated in advance for each event based on anticipated consumption, past experience, and a safety allowance, and are issued to the person in charge of the event, using a special requisition form. The person in charge has the responsibility for returning all bottles, empty or full, to the storeroom when the event is over. The liquor used, represented by the empties, is charged to the cost of the event, and the remaining stock is integrated into the storeroom inventory.

Sometimes the supply system breaks down during an unexpected rush or an emergency. If the storeroom is closed, liquor is likely to be borrowed by one bar from another so as not to lose sales, and no one thinks to keep track. Eventually empties end up in the wrong place, the records are scrambled, and sooner or later something disappears. It is the manager's responsibility to avoid such emergencies and keep the supply and record system on track.

Inventory . . .

The beverage ***inventory***—that is, the amount on hand at any given time—is of central importance to the purchasing function. The buyer must be able to determine exactly what is on hand, as well as the rate at which it is being used, in order to make intelligent purchasing decisions. In addition to storeroom records, other ways of keeping track of total inventory are needed.

There are two reasons for keeping a constant check on inventory. One is to pinpoint losses quickly in order to put a stop to them. This has to do with controls, a subject discussed more fully in the next chapter. The other reason has to do with purchasing. If you have lost stock due to theft, breakage, error, or whatever, you must buy stock to replace it, so that you can serve your customers. Therefore you have to know what you really have on hand in order to plan each purchase.

Physical inventory. The only accurate way to know what you have on hand is to take a complete ***physical inventory***—that is, to count each bottle and keg—on a regular basis. Ideally you will do it weekly and again at the end of the accounting period.

If possible the inventory should be taken by persons who do not buy liquor or handle it on the job. As a double check, it is best to have two people working together, one counting and the other writing down the count. Both people initial each page as it is completed. The inventory record should follow the arrangement of the storeroom, grouping items by category, and within categories by brand, and within brand by size. An inventory form used by a multiple-bar facility is shown in Figure 11.6.

The liquor at the bar is also part of inventory until it is sold, and it too is counted in an end-of-the-month physical inventory. It is done in the same way as the storeroom inventory except that there are opened bottles to be counted.

The simplest way to measure the contents of opened bottles is to estimate each bottle by sight and count the contents in tenths. Thus a full bottle is 10/10, a half bottle is 5/10, while an empty bottle is 0/10. This gives you an approximate amount, of course, but it is close enough.

If you have a metered pouring system you have a very accurate way of counting. The system counts the drinks it pours. You multiply the count by the size of the drink and compare this with what is left in the partly used bottle. You have to count the liquor in each line from bottle to dispensing head as part of the inventory. The manufacturer's representative should give you this capacity at the time the equipment is installed.

All inventory must be taken at the same time and it must be done when the bar is closed, so that nothing changes while the count is being taken. Weekly inventories of individual bars or of the storeroom may supplement the overall end-of-the-month count.

Perpetual inventory. Another way of providing inventory information is to compile ongoing daily records from invoices and requisitions, adding each day's purchases and subtracting each day's issues for each item in stock. This task is typically performed by the accounting department and the product is known as a ***perpetual inventory***. It is kept by hand on cards —a separate card for each item—or by computer, which can report the stock at any given moment with pushbutton ease.

A perpetual inventory is a paper record that tells you only what you are supposed to have on hand in the storeroom. It does not tell you what you really have—only a physical inventory can do that. Its primary function is to provide a standard against which a physical count can be measured, item for item, at any given time. If everything is in order the two inventories should agree. If separate records are kept on bin cards, they should also agree—they too are a form of perpetual inventory.

If the count and the records don't agree, you are faced with the dilemma of determining whether there are errors in the records or the count or

SKY CHEFS

LIQUOR INVENTORY

FORM 10B REV 4 71

*LOCATION

1 ____________ 5 ____________
2 ____________ 6 ____________
3 ____________ 7 ____________
4 ____________ 8 ____________

UNIT ____________

DATE ____________ 19____

PAGE NO. ______ OF ______ PAGES

DESCRIPTION OF ITEM (BE SPECIFIC)	UNIT	*LOCATION								TOTAL ON HAND	UNIT PRICE	EXTENSION
		1	2	3	4	5	6	7	8			

TOTAL THIS PAGE ______

MANAGER'S SIGNATURE ____________

figure 11.6 Liquor inventory form used in a multiple-bar facility. (Courtesy Sky Chefs)

whether the items themselves have disappeared. You can trace errors in the record by going back to the invoices and requisitions, and errors in the count by recounting. If you can't find any mistakes you may as well assume theft, and adjust your perpetual inventory record accordingly.

A perpetual inventory is very time-consuming, expensive, and subject to error, and there are other ways of measuring discrepancies that are accurate enough for everyday use. We explore them in the next chapters. Whether you use a perpetual inventory or not, it is an important concept. If you don't have a way of knowing where you ought to be, you don't have a measure of where you are.

Figuring inventory value, bar cost, and inventory turnover rate. When you have completed a physical inventory for an accounting period, you must figure the dollar value of the total stock. To do this you begin by entering the unit cost of each item on the inventory sheet on which you have recorded the count (refer to Figure 11.6).

The next step is to multiply the unit cost by the number of units to find the dollar value of the stock for each item. Enter each total on the form in the last column. Totaling the values of all the items then gives you the value of your entire inventory. This figure is known as the ***closing inventory (ending inventory)*** for the accounting period. The same figure becomes the ***opening inventory (beginning inventory)*** for the next accounting period.

Now you can use this figure to determine the value of all the liquor used to produce your sales for the period, or in other words your ***beverage cost***. Here is the way to do it:

TO:	Value of opening inventory (OI)
ADD:	Value of all purchases made during period (P)
EQUALS:	Value of total liquor available during period
SUBTRACT:	Value of closing inventory (CI)
EQUALS:	Value of total liquor used during period, or cost (C)

Or, to put this calculation in equation form:

$$OI + P - CI = C$$

The dollar cost for liquor used in a given period is usually expressed as a percentage of the sales for the same period. To figure this percentage you divide cost by sales:

$$\frac{C}{S} = C\%$$

This percentage figure is often referred to simply as ***bar cost***. We will discuss the significance and uses of this figure in later chapters.

You can also use the values for closing inventory and purchases for the period to determine your ***inventory turnover rate***. This rate will help you decide whether you are keeping too much or too little in inventory. To figure this rate you divide purchases by inventory:

$$\frac{\text{Purchases for the period}}{\text{Closing inventory}} = \text{Inventory turnover rate}$$

For example, if purchases were $1500 and closing inventory is $1500, the turnover rate is 1; you have turned over your inventory once during the period. If purchases were $1500 and closing inventory is $2000, your turnover rate is 0.75, while if closing inventory is $1000 with $1500 in purchases, your turnover rate is 1.5.

What is the significance of the turnover rate? Generally speaking, if your turnover rate is consistently below 1, you are probably stocking more than you need. If it runs above 2, you probably often run out of things you need and you might increase your sales if you increased your rate of purchase. However, each enterprise has its own optimum purchase rate which may vary from industry averages. You really should figure the turnover rates for beer, wine, and liquor separately, because they are not necessarily the same, the rate for wine in particular being slower.

Purchasing bar supplies . . .

Purchasing supplies for the bar follows the same cycle as beverage buying —purchasing, receiving, storing, and issuing. However, buying supplies is a good deal simpler because the products are not regulated by state and federal laws.

You can buy grocery items from grocery wholesalers and cocktail napkins from wholesalers of paper goods. But if you find beverage wholesalers who carry drink-related items, check them out. (Some state codes do not allow liquor dealers to sell anything but liquor.) The prices may be better, since the liquor dealer may get better quantity discounts on bar items such as maraschino cherries and cocktail onions than the grocery dealer does. Service may be better too. The food wholesaler may pay little attention to your small order, whereas the liquor purveyor wants to keep your beverage business. The liquor dealer may also sell in smaller quantities than the grocery wholesaler, and if you have a small enterprise it may take forever to use up a case of olives or a gross of napkins.

But liquor purveyors don't handle such items as lemons, limes, oranges, celery, eggs, milk, cream, ice cream. You will buy your produce from a produce dealer and your dairy supplies and ice cream from a wholesale dairy. Such items must be refrigerated in the storeroom or go straight to the bar. Ice cream must go straight into a freezer.

In working out your orders you will have a choice of can or bottle sizes in many items. The cost per olive or per ounce is nearly always cheaper in the larger container sizes. However, consider deterioration once the container is opened. If you use only half a can of something and have to throw out the rest, you haven't saved any money.

Receiving, storing, and issuing follow the same procedures used for alcoholic beverages. Like items must be stored together, and the stock must be rotated at each delivery. All types of bar supplies are counted on the regular physical inventory. However, once a container is opened, it is considered used and is not counted.

Summing up . . .

A good purchasing manager provides an adequate supply of beverages at all times without overinvesting in idle inventory. A careful system of records is maintained at all phases of the purchasing cycle. Through such records management can keep track of supply and can pinpoint responsibility at any stage along the way from purchase to sale.

Physical inventory at regular intervals provides the basis for figuring needs, costs, and losses. Written or computerized inventory records provide standards against which physical inventory can be measured.

A well-organized, well-managed purchasing system can contribute to profits by keeping costs down, efficiency up, and supplies flowing.

chapter 12 . . .

PLANNING FOR PROFIT

In a way, profit is what is left over after all expenses are paid. But this is an almost fatalistic point of view and it does not give you the best results. If you look at profit as a goal to be achieved instead of as a residue of operation, you can chart your course and control the process.

This chapter shows how the principle of planning for profit works to structure the financial side of an operation toward reaching a profit goal. It discusses the budget as a profit plan, the systematic pricing of drinks to maximize profits, and the setting up of controls to assure that the beverages go into drinks and the money goes into the cash register and thence into profit as planned.

How often have you been shocked to hear that a popular bar or restaurant was going out of business? If you ask what went wrong, the usual reply is, "Poor management!"

It is a phrase that could mean almost anything, but most likely money management was involved, and more than likely the failure also involved poor planning. A great many enterprises, especially small ones, operate with more enthusiasm than foresight and count on advertising or gimmickry or entertainment or somebody else's successful formula to generate profits. The accountant is expected to take care of the financial end of things.

But the shrewd entrepreneur knows that profit can and should be systematically planned for, budgeted for, and watched over like a hawk from one day to the next. Let us begin with the budget as a profit plan.

Budgeting for profit . . .

Profit is the primary goal of a beverage enterprise, whether that goal is a hope or a dream or a very specific target. A budget provides a strategy for reaching a target and a major tool for measuring progress along the way.

What is a budget? A **budget** is a financial plan for a given period of time coordinating anticipated income and expenditures to ensure solvency and yield a profit.

The budget is a document with a dual personality. In the beginning it is a plan detailing in dollar terms the *anticipated performance* of the bar during the period to come—the expected sales, the expenses that will be incurred in order to achieve the sales, and the desired profit. As the budget period arrives, the budget ceases to be a plan for the future and becomes a tool of control for the present. It now provides a basis for measuring *actual performance and results* against the performance goals of the plan.

Or perhaps it is more accurate to think of the budget as a continuing two-phase process or system, rather than a static or inflexible set of figures. The planning phase requires the manager to project the financial future in detail and make realistic forecasts of income and outgo in relation to the ultimate profit goal. This very process of thinking ahead, gathering data for forecasting, figuring one's way through the facts and numbers, and applying them to a future that may change before it arrives may suggest operating adjustments that will help to achieve the profit goal.

In the second or control phase, the frequent measuring of actual results against a budget plan can signal threats to profits and indicate the areas of trouble. Quick action may put the business back on course. On the other hand, sometimes the original plan is unrealistic, or circumstances change.

Then the budget can be modified to bring it into line with present reality. In this sense the planning process continues during the operating period.

The planning process. The first step in developing a budget is setting your profit goal—a specific, realistic goal, not just "as much profit as possible." How do you do this? You think of it in terms of return on money invested. Let's say that you have $100,000 invested in a going business or that you are going to invest that amount in a new project. What would this $100,000 bring you in a year if you put it in a savings certificate or mutual fund or invested it in real estate or some other not too risky venture? Let's say it would bring you $10,000 a year. Now double this figure (a rule-of-thumb safety margin because of the risks you are taking) for a total of $20,000. This is your projected return on investment (ROI) and your profit goal for the year.

Profit is the difference between income and expenses. Since your income comes from sales, your next step is to forecast your sales for the budget period. The accuracy of this forecast will determine how realistic the rest of your budget is and whether or not you will reach your profit goal or indeed make a profit at all.

To forecast your sales you must estimate the number of drinks you expect to sell in each period of each day, week, and month of the year. Figure 12.1 is a useful form for such a forecast. If you are already in business, draw on your past sales data. Break the data down in terms of number of drinks sold at noon, in the early afternoon, at happy hour, in the evening, and by day of the week, weekday, weekend. Note the effect of price changes for happy hours or specials as well as general price changes and the influence of holidays, conventions, football games. Make notes on high and low sales periods, seasonal variations. Then use this data to decide whether the upcoming year will be a rerun of the last, or whether you want to make it different and whether there will be external factors that will affect the forecast.

For example, has your competitive position changed—a new bar across the street, maybe? Is the general economy affecting your business? Are you going to make changes such as renovating your facility or adding live entertainment and dancing? Do you have plans for special promotions? Adapt your historical figures to accommodate all the changes you foresee.

If you are just starting out in the bar business, make the same estimates on the basis of what you have found out about your chosen clientele, your market area, your competition, and your capacity. And estimate conservatively!

Finally, multiply the numbers of drinks you forecast by the prices you plan to charge in the upcoming year. This is your sales forecast and your anticipated income. For the sake of argument, let's place this figure at $100,000 for the year (see Figure 12.2).

Your next step is to estimate all your expenses for the same period. You

SALES FORECAST FORM

Period: *From* ____________ *To* ____________

Date	*Mon*	*Tue*	*Wed*	*Thu*	*Fri*	*Sat*	*Sun*	*Total*	*Average*
Shift 11–3									
Week ending ____									
Week ending ____									
Week ending ____									
Week ending ____									
4-week totals ____									
Shift 3–7									
Week ending ____									
Week ending ____									
Week ending ____									
Week ending ____									
4-week totals ____									
Shift 7–closing									
Week ending ____									
Week ending ____									
Week ending ____									
Week ending ____									
4-week totals ____									

figure 12.1

TENTATIVE BUDGET	Dollars	Percent of sales
Projected Sales	$100,000	100%
Variable Expenses		
Beverage costs	$ 24,000	24.0%
Payroll costs	$ 22,500	22.5%
Administrative expenses	$ 4,000	4.0%
Laundry and supplies	$ 3,600	3.6%
Utilities	$ 3,000	3.0%
Advertising and promotion	$ 3,500	3.5%
Repairs	$ 1,500	1.5%
Maintenance	$ 2,000	2.0%
Miscellaneous operating expenses	$ 700	0.7%
Fixed Expenses		
Rent	$ 6,000	6.0%
Taxes	$ 1,160	1.16%
Insurance	$ 2,400	2.4%
Interest	$ 800	0.8%
Licenses and fees	$ 2,200	2.2%
Depreciation	$ 1,800	1.8%
Amortization	$ 840	0.84%
Profit Goal	$ 20,000	20.0%
	$100,000	100.0%

figure 12.2

have two sources of information for these—the past year's history and any commitments you have made for the period ahead such as a lease or a loan from the bank. Again, gather your data and make your estimated expense figures as precise as possible.

Expenses are usually grouped into two categories: fixed and variable. ***Fixed expenses*** are those that are not related in any way to sales volume but are fixed by contract or simply by being in business: that is, you could not operate at all without them, yet they go on whether or not you sell a single drink. Your rent, for example, remains the same whether your sales are zero or a million dollars, and any increase in rent comes from your landlord and not from your volume of business. Insurance, licenses and fees, taxes (except sales taxes), interest on loans, depreciation, and so on all continue independently of sales volume.

Fixed expenses are easily predictable. Write them down in dollar figures as in the sample budget (Figure 12.2)—for example, rent at $500 a month or

$6000 a year, and so on for taxes, fees, interest, insurance, depreciation, and amortization. Then add them all up for a total for the year.

Variable expenses are those that move up and down with sales volume. Chief among these are beverage and payroll costs. Itemize these too in dollars for the coming year, going back to your drinks-per-day figures to estimate the beverages and staff you will need to produce these drinks.

Actually, a certain portion of the payroll is for all practical purposes a fixed cost. Salaried employees—a manager, for example—must be paid whether you sell one drink or one thousand. A certain number of hourly employees also represent fixed costs because they must be there when the doors open for business. Some enterprises, in budgeting, divide payroll costs between fixed and variable expenses, identifying the cost of a skeleton crew as a fixed expense and the cost of additional staff needed for the estimated sales volume as a variable expense. This would give you a more finely tuned budget than lumping payroll all together. But for simplicity's sake we are following the traditional classification of all payroll as a variable cost.

A third category of expenses, ***unallocable expenses***, is sometimes listed separately because it consists of general expenses that are neither fixed nor directly tied to sales. They include promotion and advertising, and again for simplicity's sake we include them under variable expenses. These dollar figures depend on your plans, which in turn depend on how much you think such expenditures will influence sales. A certain amount of advertising and promotion is often necessary just to keep your enterprise in the public consciousness and maintain a certain sales level. Enter your planned dollar figures on your budget.

When you have estimated all your expenses for the year, add them all up. Then go back and figure each individual expense item as a percentage of your expected sales for the year, as shown in Figure 12.2.

The percent-of-sales figure is a precise way of expressing the relationship of a given expense to the sales it will help produce. This is a critical relationship, one that is watched closely in operation as a measure of performance against plan. The percent-of-sales figure is referred to in the industry as ***percentage cost***, ***cost percentage***, or ***cost/sales ratio***.

When you have completed all your percentage figures, enter your profit goal in dollar figures and again as a percentage of sales. Add the dollar profit goal to the total expenses, and compare this sum with your sales forecast. Next add up all the percent-of-sales figures for expenses and profit.

If the dollar figure for expenses-plus-profit is the same as the sales forecast or less, and the percent-of-sales figure is 100% or less, you are in good shape. If expenses-plus-profit dollars are more than the dollar sales forecast and the percentage figure is more than 100%, you have some readjusting to do. To reach your profit goal you have two choices: to cut costs or to increase sales (raise volume or prices or both volume and prices). Whatever you do, you must reach a balance. If all else fails, you must reduce your profit goal.

Your percentage figures can be very helpful in establishing a profit-yielding budget. You can see at a glance if any one category of expense is out of line. The percentages in Figure 12.2 are generally accepted industry norms, give or take a few points. However, don't look at them as hard-and-fast goals for your budget; your plan must reflect your own realities.

Figure 12.3 is a more sophisticated budget form. Filling in figures for last year and this year will give you helpful insights as you make your projections for next year.

For the sake of simplicity, in discussing the budgeting process, we have ignored the fact that the majority of enterprises combine some form of food service with beverage service. Wherever this is true, most fixed expenses and some others are shared with the food-service side of the enterprise. In allocating such expenses in a separate budget for the bar, you would figure the bar's share in proportion to its share of total sales. Thus, if beverage sales are projected at 35% of a restaurant's total sales, you would budget the shared expenses at 35% of the total projection for those expenses.

The control phase. Your paper plan is a numerical blueprint for your operations during the coming year. To turn this paper plan into reality you make whatever operating changes you have decided upon to reduce costs and increase sales to match the numbers on the plan. From this point on you monitor the actual performance of the bar, using the budget as a standard of measurement, a tool of control.

This control process consists of the following three steps:

- Comparing performance with plan to discover variations.
- Analyzing operations to track the causes of the variations.
- Acting promptly to solve the problem.

These three steps form the link that connects the plan with the profit.

Naturally the budget and the actual results are not going to correlate number for number every day or even every week or month. For example, you may put out a lot of money to buy liquor to cover the American Legion convention, and the sales won't show up until the next monitoring period. So in one period both your dollar costs and your percentage costs rise, and then in the next period your dollar costs are back to normal but your percentage cost drops because your sales went up during the convention. So how do you judge when a variation is significant?

To some extent it depends on what the variation is—whether sales, or labor cost, or beverage cost, or something else. It also makes a difference how large it is and how persistent. You can set a permissible variation limit based on past history for several periods. There are also a few rules of thumb. Generally a variation is significant if it shows up in both dollars and percent of sales. Less significant is a dollar variation by itself. Least significant

BUDGET Comparative format, three periods						
This year		*Last year*		*Next year*		
Dollars	*Percent*	*Dollars*	*Percent*	*Dollars*	*Percent*	
						Sales
						Variable Expenses
						Beverage costs
						Payroll costs
						Administrative expenses
						Laundry and supplies
						Utilities
						Advertising/Promotion
						Repairs
						Maintenance
						Misc. operating expenses
						Total Variable Expenses
						Fixed Expenses
						Rent
						Taxes
						Insurance
						Interest
						Licenses and fees
						Depreciation
						Amortization
						Total Fixed Expenses
						Profit Before Taxes

figure 12.3

is a percentage variation by itself. However, you ought to look into any persistent variation or one that is over your permissible limit.

Assessing the achievement. The control phase of any budget plan should conclude with several comparisons:

- *With past performance:* How did you do this year compared with last year?
- *With industry performance:* How did you do compared to your competitors and to the industry at large?
- *With your own goals:* How close did you come to reaching your profit goal?

The final and most important question you should ask yourself is: How can you use your experience with the past year's budget in structuring next year's budget to meet your profit goal?

Pricing for profit . . .

Profit is the difference been total sales and total costs. For a bar, total sales are the number of drinks sold multiplied by the selling prices. The essence of profitable pricing is to set the individual prices to produce the maximum difference between total sales and total costs. Several things are involved: the cost of each drink, the effect of price on demand for the drink, the contribution of each drink to total sales, and the effect of the sales mix on profits.

There are many different ways of pricing in the bar business, used consciously or unconsciously, sometimes simultaneously, often inconsistently. Some managers set prices by a strict cost/price formula. Some set them by trial and error. Some charge what the traffic will bear. Many price according to neighborhood competition. Some copy the bar with the biggest business. Some set prices low, hoping to undercut their competitors; others set prices high to limit their clientele to a certain income level. A few lucky ones, in command of a unique facility such as a revolving skyscraper lounge, set skyscraper prices because they assume the customer expects them. Some use an impressive price/value combination as a basic merchandising strategy.

Whatever the pricing policy, the goal should be to set prices that will maximize the ***gross margin*** or ***gross profit*** (sales less product cost). Each of the many drinks of the bar should be priced to achieve this total outcome.

The cost/price relationship. Any sound method of pricing a drink should start with its cost. You can establish a percentage relationship between cost and price that will give you a simple pricing formula to work with. The cost percentage of the price pays for the ingredients in the drink, and the remaining percentage of the price (gross margin) goes to pay that drink's share of all your other costs and your profit. The cost percentage will be in the neighborhood of the beverage cost percentage of your budget (Figure 12.2), though it will not correspond exactly since the latter represents the combined percentages of liquor, wine, and beer. Generally your liquor cost percentage will be 20 to 25%, while wine and beer will run higher—33 to 50%. Each enterprise must determine its own cost percentage to produce the profit needed.

To find the price, you divide the cost of the ingredients by the cost percentage:

$$\frac{\text{Cost}}{\text{Cost \%}} = \text{Sales price}$$

For example:

$$\frac{\$0.45 \text{ Cost of } 1\frac{1}{4} \text{ oz scotch}}{0.25 \text{ Cost percentage}} = \$1.80 \text{ Price for Scotch Mist}$$

You can turn this formula around and convert the cost percentage into a multiplier by dividing it into 100%, which represents the sales price. For example:

$$\frac{100\% \text{ (Sales price)}}{25\% \text{ (Cost percentage)}} = 4 \text{ (Multiplier)}$$

$$4 \times \$0.45 \text{ (Cost)} = \$1.80 \text{ (Sales price)}$$

In pricing mixed drinks, the simplest version of this cost/price formula is to use the cost of the beverage alone. This is known as the ***beverage cost method***. In the Scotch Mist this cost figure is very accurate, since scotch is the only ingredient.

However, most drinks have more than one ingredient, many of them nonalcoholic, such as mixers, garnishes, cream. A truly accurate cost/price relationship must be based on the cost of all the ingredients. The technique of determining total cost is based on the drink's recipe and is known as ***costing the recipe***.

Here is a simple method of costing a recipe (below left), with a Martini as an example (below right):

1. We start with the standard house recipe.	1. Recipe: 2 oz gin ½ oz dry vermouth 1 pitted olive

2. We determine the cost of the gin. To do this, we find:
   (a) The size bottle poured and its ounce capacity
   (b) Bottle cost (from invoice)
   (c) Cost per ounce (divide bottle cost by oz per bottle)
   (d) Recipe cost of gin (multiply ounce cost by number of ounces in recipe)

2. Cost of gin:
   (a) Liter = 33.8 oz
   (b) \$8.12
   (c) $\frac{\$8.12 \text{ bottle cost}}{33.8 \text{ oz per bottle}} = \$0.24 \text{ per oz}$
   (d) \$0.24 cost per oz
   × 2 oz in recipe
   \$0.48 recipe cost of gin

3. We determine the cost of the vermouth in the same way:
   (a) Size bottle and ounce capacity
   (b) Bottle cost (from invoice)
   (c) Cost per ounce
   (d) Recipe cost of vermouth (ounce cost × ounces)

3. Cost of vermouth:
   (a) 750 ml = 25.4 oz
   (b) \$3.25
   (c) $\frac{\$3.25 \text{ bottle cost}}{25.4 \text{ oz per bottle}} = \$0.128 \text{ per oz}$
   (d) \$0.128 cost per oz
   × 0.5 oz in recipe
   \$0.064 recipe cost of vermouth

4. Then we determine the cost of the olive:
   (a) Jar size and olive count (from invoice or label)
   (b) Cost of jar (from invoice)
   (c) Cost per olive
   (d) Recipe cost of olive

4. Cost of olive:
   (a) Quart = 80 olives
   (b) \$2.20
   (c) $\frac{\$2.20 \text{ container cost}}{80 \text{ number contained}} = \$0.027 \text{ per olive}$
   (d) 1 olive = \$0.027 recipe cost of olive

5. Now we total the costs. This gives us our standard recipe cost as of the date of costing.

5. Total costs:

\$0.480	cost of gin
0.064	cost of vermouth
0.027	cost of olive
\$0.571	standard recipe cost

To arrive at the selling price of the drink we divide the cost by the cost percentage. This gives us our sales price:

$$\frac{\$0.571 \text{ Recipe cost}}{0.25 \text{ Cost percentage}} = \$2.284 \text{ Sales price, which we round off to } \$2.30$$

A quicker though less accurate way to arrive at a selling price is to base the price on the cost of the ***prime ingredient***—that is, the base liquor—instead of the actual recipe cost, adding a certain percentage for the extra ingredients. This is a refinement of the beverage cost method. Using our Martini as an example, it works like this:

$0.48	Prime ingredient cost
+ 0.048	10% allowance for additional ingredients
$0.528	Cost base
× 4	Multiplier
$2.112	Sales price, rounded off to $2.15

Pricing mixed drinks on the basis of prime ingredient cost is a quick, easy, and widely used method. It is typically used to establish a base price for all drinks having the same amount of the same prime ingredient; then the percentage is added for drinks having extra ingredients. Obviously the method is not as accurate as costing the recipe; you can see that the added percentage in the example does not reflect true cost. A single percentage figure will never be accurate for a whole spectrum of drinks.

Costing takes a great deal of time. But to arrive at a sound price structure you have to establish the cost of each drink, one way or another. Only if you look at each drink by itself can you establish sound cost/price relationships and build a coherent overall price structure.

The demand/price relationship. But costs and percentages are not the whole answer to pricing. Price is only one of two factors in total sales; the other, as you know, is the number of drinks sold. And the price affects the number. It is the effect of prices on numbers that is the elusive secret of successful pricing—elusive because no one ever knows precisely what effect a change of price will have on demand in any given situation.

If you increase the price of a given drink, you increase the gross margin per drink sold, since cost remains the same. But sales volume is typically sensitive to changes in prices. In a normal situation (whatever that is) there is a seesaw relationship: when the price goes up, the gross margin per drink goes up but the number of drinks sold goes down (demand falls). When the price goes down, the gross margin per drink goes down but the number sold goes up (demand rises).

However, one seldom has a totally "normal" situation. In any given bar, more than price is influencing demand—drink quality, ambience, food, entertainment, individual capacity, other drinks, other bars. How much of a price increase for a given drink will cause a customer to order fewer drinks or switch to a lower-priced drink? At what lower overall price level will customers choose you over your competitors? At what higher price levels will they start to patronize the bar next door instead? Where does the seesaw balance?

Most bars operate in a competitive environment in which demand is fairly responsive to price changes. The trick is to know your competition and especially your customers, make shrewd guesses on demand/price sensitivity, and watch your sales to see how they respond. Your objective is to balance the seesaw for all the drinks taken together in a way that will bring the largest overall gross margin for the enterprise.

Coordinating prices to maximize profits. When you have the whole detailed picture of cost-based prices, you can round off and finalize drink prices with your eye on the seesaw of demand. The first modification of cost/price ratios is to simplify the picture by establishing several broad categories of drinks and setting uniform prices within each category. This makes it easier for everybody and enables bartenders and servers to complete sales transactions with efficiency.

The following price categories may be useful, or at least give you a point of departure for your own scheme:

- *Highballs:* Two prices, one for drinks poured with well brands and one for call-brand drinks. Generally this is the lowest price category.
- *Cocktails:* Two prices, one for cocktails made with well brands and one for those made with call brands.
- *Frozen drinks and ice cream drinks.*
- *After-dinner liqueurs and brandies:* Two prices, one for ordinary liqueurs and one for premium liqueurs and French brandies.
- *Specialty drinks.*

All pricing decisions involve interdependent relationships among drinks and drink prices. The price you charge for one kind of drink can affect the demand for others. For example, if you raise your cocktail prices, you may sell fewer cocktails and more highballs for fewer total dollars. On the other hand, if you introduce a high-price frozen drink with customer appeal you may sell more of them and fewer of your lower-priced highballs and cocktails. A high-profit coffee specialty at a premium price may not affect demand for predinner drinks at all. You have to keep potential interrelationships in mind in order to arrive at a combination that maximizes the sum total of all drink profits. This is known as ***total business pricing***.

Some drinks always have a lower gross margin because demand drops sharply as the price goes up. Among these are beer and wines by the bottle. Few bottles of wine are sold when they are priced at three or four times the cost. On the other hand, it is usually easy to get a high gross margin for specialty drinks and house wine by the glass.

You can use a lower-gross-margin price as a way to increase volume by bringing in new customers. Bars do this with happy hours, 2-for-1 specials, and oversize drinks. You can also reduce prices selectively to create a mood for buying other drinks; you might, for example, have a special low price on frozen Strawberry Daiquiris on a hot day and sell many other frozen drinks at their regular price just by planting the frozen-drink idea. Or you can reduce your standard prices on higher-priced drinks and raise them on lower-priced drinks and you *may* come out ahead.

All these price relationships must be worked out for your own enterprise. Some work well in some places and others don't. Whatever combination of prices gives you the best overall price-times-numbers-less-cost is the best set of prices for you. For the moment, at least. Things change, and you have to watch sales closely. It is a fascinating game.

It is a good idea to provide a list of standard prices, both for the customer and for service personnel. You can post prices at the bar (see Figure 12.4) or print them on a menu. They will eliminate arguments between employees and customers, and they will make it harder for unscrupulous employees to overcharge and pocket the difference.

Establishing product controls . . .

In Chapter 9 we talked about what makes a successful mixed drink and about developing recipes for your drinks so that each drink served from your bar will taste the way you want it to. How can you be sure the recipes you have carefully developed will be faithfully executed by each bartender if they all have their personal ways of pouring and they come and go at the normal rate for the species?

Not only do you want the drinks served at your bar to be your recipes, you want them to be the same every time. Product consistency is very important in building a clientele. New customers expect your Bloody Mary to taste the way a Bloody Mary should, and repeat customers expect your Bloody Mary to taste the way it did the last time. Meeting customer expectations may be even more important to profit than setting your drink prices correctly. It is sales price multiplied by sales volume that produces the profit your budget contemplates, and you can't build volume on the basis of drinks that don't meet customer needs and desires *consistently*.

To achieve this consistency you need to establish standards for all drinks —standard ingredients, quantities, portion sizes, and procedures for making them. When these standards are put into practice and everyone is required to follow them, the customer will get the same drink every time no matter who makes it.

In addition to producing consistent drinks, these standards give you ways of controlling the quantities of liquor used. If you can control the quantities, you also control the costs. And if you control the beverage costs you can maintain your projected cost/sales ratio and protect your profit. To achieve all this you must standardize three major elements for each drink —size, recipe, and glass.

Standard drink size. In the vocabulary of the bar the term ***drink size*** refers to the amount of the prime ingredient used per drink poured, not the size of the finished drink. In each bar this amount is the same for most

PRICE LIST
Crackers Restaurant & Bar

Glass	Drink category	Drink size	Regular price	Happy Hour
8-oz highball	HIGHBALL			
	Well	1¼-oz	$1.95	$1.00
	Call	1¼ oz	2.25	1.50
5-oz rocks	ROCKS			
	Well	1½ oz	2.25	1.00
	Call	1½ oz	2.50	1.50
1½-oz jigger, lined	SHOTS			
	Well	1 oz	1.50	1.50
	Call	1 oz	1.75	1.75
4-oz cocktail 5-oz rocks	MARTINI/MANHATTAN			
	Well	2 oz/¼ oz	2.25	1.50
	Call	2 oz/¼ oz	2.50	2.50
10-oz wine	JUICE DRINKS Mary/Driver/Collins/Sling			
	Well	1½ oz	2.25	1.50
	Call	1½ oz	2.50	2.50
7-oz rocks 10-oz wine	DAIQUIRI/MARGARITA/SOUR			
	Well - Rocks	1½ oz	2.25	1.50
	Well - Frozen	1½ oz	2.50	1.50
	Call - Rocks	1½ oz	2.50	2.50
	Call - Frozen	1½ oz	2.75	2.75
12-oz snifter	BRANDIES and LIQUEURS			
	Well	1½ oz	2.25	1.50
	Call	1½ oz	2.50	1.75
	Rémy Martin	1½ oz	3.00	3.00
5-oz rocks	CREAM & TWO LIQUORS			
	Well	1¼ oz ea	2.50	2.00
	Call	1¼ oz ea	3.00	3.00
8-oz cup	COFFEE DRINKS			
	With 1 liquor	1¼ oz	2.50	1.50
	With 2 liquors	1 oz ea	3.00	3.00
10-oz wine	WINE			
	By the glass	8½ oz	1.50	1.00
	Spritzer/Cooler	6 oz	1.50	1.00
10-oz wine	CHAMPAGNE	8½ oz	2.50	2.50
12-oz mug	BEER			
	Domestic	Bottle	1.25	1.00
	Imported	Bottle	1.50	1.50
10-oz wine	PERRIER	6 oz	1.25	1.00
12-oz mug	SOFT DRINKS	8 oz	.50	.50
10-oz wine	VIRGIN DRINKS	No liquor	1.95	1.95
10-oz wine	JUICES	No liquor	1.00	1.00

figure 12.4 Sample list of drink standards showing sizes, glasses, and prices. (Courtesy Crackers, Dallas)

spirited-based drinks, with different standard amounts for a few drink types and special drinks. Thus, if your drink size is 1½ ounces, you pour 1½ ounces of the base liquor in each drink, whether it is a gin drink or vodka or scotch or whatever. This is your ***standard drink size***.

Each bar has its own standard drink size. Most bars pour a 1-ounce or 1¼-ounce standard drink, but it varies across the industry from ¾ ounce to 2 ounces, depending on the nature of the enterprise and the clientele. In each bar special drink types that vary from the standard will have their own standard drink size. You can see how this works in one bar by studying the *Drink Size* column in Figure 12.4.

To pour a standard drink, the bartender must have an accurate means of measuring the liquor. As discussed at length in Chapter 9, there are several ways of measuring, notably hand measuring, automated pouring, and free-pouring.

In a bar concerned with consistency and tight liquor controls, free-pouring is out of the question, no matter how skilled the bartender. There is too much potential for variation. The most common way of measuring is to use a shot glass and stainless-steel jiggers of various sizes. All shot glasses should be the standard drink size. They will function as the jigger for the base liquor. Different sizes of stainless-steel jiggers will provide the means of measuring the smaller amounts called for in drinks containing more than one liquor as well as larger amounts for brandies and oversize drinks. If the bartender is well trained, skillful, and honest, measuring with these hand measures will produce controlled, consistent drinks.

Even so, overpouring and underpouring are hazards of the hand method, whether intentional or caused by the rush of business. In a bar with annual sales of $100,000, a steady overpouring of ¼ ounce per drink could add $2000 to $3000 a year to your costs. Many enterprises, large and small, are using some method of controlled pour, either pourers that measure or an automated pouring system.

Several types of automated pouring systems were described in Chapter 3. Preset to pour the standard-size drink and to record each drink poured, they eliminate overpouring, underpouring, and spillage, and help to provide a consistent drink. Most systems pour only the major liquors, however, so there is still room for error, inconsistency, and loss.

Whatever the measuring method, a list of drink sizes should be posted at the bar showing mixed drinks, wine, beer, and special drinks. Figure 12.4 is a comprehensive list showing drink sizes, glass sizes, and prices.

Standard drink recipe. In our discussion of mixology we had a good deal to say about the importance of proportions in making a good drink and the wisdom of writing them down in recipes. A recipe that specifies exactly how a given drink is made at a given bar is known as a ***standardized recipe*** (Figure 12.5). It specifies the exact quantity of each ingredient, the size glass to be used, and the exact procedure for preparing the drink. The garnish is

INGREDIENTS

2 oz gin

½ oz dry vermouth

1 olive

DRINK: Martini

GLASS: 4-oz cocktail
5-oz rocks

PROCEDURE

Stir gin/vermouth with 1/3 mixing glass cube ice.

Strain into prechilled cocktail glass or over cube ice in rocks glass.

Garnish with olive.

INGREDIENTS		BOTTLE	BOTTLE COST				DRINK	DRINK COST			
	DATE:		9/31			DATE:		9/31			
Gin		L	8.12				2 oz 60 ml	0.480			
Vermoutn		750 ml	3.25				½ oz 15 ml	0.064			
Olive							1	0.027			
						DRINK TOTALS	2½ oz 75 ml	0.571			

DATE	9/31/81
COST	0.571
PRICE	2.284 = $2.30
COST%	25%

figure 12.5 Standardized recipe card with cost and price data.

included, and anything else that is necessary to the drink. If there is a picture of the finished drink in its assigned glass, so much the better.

You should have a standardized recipe for every drink you serve. Keep the recipes together in a loose-leaf notebook with plastic page covers. Use them to train new bartenders, and see that every bartender, old and new, follows the house recipes to the letter. If you can make this happen you will really have achieved control of quality and quantity.

The standardized recipe is the basis for the costing and pricing described earlier. Every recipe should be costed periodically, and the cost of each ingredient should be recorded on the recipe card along with the date of costing. It is a good idea to review the whole recipe at this time, to be sure you haven't changed a garnish or a glass or changed Cointreau to triple sec in your Side Car without writing it down.

A file or book of standardized recipes also provides data for certain types of beverage control. We'll be discussing this shortly.

Standard glassware. Each standard drink should be served in a ***standard glass***—a glass of specified size and shape that is used every time that drink is poured. The size is the most important, since it controls the quantity of the ingredients that must fill it as well as the taste, as explained in Chapter 9. But a standard, distinctive shape is important too. It makes a drink look the same every time, as well as tasting the same, and it reduces the possibility of using the wrong-size glass. If you have two glasses of different sizes with the same shape, it is easy to get them mixed up, and if you use the wrong one it will alter the drink considerably. (A word to the wise: order replacements by both size and catalog number; it is difficult to judge capacity by appearance.)

Some people equate glass size with portion size, and this has a superficial validity. But the real portion size is the drink size—the base liquor—controlled by the shot glass or jigger. In most cases the drink glass has more to do with proportion than portion.

In standardizing your glassware, choose whatever size and shape will give each drink the most appeal. This doesn't mean the glassware itself has to be special; it means that *the drink in the glass* should look appealing. For example, a drink on the rocks looks skimpy if the glass is either too big or too small. A straight-up cocktail looks undersize in too large a glass. Shapes can make a big difference. A footed glass can make a drink look bigger than it does in a tumbler of the same capacity. Rounded shapes make a difference in the way the ice fits in, and that changes the way the liquid fills in around the ice. You really have to experiment. You must also allow for what the garnish will do to the drink. No glass should be so small that the measured drink fills it to the brim.

You can manage nicely with very few styles and sizes of glass if you wish—one glass for cocktails and champagne, one for rocks drinks, one for highballs, one for tall drinks and beer, one for wine, and one for brandy and after-dinner liqueurs. The fewer the glass types, the fewer the mistakes in service. Chili's Restaurant in Dallas has simplified its glassware to a single-size mug, even for straight-up cocktails. Your choices depend on your clientele, your drinks, your image. Chili's clientele is not the straight-up cocktail type. In a casual establishment like Chili's, where the food is burgers and fries served in baskets and the lines are long every night, the one-mug concept works.

The glass for each drink should be specified on its written standard recipe. It should also be on your drink list posted at the bar.

Establishing beverage controls . . .

Ideally, all beverages "used" or "consumed" should be used to make drinks for which payment is collected, which are consumed by the paying customer. But not all liquor is so used or consumed. It may be overpoured, spilled, improperly mixed, or otherwise wasted. It may cling to the sides of empty bottles. It may be rung up incorrectly. Customers may refuse drinks, or avoid paying, or pay too little. It may be pilfered by your own employees. One way or another it is no longer available for sale. This kind of use, unchecked, can quickly consume your profit margin.

If you have standardized your cost percentages and your drink sizes and your recipes and your prices and your inventory procedures and your par stock, you have already minimized certain kinds of losses. You have also provided tools for measuring loss. If measuring reveals that losses are indeed threatening profits, you can track down the causes and plug the leaks.

Three techniques of measuring are commonly used. One is based on cost percentage, a second compares ounces used with ounces sold, and the third measures potential sales value against actual sales. Each in its own way measures what you have to sell against what you do sell—the "slip twixt the cup and the lip."

The cost percentage method. The first technique of measurement compares the cost of liquor used during a given period to the sales of the same period. The resulting percentage figure is then compared to the standard cost percentage. This method requires a physical inventory of storeroom and bar at the beginning and end of the period, plus purchase and sales figures for the period. With this information in hand, you make the following calculations:

step 1: To find the value of the liquor available for sale during the period, add the opening inventory value to the value of the purchases:

  Value of opening inventory
+ Value of liquor purchases
  Value of liquor available for sales

step 2: To find the cost of the liquor used during the period, subtract the value of the closing inventory from the value of the liquor available.

Value of liquor available
−Value of closing inventory
Cost (value) of liquor used

step 3: To find the bar cost percentage, divide the cost of liquor used by the total sales for the period:

$$\frac{\text{Cost of liquor used}}{\text{Total dollar sales}} = \text{Bar cost percentage}$$

The percentages for beer, wine, and spirits must be figured separately because they are all different.

This percentage figure can now be compared to the planned, or standard, bar cost percentage (your budget figure) as well as to the percentages of previous periods. If your bar cost for the current period is more than 1½ percentage points higher than your planned bar cost or the average bar cost of several periods, you should start looking for the reasons.

You probably recognize this procedure as part of the end-of-the-month inventory routine described in Chapter 11. But if you wait till the end of the month, the damage is done and you may not be able to trace the cause. It is better to take weekly or even daily inventories to keep track of your bar costs. Where only beverage is concerned, physical inventories are usually not major undertakings.

The ounce method. The second technique of measurement compares the ounces of liquor used with the ounces of liquor sold. In this method, opening and closing inventories at the bar each day reveal the number of ounces sold. You take the opening inventory before the bar opens but after the par stock has been replenished, and the closing inventory after the bar closes. Then you use your data as follows:

step 1: Subtract the ounces of each liquor in the closing inventory from the ounces of liquor in the opening inventory. This gives you ounces used.

step 2: From your guest checks, tally the drinks sold by numbers of each type. For each type multiply the ounces of liquor in one drink by the number of drinks sold, giving you the number of ounces per drink type. Then add them all up to find the number of ounces sold.

step 3: Subtract the number of ounces sold from the number of ounces used.

This is a daily measure, and it measures only what is happening at the bar, but it is accurate and it pinpoints this area of loss. Unfortunately it takes time, unless you have a computerized register that can replace the guest check and the hand tally.

The potential sales value method. The third method of measurement compares actual dollar sales with potential sales value. Each bottle you buy represents potential dollars in the register. We can determine its potential sales value by using standard drink sizes, standard drink selling prices, and the number of drinks that can be served from each bottle. For example, if you pour 1-ounce drinks, sell each drink for $1.00, and use only liter bottles, you should have $33.80 in the register for each bottle you use.

In real life things are more complex. Most bars have more than one drink size and more than one drink price. Therefore the potential sales value of each bottle must be adjusted for these variations. One method of doing this, the weighted average method, is based on averages of drink sizes and prices for drinks actually sold over a period of time.

For the sake of illustration let us telescope our time period into one typical day:

step 1: We find the average drink size for all the drinks made with one kind of liquor—gin, for example:

drink	*drinks sold*	*ounces sold*
Gimlet (1.5 oz)	10	15
Martini (2 oz)	40	80
Gin/Tonic (1.5 oz)	12	18
	62	113

To complete this step we divide the total ounces sold (113) by total drinks sold (62) to obtain an average drink size:

$$\frac{113 \text{ oz}}{62 \text{ oz}} = 1.82 \text{ oz Average gin drink size}$$

step 2: Now we find the number of drinks per bottle. If we divide our liter of gin (33.8 oz) by our average drink size (1.82 oz), we obtain the number of drinks per liter:

$$\frac{33.8 \text{ oz}}{1.82 \text{ oz}} = 18.57 \text{ Average drinks per liter of gin}$$

step 3: Next we find the average selling price of these drinks:

drink	*number sold*	*total sales*
Gimlet @ $2.00	10	$20.00
Martini @ $2.00	40	80.00
Gin/Tonic @ $1.75	12	21.00
	62	$121.00

To complete this step we divide total sales ($121.00) by total drinks sold (62) to obtain the average drink selling price:

$$\frac{\$121.00}{62} = \$1.95 \text{ Average selling price}$$

step 4: Finally we find the potential sales value of each quart of gin by multiplying the average drinks per liter (18.57) by the average selling price ($1.95):

$$
\begin{array}{r}
18.57 \\
\times \ \$1.95 \\
\hline
\$36.21
\end{array}
\text{ Potential sales value of 1 liter of gin}
$$

In the same way we can figure the potential sales value of every bottle of liquor used in a given period. We can then compare the actual sales dollars with the potential sales value of the bottles to measure discrepancies. Ideally the actual sales should equal the potential sales value.

The weighted average method assumes that the sales mix, liquor costs, and sales prices remain constant. When anything changes everything must be refigured.

This method consumes many hours with a calculator in hand in order to approximate accuracy, and it requires further refinements to take into account drinks having more than one liquor. There is another, simpler method of approximating potential sales value.

This method requires a test period of 45 to 60 days during which you enforce all standards to the letter and observe all bar operations continuously. At the end of the period you carefully determine bottle consumption from purchases, issues, and inventories, and translate it into potential sales value as though all liquor had been sold by the straight drink. Then you compare this value with actual sales during the period. The difference represents the varying amounts of liquor in your drinks plus an inescapable minimum of waste and inefficiency. You convert this difference into a percentage figure and accept it as your ***standard difference***. You can use it from then on to compare potential sales with actual sales. Any time the percentage of difference is higher than the standard difference, investigation is in order.

If costs and sales prices change, the percentage figure is still applicable. If the sales mix changes, a new test period is needed to determine a new standard difference.

The role of par stock. In earlier chapters we have talked about par stock as a means of assuring a full supply of liquor at the bar and as an inventory tool. Now we can see par stock in still another role: it is a key tool of control at the bar, where it is a standard against which consumption can be measured *at any time.*

With the one-empty-for-one-full system of requisitioning, a manager has only to count the bottles at the bar and compare the number with the par stock form to determine whether anything is missing. This can be done as part of a full-bar inventory, or the manager can spot-check any given brand at any time.

And it is a wise manager who does so from time to time. Frequent checking is sometimes enough to prevent pilferage entirely. The oftener you check, the more closely you can pinpoint the possible culprits and the less likely it is they will be tempted. If you are not actually preventing losses you have a measure of what is missing, how serious it is, and most likely who is responsible.

Par stock is the last step in the series of controls over the liquor itself. In Chapter 11 we followed the liquor controls in a large operation from receiving to sale. Liquor was counted by the receiving agent, who assumed responsibility by signing the invoice. It then moved to the storeroom, where it was again counted and responsibility transferred to the storeroom supervisor. The requisition, signed by the storeroom supervisor and the receiver, transferred the responsibility to the receiver, usually the bartender, who again counted the bottles. There the responsibility lies until it is sold.

This chain of responsibility resting in the hands of a single person at all times is a system every manager would do well to follow. It still does not guarantee security, but it discourages theft and it facilitates finding the leaks and plugging them. In a small enterprise, of course, the owner/manager carries the responsibilities that are divided among several people in the large establishment.

Establishing cash controls . . .

When a customer buys a drink, there has to be a way of making sure that the sale is recorded and the money finds its way into the cash register.

There are many systems of paying for drinks. The simplest procedure, and the riskiest, is for the bartender to pick up the cash laid out on the bar and ring it up on the register without benefit of guest check. The most complicated—and probably the safest though most expensive—is a computerized system that prerecords the sale, pours the liquor, and rings up the sale on both register and guest check, with a receipt for the customer. In between are many different systems. Some bars may run on a pay-as-you-go system; others may use a running-tab system in which drinks are entered on the guest check but not paid for until the customer is ready to leave. In some places the customer pays the bartender. In others he or she pays the server, who pays the bartender or a cashier. In still others the customer pays the cashier directly. In busy bars the server may carry a bank, paying the bartender for the drinks when they are ordered, and collecting and making change on the spot as each round of drinks is served.

Whatever the system, you need standard procedures for handling cash and some form of guest check for the record. You need the record. You want the cash. There may be others among your personnel and your customers who want the cash too.

When payment is routed through several persons it is usually intended as a system of checks and balances, but sometimes it works the other way. More people are exposed to temptation and opportunity, and losses are more complicated to track to the right source. On the other hand, at the bar you have one person taking the order, filling the order, recording the sale, and collecting the cash, uninhibited by the controls a division of responsibility might provide, and surrounded by opportunity. For you it is a Catch-22 situation, magnified by the large number (let's hope) of fast-paced individual transactions and an environment permeated with liquid gold and money changing hands.

Here are some common practices that make cash disappear. Most are at your expense, but sometimes it is the customer who is out of pocket, and that can hurt your business too.

The bartender:

- Fails to ring up sales and pockets money.
- Overcharges and pockets the difference.
- Short-changes customer and keeps change.
- Brings in own liquor and sells it (using house mixes and garnishes).
- Brings in empty bottle, turns it in to storeroom, then sells from bottle that replaces it and pockets money.
- Short-pours a series of drinks, then sells others from same bottle, keeping money.
- Sells liquor from one bottle without ringing up, then waters remaining booze to cover theft.
- Substitutes well liquor for call brand, collects for call brand but rings up well price, keeps change.
- Smuggles out full bottles.

The server:

- "Loses" guest check after collecting and pockets money.
- Reuses a check and keeps money for one use.
- Overcharges for drinks and pockets difference.
- Makes intentional mistakes in totaling check and keeps overage.
- Intentionally omits items from check to increase tip.
- Changes items and prices on check after customer pays.
- Gives too little change and pockets balance.

The cashier:

- Gives too little change and pockets balance.

- Fails to ring up check, pockets money, and "loses" check (blame falls on server).

The customer:

- Walks out without paying.
- Sends back drink after half-emptying glass.
- Uses expired credit card.
- Pays with hot check.

The cash-control system you devise to forestall all these little tricks should both reduce opportunity and pinpoint responsibility. You need a system that you can enforce that also leaves a trail behind when it is evaded. Then you must keep after it. Wherever it is not foolproof, you must keep checking up. If your employees know you are policing the system they are likely to remain honest. Lax enforcement invites pilferage and almost seems to condone it.

Here is a suggested system:

1. To start with, you should have numbered guest checks with your bar's name or logo. Each bartender or server signs out a sequence of numbers for the shift and turns in the unused checks at the end of the shift. You keep the master list and check the used and unused checks. The employee must pay for any guest checks that are not used or turned in.
2. All guest checks must be written clearly in ink. Alterations are not allowed unless they are initialed by a responsible person. Drinks should be machine-priced.
3. If possible, use a precheck method of registering drinks. In this system the order is rung up before the drinks are poured. When payment is made, another register, or another section of the same register, is used to ring up the same sale, on the same check but a different register tape. The totals on the two tapes should be the same.
4. All items on the check are individually rung up and totaled by the register or else added on an adding machine before the check is presented to the customer for payment. Do not rely on the server's mathematics.
5. The server must print the amount received from the customer on the bottom of the check. A box for this purpose is useful.
6. Each check must be rung up individually when paid, and the register drawer must be closed after each transaction. The paid check is filed in an assigned place, even in a locked box.
7. Receipts are given to customers along with their change.
8. Only one person at a time operates the cash register and is responsible for the cash in the drawer. Train that person thoroughly in the register function and in your opening and closing routines (Chapter 8).

Some other suggestions: Post prices for all drinks so that both customers and servers are informed. Position the register so that the customer can see the amount being rung up. Make access to the bar interior difficult and lock up the liquor when the bar is closed. Check credit cards against lists of invalid cards provided by the card companies.

A register that provides sales breakdown by drink categories, sales periods, salespersons, and departments can help you to pinpoint discrepancies quickly and determine their origin.

Remember that no records of any kind actually control costs; they only pinpoint losses. The flip side of controls—the effective side—is doing something to stop the losses.

Summing up . . .

Setting up a specific profit goal and preparing a financial plan for reaching it can start the business side of an enterprise on the path to success. A financial plan also sets up a means of continuously measuring the degree of success the bar is achieving in its operations.

Since profit is the margin of sales over costs, the profit plan sets up a two-pronged effort—to maximize sales and minimize costs. On the sales side, pricing is of strategic importance in maximizing profit per drink without inhibiting demand and reducing volume. On the cost side, while every expense must be watched, the primary focus is on controlling beverage losses. There are two major ways to do this. One is to minimize the opportunities for pilferage. The other is to set up systems for measuring and pinpointing losses and take prompt action to stop them.

YOUR CHECKLIST FOR PLANNING AND CONTROLS

Budgeting

Profit goal $ ________________

Less beverage cost $ ________________

Gross profit $ ________________

Less operating expenses $ ________________

Net profit $ ________________

Pricing

Overall cost/price ratio desired ________%

Pricing considerations (check those that apply)

Cost/price ratio _____

What traffic will bear _____

Neighborhood competition _____ Meet it _____ Beat it _____

Limit clientele _____

Customer expectations _____

Price/value combination _____

Trial and error _____

Special pricing devices (check those you will use, with tentative prices)

Price categories: Well drinks ________ Call drinks ________

Frozen/ice cream drinks ________ Liqueurs and brandies ______________

Specialty drinks (types and prices) ____________________________________

Happy hour ________ Promotional specials ____________________________

Method of figuring cost: Beverage cost ________ Prime ingredient cost ________

Controls

Standard drink sizes: Highball ________ Cocktail ________ On the rocks ________

Standard glass sizes: Highball __________ Cocktail __________ Rocks __________

Beer ________ Champagne ________ Wine ________ Collins/cooler ________

Beverage control methods you will use:

Cost percentage _____ Ounce _____ Potential sales value _____ Par stock _____

Cash controls ____

chapter 13 . . .

TRACKING THE BUSINESS

This chapter is about accounting, but it is not intended to teach accounting. Its purpose is to provide an understanding of what the accountant does, why it is important to the bar, and how to use the accountant's information in pursuit of profit.

We will look at the entire accounting activity as a financial information service that provides the kinds of financial data a bar needs in its dealings with others and for its own use. We'll see how the raw financial data of daily operation is organized into the kinds of records and reports that show how the business is doing. Finally, we will examine some ways in which accounting data and techniques can be applied to daily operations to track the business and improve results.

You've pulled together some hard-earned money, borrowed some from your father-in-law and some from the bank, and the bar you have been dreaming about for years is about to open. Tomorrow, and every day thereafter, that money will be in motion, flowing in and out of your business. How will you keep track of the flow, and how can you make it work to yield your carefully projected profit? For this you will need a corresponding flow of information about your business, organized information kept continually up to date. This is provided by ***accounting***, both the formal accounting of the professional accountant and the recordkeeping of the bar itself.

Functions of accounting . . .

Accounting has two major functions. It *collects* the financial data of a business and transcribes it in an orderly historical record, and it *analyzes* the data by arranging it in financial reports that reveal what is going on and what it means for the enterprise.

The historical record chronicles the daily flow of money in and out of the business—where it comes from, where it goes. *Money coming in* is called ***income*** or ***revenue*** or ***sales***. *Money going out* is called ***expenses***. The accountant organizes this information into a series of ongoing ***accounts*** that parallel the categories of information in your budget, as well as others for special purposes. At the end of each accounting period (usually a month or four weeks), the accountant draws up an ***income statement (profit-and-loss statement)*** summarizing the income and outgo and the profit or loss for the period. This statement is arranged to show the relationships between incomes and outgos so that you can see what has happened during the period, and whether you are ahead or behind where you want to be and what, if anything, is out of line.

Periodically—at least once a year—the accountant compiles a ***balance sheet*** showing the state of business at that point in time. This lists the ***assets*** of the business (what it owns), its ***liabilities*** (what it owes), and the ***equity*** (what belongs to the stockholders or to you as owner).

The income statements and balance sheets are formal analyses of your financial records. In addition—for an extra fee—the accountant can explain what the data mean for your business and how you can apply them to improve your financial performance in the months to come.

In a small enterprise the accountant is not part of the staff but sells his or her services to the bar on a regular basis. Usually the accountant is selected by the owner rather than by a hired manager. In a large corporation there will be an accounting department separate from the operations side. In both situations the hired manager is one step removed from the accounts, which acts as a form of control.

All accounting has a fat price tag because of the amount of work involved. If you understand what the accountant is doing and how it can be useful to you, you can really get your money's worth. Here is the indispensable minimum of knowledge that you should have:

- You should be familiar with basic accounting terms so that you know what your accountant is talking about.
- You should know what kinds of financial information you need so that you can discuss with the accountant what accounts should be set up. (The more separate accounts, the higher the cost. The more you understand about accounting, the better your judgment will be.)
- You should be able to understand an income statement and a balance sheet and recognize what these financial reports mean for your business.
- You should know how to use accounting data as tools for controlling costs and increasing profits.

You should also know what accounting does *not* do for you. Some people believe that a good accountant can create profit and control costs through financial wizardry of some sort. No amount of recordkeeping or analysis can do this, of course. It is the decisions made and actions taken *in response to* accounting data and analysis that influence costs and profits, and it is the manager or owner who does this. You can pay an accountant to give you advice and point the way, but the responsibility for action is yours.

You should review all accounting statements promptly and thoroughly. In addition, if you understand accounting techniques, you can use these techniques yourself on a day-to-day, on-the-spot basis without waiting for a monthly income statement or a meeting with the accountant to see how things stand.

But first let's look at accounting as a financial information service. When you list out the kinds of information you need, you will have a better idea of the accounts you want to set up and how you will use the information.

Financial information a bar must have . . .

A surprising amount of financial information is necessary to run a bar efficiently, and a surprising number of people require information from even the most modest of enterprises. Here are some kinds of information an efficient accounting system must provide:

- Information required by law.
- Information creditors want.
- Information owners want.
- Information the manager needs to run the business successfully.

Information required by law. The new bar owner or manager may be astonished at the kind and amount of financial data needed for reports to various government entities.

One such report is the annual federal income-tax return required of every business in the United States. This report, filed with the Internal Revenue Service (IRS), requires a listing of major sources of income and major categories of expenses, a schedule of inventory, a profit or loss figure, and a tax calculation. Many states also have an income tax. This generally follows the same outline as the federal tax return. In most instances the income, expense, and inventory records you keep for the bar's own use will correspond with the questions on the various tax forms. Your accountant will fill out the forms, figure the profit/loss, and calculate the taxes.

Another kind of information is needed for property-tax returns. Your city or county may require an annual report showing the value of your real estate and business property. This is usually a simple summary statement, and the figures are available on your balance sheet.

Still another call from one government or another is for sales-tax information. Most states and cities have a sales tax on drinks, which the bar must collect from the customers. Each enterprise must keep a record of all sales and submit a monthly, quarterly, or yearly report listing total income for the period and tax collected, along with a check for the amount of the tax. The reports require only summary figures and, since the taxes are a percentage of sales, no expense records are involved. But though the return is short and sweet, the recordkeeping is long and detailed (usually done by the cash-register tape), and the records must be retained for a specified time to back up the tax return. In the accountant's books the sales-tax record is kept in a special account.

The greatest demand for financial information relates to payroll. Federal, state, and even local governments have many requirements that affect payroll. You must withhold federal income tax, social-security tax, and perhaps state and local income taxes from employee paychecks. For tipped employees you must report tips and withhold taxes on tip income. You must make specified contributions to federal and state unemployment-insurance funds, and you must keep records and file reports on everything.

All this requires an enormous amount of detailed accounting work. Records for each employee must be kept from the first day on the job. These records must include base pay, overtime, tips, federal income tax withheld, and amounts withheld for social security (FICA). Monthly, quarterly, and annual reports must be filed with the IRS, with payment of amounts withheld plus the employer's share of FICA taxes. To your federal employee records you must add similar information and withholding required by state insurance and tax programs, and you must file state returns and make the required payments.

You must give each employee a year-end statement of base pay and deductions for the year. (The term base pay in this context refers to the total wages or salary paid, not the base rate of pay on which overtime is figured.)

The accountant sets up payroll accounts, with a separate record for each employee, to provide the huge amount of detail needed to fulfill federal, state, and local requirements. Usually payroll accounts are kept in a separate book. In addition to keeping the records, the accountant typically completes the withholding reports and payments for the various governments and the employees.

Information creditors want. Creditors are businesses or people to whom you owe money. You may have borrowed money from them or you may have purchased goods from them without paying cash. In either case there is a promise to pay at a certain time, or in regularly scheduled increments over a certain period with interest added at a certain rate.

Suppose you borrow money from a bank. When a bank considers an investment in any business it weighs the potential earnings for the bank against the degree of risk. An investment with a high degree of risk—your bar, perhaps—should give the bank a high return—that is, command a high rate of interest. In order to assess the risk the bank will require a detailed financial statement showing the borrower's financial standing at that point in time. It will include how much is owned and how much is owed, with the result showing that the applicant has a net worth (owns more than is owed), which will be used to repay the bank if the business fails.

If your enterprise is a going concern and you are borrowing to renovate or enlarge your operation, this financial statement for the bank will be similar to your regular company balance sheet and can be prepared from the records you keep for your own purposes. If you are a first-time entrepreneur the financial statement will deal with your personal financial condition. Besides the financial statement a bank will want to know your character and track record in the bar business. This last is something a new owner often does not have, and is the one thing that causes most first-time applications to be rejected.

But let us assume that you receive your loan. As long as you make your payments regularly and on time, the bank is not interested in any further information about your business. But if you cannot make a payment you may have to go and see the bank and provide current information on your business so that the bank can assess your financial future. Though it may have a mortgage on your property as security for its loan, it will not want to foreclose unless it is obvious you are not going to make it; foreclosures tend to frighten new business away from the bank. So it will want to look at your recent income statements, and it may want more detail than these show—the kind of detail your income-tax return asks for. From this data the bank will decide whether your profit potential is enough to repay your indebtedness.

A loan guaranteed by the Small Business Administration (SBA) requires much more detail, both initially and in cases of trouble. It usually requires monthly income statements and a yearly independent audit.

Trade creditors usually require much less information than a bank does. Suppliers of furniture and fixtures or other one-time purchases may ask for an abbreviated financial statement as well as several business references. One reference is usually the bank. It is a good idea to give the name of the bank officer with whom you are dealing, as someone who knows your financial position and can discuss credit.

Your relationships with beverage suppliers will be somewhat different. They are not interested in the details of your business except in bankruptcy proceedings. What they are interested in is new customers, and they will generally extend credit up to 30 days (if the state allows it) without requiring any special information. A new bar may have to begin on a cash basis, moving to a credit arrangement as business improves.

All credit arrangements involving alcoholic beverages are subject to federal and state regulations. Many states restrict the credit period, and some do not allow credit at all. The federal government limits credit to 30 days on all alcoholic beverages figuring in interstate commerce.

Information investor/owners want. A bar may be owned by one person who also manages the enterprise, or there may be one or more investor/owners who do not take part in its management—often referred to as absentee owners. Their financial relationship to the enterprise is different from that of the creditors. There is no agreement on the part of the business to pay back the money invested or to pay interest on it. The investors simply put their money into the business in exchange for a share of its ownership, and they take the risks that go along with it. If there is a profit, they receive their share. If there is a loss, they lose some or all of their investment.

The information absentee owners want before investing has to do with the expected profit in relation to the funds to be invested, or in other words the ***return on investment***, the ***ROI***. For a going concern the investor can look at the current dollar profit (the bottom line on the income statement). For a new enterprise the critical figure will be the anticipated sales and the expected dollar profit on the bottom line of the budget. The prospective owner then calculates the percentage share of the profit in relation to the amount to be invested and compares it to what could be earned by investing this money somewhere else.

During the year the manager or the accountant sends absentee owners monthly reports: income statements and balance sheets. These do not show the ROI as such; the investors figure that for themselves.

As an example, let's look at expectations and performance as an absentee owner sees them. Let's say the investment of a sole owner in a new enterprise is $80,000. If this owner wants a 25% return, he or she is looking for $20,000 a year. This amount becomes the profit goal of the business, and its budget is prepared accordingly. Anticipated sales are set at $100,000.

At the end of the year the accountant's income statement shows that both goals have been achieved:

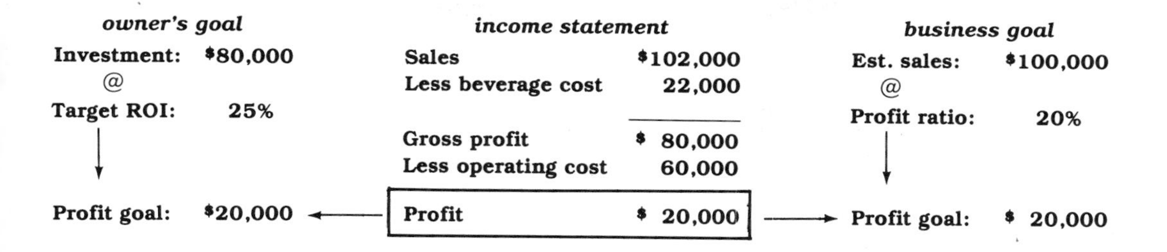

Information the manager needs to run the business. The financial information a manager needs is far more extensive than that required by governments, creditors, and investors.

In particular, a manager needs ongoing information about the flow of money in and out, the effects of outgo on income and income on outgo, and the spread between the two. This means records of sales over successive periods of time, records of various categories of expenses over the same periods of time, and a way to measure the relationships between the expenses and the income.

If you figure your expenses as percentages of sales, as in your budget calculations, you have the tool for measuring the relationships. And if you set up your records using the same expense categories you used in your budget, you can compare your cost/sales percentages to measure your performance against your plan. You can also compare your current period with previous periods, to see if you are doing better or not as well. Then you can use the information to figure out why, and decide what to do about it. We have already seen how this system works for controlling beverage costs (Chapter 12).

Your accountant organizes your financial information in accounts that make these comparisons possible. Each income statement summarizes the information for the preceding accounting period. You can use the statement to calculate your percentages, or the accountant will calculate and interpret them for you, for an extra fee.

In addition to the regular accounting records, a manager needs daily information: daily reports on sales volume, daily inventory records showing purchases and issues. These records must be organized and recorded each day. They are used in day-by-day operations and they also function as raw materials for the accountant's formal records. The manager also needs as much detail as possible on sales—volume per time period, what is selling and what isn't, when it sells best, at what price, and so on. This kind of comparative detail becomes very expensive to gather and record.

Financial records and reports . . .

The heart of an enterprise's financial information service is the accountant's ***books***—the formal financial records of the business. They chronicle the day-by-day financial history of the enterprise and organize it in records that can supply the information demanded by governments, creditors, investors, and management.

A bar may employ a CPA (certified public accountant) to keep its books, but the manager is responsible for supplying the raw data and ought to understand the way the books are set up. It is even better, as we have said before, if the manager (or owner) knows enough about the subject to have a hand in setting them up. This simply means deciding what types of information should be grouped together in one account and what should be set up in separate accounts to meet special information needs.

The raw materials. Recordkeeping starts at the bar with the written and printed records that note each transaction as it takes place. Among these are register tapes, guest checks, credit-card slips, invoices, cash payout slips, banking records, check vouchers. They are the raw financial data that must be organized each day into records that make sense, both for your own immediate use and for the formal books. Since you want your records to document the *flow of money in and out*, let's look at what is involved in sorting it all out.

Money coming in in the form of cash should be itemized in a daily summary such as the one in Figure 13.1. This summary is compiled from the day's individual records—register tape, guest checks, credit-card charges, credit sales (signed guest checks). This summary form enables the manager to keep track of the volume of daily sales, as well as to check the record of cash receipts against the cash on hand. (Cash payout slips are included in this summary in order to balance the cash.) Copies of the daily summaries are given to the accountant, providing data to be entered in the books.

In addition to the day's sales the accountant must be given a record of receipts for previous sales, such as checks from credit-card companies or individual credit customers. Records of such receipts may be found in your bank records (check stubs and deposit slips—which should itemize each transaction separately). As an alternative, the bar may keep a cash receipts journal, combining the information from the sales summary with all other daily receipts.

Money going out should all go out by check. Nothing should be paid out without a written record. If anything must be paid in cash, such as a COD beer delivery, a cash payout slip should be made out recording *date, amount, payee,* and *purpose.* (This slip is included in preparing the daily sales summary.) There should be a written document for every check—an employee time card, an invoice, a bill, a tax statement. As a form of control

DAILY SALES SUMMARY

Date____________ Day of week____________

Sales by Register

	11 AM - 3 PM	3 PM - 7 PM (Happy Hour)	7 PM - Close	Total
Liquor				
Beer				
Wine				
Tax				
Total				

Cash Count

	Sales	Deposit
Currency		
Cash payouts		XXXXXXXXXXXXXXXXXXXX
Checks		
Accounts receivable		XXXXXXXXXXXXXXXXXXXX
Total sales/deposit		
Cash/short or over	±	±
Adjusted deposit	XXXXXXXXXXXXXXXXXX	

Accounts Receivable

	Sales	Discount	Accts receivable
American Express		–	
Visa/MasterCard		–	
Other		–	
Other		–	
Total			

Reconciled

Deposits + Accts receivable = Total sales

________ + ________ = ________

OVER	SHORT

figure 13.1 A daily sales summary form used by a small operation.

at the time of payment, invoices and statements should be checked to be sure that they reflect the terms and prices agreed upon and that the goods and services were actually received. All invoices should be approved for payment by the manager or controller, and paid using the original signed invoice copy.

There should be a systematic written record of money paid out. There are several ways of doing this. A small enterprise can keep the record in its checkbook. Each check should be recorded on the check stub by *date, amount, payee,* and *purpose.*

A more common way to keep a record of money going out is to make payments with multiple-copy ***voucher checks***. The top of the voucher is the check itself and the bottom portion provides space for a notation of what the money went for. A carbon copy of both parts becomes the written record of the disbursement.

A third way is to keep a cash disbursements journal, in which all payouts are entered daily. The two cash journals together can give you current readings on your cash flow.

The checkbook stubs, voucher carbons, or cash disbursements journal, along with the supporting documents, supply the accountant's raw data for money going out. In addition the accountant needs the invoices for credit purchases not yet paid for (known in accounting jargon as ***accounts payable***) and invoices that have been paid in cash—in short, all invoices for the period.

It goes without saying that the bank account for the bar must be used for bar business only. If your bar is set up as a sole proprietorship and you combine business and personal assets for tax purposes, you have to be extra careful to keep your money records straight—business funds in one bank account and one checkbook, personal funds in a separate account.

The final category of raw data is your closing inventory, which you figure after the close of business on the last day of the accounting period (Chapter 11).

In addition to the data that go into the accountant's books, a manager needs several types of ongoing records kept on the premises:

- A separate inventory record showing purchases and issues—essential to making purchasing and pricing decisions, as well as providing income-tax data.
- A separate record for each employee showing rate of pay, hours worked, overtime, tips, bonuses, deductions for social security and withholding, and amounts paid—essential not only for payment purposes but for social security and withholding payments and reports to the government.
- A separate sales-tax record—essential for reporting the tax and making payments.
- A separate record for each credit customer showing charges and payments—essential for collection purposes.

- Separate records of credit-card charges and payments, itemized by card company.

The accountant's books. The books of a business consist of two parts. The ***journal*** is the financial chronology of the enterprise in which are entered all the business transactions day by day. It is also an index to its companion book, the ***ledger***, in which the individual accounts are kept.

The ledger contains a separate account for each category of assets and each category of liabilities, an equity account, and a separate account for each of the various subdivisions of the equity. These subdivisions include the various categories of revenues, which increase the equity, and the various categories of expenses, which decrease the equity. The total of the assets accounts will always equal the total of the liability-plus-equity accounts. This is the fundamental equation of bookkeeping:

Assets = Liabilities + Equity

The accountant analyzes and records each transaction in the journal in terms of its effects on the various asset, liability, and equity accounts. Each transaction has a double effect: it increases one or more accounts and decreases one or more other accounts by the same total amount. Increases in assets and expenses and decreases in liability, revenue, and equity are entered in the left-hand column of the journal and are known as ***debits***. Decreases in asets and expenses and increases in liability, revenue, and equity are entered in the right-hand column and are known as ***credits***. To the layman (who expects all debits to be decreases and all credits to be increases) this system seems to fly in the face of all reason, but accountants have been doing things this way since Italian merchants invented the system in the thirteenth century—and it works. Just suspend belief and remember that, in accounting books, debit means the left side and credit means the right side.

The debits and credits from the journal are then entered as debits and credits in the specified ledger accounts, a process known as ***posting***. This system of entering each transaction in different accounts in equal amounts on the left and right is known as ***double-entry bookkeeping***.

At the end of the accounting period the books are balanced. For each account in the ledger the debits and credits are added up and the smaller total is subtracted from the larger, resulting in a debit or credit balance. The balances for asset accounts are totaled. The balances for liability and equity accounts are totaled. If the two totals are equal, the books are in balance. If they are not, errors must be sought out and corrections made.

When the books are balanced, the accountant uses the ledger data to prepare the two basic accounting reports: an ***income statement*** analyzing and summarizing the flow of money in and out during the accounting period, and a ***balance sheet*** showing the financial state of the business at that point in time.

The double-entry system provides a nearly infallible check on the accuracy of the records. It has the other major advantage we have mentioned: it focuses the effects of the money flow on the different aspects of the business, which makes possible the kind of analysis and comparison you need.

A small enterprise may start out with a single journal and a dozen or so accounts. As the business becomes more complex it may need several separate journals such as cash receipts, cash disbursements, payroll. A large operation may want separate ledger accounts for sales of mixed drinks, wine, and beer, or even for specific drinks, and separate accounts for specific categories of expenses. Big corporations will have books running to many volumes, or more likely will have them all computerized. Many CPAs also use computers for their clients' books.

Each enterprise should design its information system to fulfill its own goals and fit its own budget. The more detail, remember, the higher the cost of the service.

You should also discuss with your accountant how much of the recordkeeping you want the accountant to do and how much you want to do yourself. Such arrangements vary. You can have it all done professionally beginning with your cash summaries and your invoices, including figuring your payroll and writing all your checks. On the other hand, if you understand the system, you can do a good deal of the sorting out and classifying and summarizing yourself so that you turn over your raw material in organized form. You can also figure your own payroll and write your own checks. It depends on what division of labor is most cost-efficient for your enterprise.

The income statement. For the manager, the income statement is the most useful report the accountant produces. It shows the kinds and amounts of revenue, the kinds and amounts of expenses, and the resulting profit or loss over a given period of time. It is routinely prepared at the end of each accounting period.

The income statement is compiled from the various revenue and expense accounts for the period. It shows a summary figure for each account—the account balance. The way in which the income statement groups and presents these summary figures has a lot to do with its usefulness to the manager of the bar.

Figure 13.2 shows you an income statement structured for maximum usefulness. There are two principles behind this arrangement:

- ***Reponsibility.*** Revenues and expenses are grouped according to the responsibilities involved. Thus for each grouping the performance of the person responsible can be measured and evaluated.
- ***Controllability.*** Expenses are classified and grouped according to the extent to which they can be controlled. This income statement shows three levels of controllability: beverage expenses (directly related to sales

INCOME STATEMENT

March 1981

TOTAL SALES					
Liquor	$22,550.30				
Beer	$ 7,129.45				
Wine	$13,743.85			$43,423.60	100.00%
Less COST OF BEVERAGE SOLD					
	Liquor	Beer	Wine		
Opening inventory	$2692.70	$1620.10	$2179.45		
+ Purchases	5101.35	2130.00	3603.10		
	$7794.05	$3750.10	$5782.55		
- Closing inventory	3622.24	902.10	1530.20		
Total COST	$4171.81	$2848.00	$4252.35	$11,272.16	25.96%
GROSS PROFIT				$32,151.44	
Less OPERATING (CONTROLLABLE) EXPENSES					
Payroll costs		$13,620.70			
Administrative expenses		$ 1,328.78			
Laundry and supplies		$ 1,101.20			
Utilities		$ 385.26			
Advertising and promotion		$ 275.00			
Repairs		$ 510.00			
Maintenance		$ 874.00			
Miscellaneous operating expenses		$ 162.00			
Total VARIABLE EXPENSES				$18,256.94	42.04%
GROSS OPERATING PROFIT				$13,894.50	
Less FIXED OVERHEAD (NONCONTROLLABLE) EXPENSES					
Rent		$ 850.00			
Taxes		$ 440.00			
Insurance		$ 357.00			
Interest		$ 818.00			
Licenses and fees		$ 907.00			
Depreciation		$ 774.00			
Amortization		$ 320.00			
Total FIXED EXPENSES				$ 4,466.00	10.29%
NET PROFIT BEFORE TAXES				$ 9,428.50	21.71%

figure 13.2 A typical income statement for a small operation.

Level 1	TOTAL BEVERAGE INCOME Less BEVERAGE COST (direct expenses) GROSS PROFIT
Level 2	GROSS PROFIT Less OPERATING EXPENSES (controllable) GROSS OPERATING PROFIT
Level 3	GROSS OPERATING PROFIT Less FIXED OVERHEAD EXPENSES (noncontrollable) NET PROFIT OR LOSS BEFORE TAXES

figure 13.3 Structure of Figure 13.2.

and fully controllable), operating expenses (related to sales to some extent and controllable in some degree), and fixed expenses (noncontrollable overhead expense).

According to these two principles this income statement is separated into three successive profit levels. Its structure is shown in Figure 13.3.

The first level, as you see, is ***gross profit (gross margin)***, or profit from beverage sales less beverage costs. The responsibility includes sales volume, beverage supply, and beverage cost control. This level of profit is the responsibility of the manager.

The second level is ***gross operating profit***, or gross profit less operating expenses. This responsibility includes supplying items of sales-related expense (labor, supplies, laundry, utilities, maintenance, promotion, and so on) in amounts that support and increase beverage sales, while controlling the costs of these items.

The third level is ***net income (net profit before taxes)***, or operating profit less fixed overhead expenses, such as rent, taxes, licenses and fees, insurance, interest, depreciation, amortization. The responsibility here, to the extent that these expenses can be altered, involves finance and investment policy, contracts, and high-level decision making. Those responsible for this segment of expenses and for the net profit level are the owners of small businesses or top management in a large corporation.

However, though responsibility for net profit remains at the top, as a practical matter profitability depends on the extent to which the gross operating profit (Level 2) exceeds the fixed expenses. Since the manager is responsible for this level of profit, profitability is in his or her hands.

Thus the goal should be to maximize gross operating profit, and the strategy should be to relate specific costs to the sales they produce, rather than simply thinking in general terms of maximizing sales and holding down costs. It's okay to spend money to make money. If increasing a cost increases sales by a more than proportionate amount, it is profitable to increase that cost.

Each monthly income statement tells you how you are doing when you compare it with your budget and the statements of preceding periods. It also gives you the dollar data you need to figure the cost percentages you use for analysis and control. The income statement is management's most important accounting tool.

The balance sheet. The balance sheet is the accounting report that shows the financial condition of an enterprise at a given date. It is a snapshot image of the state of the business frozen at an instant in time.

The balance sheet has three major segments: ***Assets***, ***Liabilities***, and ***Equity***. In accordance with the basic equation—Assets = Liabilities + Equity—assets are shown on the left side of the balance sheet and liabilities and equity are shown on the right. The totals for each side of the balance sheet are always equal. Figure 13.4 shows you a sample balance sheet for a small bar.

Assets, as you can see, are subdivided into several categories.

- ***Current assets.*** Anything owned that can be turned into cash within a year—inventory, for example.
- ***Fixed assets.*** Tangible permanent assets such as building, land, equipment.
- ***Other assets.*** Assets that are neither current nor fixed, such as a long-term investment.

Liabilities are subdivided into:

- ***Current liabilities.*** Anything owed that must be paid within a year, such as beverage accounts payable or tax payments due within a year.
- ***Long-term liabilities.*** Anything owed that is due a year or more in the future, such as a long-term bank note.

Equity is what is left of the assets for the owners after all the creditors' claims (liabilities) are satisfied. There are two forms of equity:

- ***Owners' equity.*** The original contribution of the owners to the establishment of the business.
- ***Retained earnings.*** Profits that have not been distributed to the owners but are kept in the business.

People you want to borrow money from look closely at the retained-earnings figure on your balance sheet. Keeping a good amount of cash in the business in case of an emergency or an extraordinary opportunity is better than distributing all the profits.

The accountant prepares a balance sheet at the end of each fiscal year and at any additional intervals you want—monthly, quarterly, whatever. Banks and investors will use your balance sheet for assessing the risks they are taking if they loan you money or invest in your business, and reports to stockholders of large corporations are centered around the balance sheet. It is very useful to you too, if you are the owner of a small business or have

BALANCE SHEET

A Record of Ownership

ASSETS: What Is Owned			LIABILITIES + EQUITY: What Is Owed		
Current Assets			Current Liabilities		
Cash	$23,500		Accounts payable	$ 700	
Receivables	1,300		Wages payable	2,300	
Inventory	2,000		Taxes payable	1,100	
Prepaid expenses	1,100		Notes payable	15,000	
Total Current Assets		$27,900	Total Current Liabilities		$19,100
Fixed Assets			Long-Term Liabilities		
Building	$180,000		Notes payable	$102,000	
Equipment	35,000		Total Long-Term Liabilities		$102,000
Furniture/fixtures	2,500				
Less Depreciation	50,000		Total Liabilities		$121,100
Total Fixed Assets		$167,500			
			Equity		
Other Assets			Owner's equity	$35,000	
			Retained earnings	39,300	
			Total Equity		$74,300
TOTAL ASSETS		$195,400	TOTAL LIABILITIES + EQUITY		$195,400

figure 13.4 A typical balance sheet for a small operation.

financial responsibility in a large one. It enables you to evaluate your current financial position and to make major plans for the future.

Using accounting to maximize profits . . .

Since profit is the difference between costs and sales, the way to maximize that difference is to increase sales, or decrease costs, or increase costs in a way that will increase sales even more. Accounting data and techniques can help you do all of these things. Here are some ways you can use them.

Data analysis. Data analysis is that part of the accounting function that goes beyond recordkeeping. It includes organizing the records so that relationships can be observed and measured and their significance interpreted. The accountant's income statement is a form of data analysis. You take the process further when you calculate relationships numerically (cost/sales percentages, for example) and use them to control costs or set prices or make out a budget or compare any body of data with any other.

Three areas of analysis are especially important in maximizing profits: sales, operating expenses as they relate to sales, and overhead expenses as they relate to sales.

All profits come from sales; therefore, increasing sales is one road to maximizing profits. For this you need all kinds of information about sales: volume per drink, per price, per spirit, per employee, per customer, per shift, per bar if you have more than one, per day of the week, per week, per month, per season. If you can gather the data and calculate the relationships, you can see where the profit is being made and where to put your merchandising emphasis. You can also measure the effects of any changes you make.

You can have your accounts set up to compile this type of data, but this procedure is expensive and when you get the information in your income statement it may already be out of date. What you need is information as it is happening. A computerized cash-register system can be programmed to give you instant feedback of this type. Or you can collect and organize the data yourself using forms of your own design, such as those illustrated.

Figure 13.5*a*, for example, lets you compare the sales appeal of your various drink specials. This in turn can help you in pricing for profit. Figure 13.5*b* organizes your total sales by period as well as giving you the daily bar cost. Using these forms daily over a week's period gives you day-of-the-week variations, which tell you something about your clientele as well as the drinks.

The other piece of the profit equation is cost: profit equals sales less cost. A bar's operating expenses are the manager's responsibility. They are tied in

SALES OF SPECIALS

	Open to 3 pm	Happy hour	7 to closing	Total
Item______				
Item______				
Item______				
Item______				
Item______				
Item______				
TOTAL SALES PERIOD				

figure 13.5a

DAILY COST/SALES ANALYSIS

DATE ______________

DAILY BAR SALES		DAILY COST	
Open to 3 p.m.	______	Opening stock	______
Happy hour (3-7)	______	Issues +	______
7 p.m. to closing	______	TOTAL	______
TOTAL SALES	______	Closing stock -	______
		TOTAL USED	______

→ ______ ←

PERCENTAGE COST

figure 13.5b

varying degrees to the rise and fall of sales, and they are always examined in relation to the sales they help produce. Keeping close track of this relationship can tell you a great deal about what is happening to profit day by day.

The two major categories of operating expenses—beverage cost and payroll expense—are called ***prime costs***, because they are the largest and most necessary items you spend your money for. Beverage cost is the cost of the alcoholic beverages sold. Since the cost of these is dictated largely by customer taste, the manager's major focus is on controlling losses. Here data analysis in the form of cost/sales ratios is a common way of detecting and measuring losses, as discussed in Chapter 12. The cost/sales ratio on the monthly income statement can tell you how well you controlled losses last month, but it doesn't help to detect losses as they happen. Here again, your own data gathering and measuring on a daily or weekly basis can adapt the accountant's techniques to pinpoint losses quickly.

You can use in-house forms such as those illustrated to figure bar cost percentages. Figure 13.5*b* gives you bar cost on a daily basis. Figure 13.5*c* gives you the percentages on a weekly basis from physical inventory, and Figure 13.5*d* gives you the same thing by a short-cut method from

WEEKLY COST/SALES ANALYSIS

Period ______ to ______ 198__	Liquor	Beer	Wine
Total sales			
Opening inventory			
Purchases			
Closing inventory			
Cost of beverage used			
Beverage cost percentage			

figure 13.5c

WEEKLY COST/SALES ANALYSIS

Period ______ to ______ 198__	Liquor	Beer	Wine
Sales			
Purchases			
Cost of beverage used			
Beverage cost percentage			

figure 13.5d

invoices—approximate, to be sure, but quick and timely. When compared to your standard cost percentage, they can help you spot discrepancies in time to do something about them.

Payroll expense, the other prime cost, basically represents the cost of preparing and serving the products sold, though it includes the pay for all employees as well as employee benefits. Although payroll costs do not rise and fall with sales as closely as beverage costs do, they should be correlated with sales for the same period in a cost percentage figure and measured against the budget goal and percentages for previous periods. If your income statement shows that your cost/sales ratio is high, you look for ways to avoid overtime or cut staff or hours worked.

Other operating expenses—utilities, supplies, laundry, and so on—are also related in one way or another to the service of customers and hence to sales. They too should be related to sales in percentage terms and compared with budget goals and past history. Promotional and advertising costs should be correlated to sales over successive periods as a measure of their effectiveness. Specific expenses should be correlated to sales they help produce. For example, the breakage cost of special glassware for specialty drinks should be correlated to the sales of those drinks.

Even the smallest category of operating expenses should be watched in terms of cost/sales ratios. The higher the percentage for costs, the lower the percentage for profits.

Overhead expenses are those that go on regardless of hours of operation, numbers of customers, or sales volume. They include rent, insurance, fees, loan payments, and other fixed costs. The manager is not responsible for this level of business management; even so, the cost/sales ratios shown by income statements are important to watch. Though the costs are fixed, any cost/sales ratio can be reduced by increasing sales. You might step up promotion or decide to stay open longer.

On the investment level, cost/sales ratios are important in determining when and whether to renovate or to renegotiate contracts such as loans and leases. It is the owner or financial manager who needs the data analysis.

The physical condition of the premises can have its own effect on sales. Innovations introduced by competing bars, changing tastes of customers in matters of decor and ambience, development of new bar equipment that outdates yours, all sooner or later have a dampening effect on sales and profits.

Up to a point the updating of paint on the walls, new carpet on the floor, and new equipment behind the bar is an ongoing expenditure for maintenance and repairs and is considered an operational expense. But there may come a point when a major renovation is necessary to remain profitable. Then the money spent can be viewed as an investment and the data analysis is from the investment point of view. Using past and present data, the owner can project how long it will take to recover the money invested with the anticipated improvement in sales and operating efficiency. An accepted standard in the industry is five years. If this does not seem likely, the alternative may be to move from the present facility and invest in a more promising place.

Forecasting cash flow. A bar can do a high-volume business yet still run into trouble if there is not enough cash in the bank to meet the payroll and pay the bills. This can happen because money does not come in and go out at the same daily rate. You may have lump-sum payments due at certain times. Customers may pay by credit card. You may put out a large sum to buy a special wine that is hard to get. Forecasting cash flow can help you avoid a crunch.

A ***cash-flow forecast***, or ***cash budget***, as it is sometimes called, is a short-term forecast of cash flowing into the bank and cash flowing out, week by week or month by month. It requires predicting each item that will put money in your bank account or take it out. You can use your monthly income statements from previous years and other records to help you estimate variables accurately. Fixed expenses such as rent, license fees, taxes, insurance and loan payments are typically due on fixed dates in fixed amounts.

Figure 13.6 is a detailed guide for setting up a cash-flow forecast.[1] It is organized for a monthly forecast but can be made to work for any interval.

Break-even point. For every business there is a level of operation at which it is neither making money nor losing money. It is a level a new bar must reach fairly quickly if it is to survive. In the first days and weeks it is the measure for the dollars in the register, to determine whether or not this new enterprise will make it.

To figure your daily break-even point, start with all fixed costs for your enterprise for a month. Add a month's wages and salaries for a skeleton crew, the people without whom you cannot open your doors. Divide this total by the number of days in the month you will be open. This is your daily cost for opening your doors. Now add the cost of a day's liquor supply and any extra staff needed to serve it. This will give you the dollars you need in the register each day to break even.

A daily break-even point is useful at any time. When you use it over a period of days it can show you when you might cut back on staff and what days it doesn't pay to open at all. Or it may indicate that you need to do something to raise your general sales level.

[1] This chart is adapted from materials published by the Small Business Administration. It uses what are called *cash accounting* procedures. If you find slight differences from procedures your accountant uses, it is because accountants generally operate on an *accrual* basis. Your accountant can explain the differences if you are interested.

CASH FLOW FORECAST

+ = Money coming in
− = Money going out

	Jan	Feb	Mar	Apr	May	Jun
+ 1 Cash sales						
+ 2 Accounts receivable collections						
+ 3 Other cash receipts						
4 TOTAL CASH AVAILABLE						
− 5 Purchase payments						
− 6 Operating expenses						
− 7 Other cash expenses						
8 TOTAL CASH PAYOUT						
9 NET CASH FROM OPERATIONS						
+ 10 Short-term financing						
+ 11 Long-term financing						
+ 12 Additional equity						
− 13 Repayment of short-term debt						
− 14 Capital expenditure						
− 15 Repayment of long-term debt						
− 16 Cash withdrawals						
17 NET MONTHLY CASH POSITION						
18 BEGINNING CASH						
19 CUMULATIVE CASH POSITION						

figure 13.6

1. **Cash sales.** Enter your estimated monthly sales based on last year's records, allowing for seasonal influences, inflation, and growth or decline in your business during the past year.
2. **Accounts receivable collections.** Enter receivables you expect to be paid during the month.
3. **Other cash receipts.** Enter anticipated cash income not directly related to sales.
4. **TOTAL CASH AVAILABLE.** Add items 1–3. This is the sum of all projected inflows of cash for the month.
5. **Purchase payments.** Enter amount to be paid out for beverages. Estimate this from last year's records, using the cost/sales ratio for the corresponding month.
6. **Operating expenses.** Enter anticipated operating expenses estimated from last year's income statement for the same month and last month's figures.
7. **Other cash expenses.** Enter all other anticipated cash expenses.
8. **TOTAL CASH PAYOUT.** Add items 5–7. This is the sum of all projected outflows of cash for the month.
9. **NET CASH FROM OPERATIONS.** Subtract item 8 from item 4.
10. **Short-term financing.** Enter cash you expect to come in from short-term loans during this month.
11. **Long-term financing.** Enter cash you expect to come in from long-term loans during this month.
12. **Additional equity.** Enter additional cash to be invested in the business by owners during this month.
13. **Repayment of short-term debt.** Enter payments of principal and interest on short-term loans due during the month.
14. **Capital expenditure.** Enter payments on permanent business investment, such as real estate mortgages or equipment purchases due during the month.
15. **Repayment of long-term debt.** Enter payments on long-term principal and interest due during the month.
16. **Cash withdrawals.** Enter owners' cash withdrawals anticipated during the month.
17. **NET MONTHLY CASH POSITION.** Add items 9–12, then subtract items 13–16.
18. **BEGINNING CASH.** Enter previous month's cumulative cash (item 19 of previous month's forecast or actual cash position at first of month).
19. **CUMULATIVE CASH POSITION.** Add beginning cash (item 18) to net monthly cash position (item 17).

figure 13.6 (continued)

Summing up . . .

Accounting is a well-organized financial information service that can be fine-tuned to your needs. It cannot in itself produce a profit or control costs. It is you who must translate the data into action.

Whether it is human hours or computer hours you are paying for, recordkeeping is a costly business. Among large enterprises, computer systems linked to the cash register are being used increasingly to gather information at the point of sale and to provide instant data analysis. But even with computers you must understand the data generated in order to put them to use. Data analysis enables you to keep your finger on the pulse and diagnose the patient, but you are the doctor. You prescribe the cure—or spot the trends, measure the risks, and seize the opportunities.

chapter 14 . . .

MARKETING

"Make a better mousetrap and the world will beat a path to your door" may have been true when Emerson said it, but it doesn't work today. You can't create a product—even a better one—and then wait for people to come and buy it. They may not want it, and even if they do, they may not know they want it, or even know that it exists. Selling something people want and getting them to buy it from you is what marketing is all about.

Marketing should begin while your enterprise is still only a gleam in your eye, and go on as long as the enterprise does. This chapter explores the various phases of marketing—setting market-oriented goals, analyzing the market, attracting customers, selling them your wares, and making them want to come back.

In a beverage service operation, you are the seller of products (alcoholic beverages) and services (the dispensing of such beverages in a congenial environment). Potential customers are out there somewhere in the area in which you operate. Since you cannot move your products/services to them, you must induce them to come to you and, once there, to buy your products/services and come back to buy again. To be successful in this you must make sure there are enough customers within reach to support a profitable enterprise and you must offer them the goods and services they want. Clearly you must know where you are going before you start.

Setting patron and product goals . . .

The starting point of any marketing plan is choosing the market—identifying the customer group—and the second step is determining what products and services that group wants to buy. When you have defined these two essentials, you can shape everything about your enterprise to attract the customers and sell the product. This is true whether you are planning a new enterprise or revamping an established one that is losing its drawing power.

The patrons. People go to bars and restaurants for many different reasons. Usually the drink is not the primary motive; they can buy the beverage at a package store and do their drinking at home. We can divide customers into several different groups according to their reasons for coming:

- Diners at restaurants where drinks are served. They come to enjoy a good meal, and the drink—whether cocktail or wine or after-dinner drink or all three—enhances the enjoyment of the total experience. Though the food may be the primary focus, these people usually want the drink too. Restaurants that do a flourishing business without alcohol are the exception rather than the rule.
- Drop-in customers who are on their way somewhere else. These usually want refreshment—a quick pick-me-up or let-me-down after a day's work. The drink is the focus here—one or two at most and then the customer is on the move again. People waiting for a plane or a train or meeting someone at the bar also belong to this group. Bars near office buildings or factories, or in train or bus stations, airports, and hotel lobbies typically serve this kind of customer.
- Entertainment seekers—leisure-time customers looking for relaxation, stimulation, change of pace, or maybe a member of the opposite sex. They frequent bars, lounges, clubs, and restaurants where there's

something going on—country and western music, games, dancing, a chance to meet new people or keep up with social trends. They may visit several places or spend a whole evening in one if the entertainment and the drinks and the company are good.

- Regular patrons of neighborhood bars or taverns where friends are likely to gather. They are interested in enjoyment and relaxation too, but their primary desire is for companionship—being with people they know and like, feeling comfortable and at home, feeling they belong.

Most customers fall into one of these groups. The moods, tastes, and interests of the groups differ, and the people tend to differ in background and lifestyle as well, although individuals do cross group lines at times. A drop-in customer on a business trip may become a diner or entertainment seeker at home, but the mood and the purpose have changed.

Generally these groups are not compatible. A customer from one group visiting a bar frequented by another group is likely to think of the others as "the wrong crowd" and feel uneasy and out of place. People who are turned off by the people around them do not experience the place as having a friendly atmosphere even if the bar's own personnel are friendly. Such customers don't stay long and don't come back.

Within the four broad groups are many subgroups. They may divide loosely according to lifestyle, interests, age, income level, family status, occupational or social level (professional, blue-collar, leaders in society). Often these groups do not mix well either. Other subgroups may be even more specialized, forming around a common interest such as football or jazz or a particular social clique.

You can cater to any one group or subgroup, but you cannot please them all. It is a common mistake in a first-time enterprise to want anybody and everybody to come, or not to think about the type of customer at all. This doesn't work. Part of the atmosphere of any bar is its customers, and if they don't have something in common—mood, attitude, reason for coming—something is missing and they don't have a good time. The experienced entrepreneur concentrates primarily on a single, definable customer group who will have similar reasons for visiting a bar, and shapes the entire enterprise to attract and please this group.

Sometimes a bar will deliberately make itself the "in" place for a special subgroup—say a fraternity group from a college campus or the fans of the local football team. This can be very profitable, or it can have certain hazards. It depends on how large the group is, and whether its leaders support the place, and whether the group has continuity—what will happen when its leaders graduate or the football season is over? It also depends on whether you already have customers that won't care for this group and will take their business elsewhere.

It is possible to serve more than one clientele by separating them. Large hotels, especially resort hotels, may have several bars for different customer types, or may serve different groups in different rooms from the same bar. The Jack Tar Hotel described in Chapter 3, for example, serves drop-ins at the bar and lounge—including a happy-hour crowd from nearby office buildings—diners in the dining room and coffee shop (two subgroups), and business groups and conventioneers from portable bars in various conference rooms. Many restaurants have bar and dining room in separate rooms with separate entrances from the street, in order to keep drop-in or leisure-time clientele separate from the diners.

The products and services. The first essential in choosing products and services to sell is that they must fulfill customers' needs and desires. In the terminology of economics, there must be a demand for them, whether expressed or potential. If customers do not need or want your product, nothing you do can make them buy it.

There is no question about the continuing demand for alcoholic beverages; one has only to look at history to be convinced. But people, as we've been saying, come to bars for other reasons. Your products/services must fill these less tangible needs as well, must satisfy their reason for coming. Once you have targeted a chosen clientele you must find out all you can about their needs and desires and tailor your products and your services accordingly. This includes your location, facility, drinks, prices, staff, hours of service, food, entertainment, ambience—the total package the customer is paying for.

Customers of all groups have certain expectations. They expect that they will receive the products and services they need and want in a manner that is worth the dollars they spend. They expect quick, accurate, courteous service for their drink orders, and they expect each drink to taste the way it should. They expect cleanliness—in the bar itself, in restrooms, in the personal appearance of the staff, in glassware. They expect a fair deal—no false-bottom jigger, no cheap liquors and wines in place of quality brands. (Once a customer begins to suspect a raw deal—even if it isn't so—he or she is likely to see imaginary evidence of it everywhere.) They expect that you are glad to have them there, and they like to have you show it.

Beyond that, expectations differ according to the type of clientele. But whatever the expectations, customers must feel they are getting their money's worth.

Since different types of customers go to bars for different reasons, you need to find out what specific kinds of products and services will satisfy your chosen customers' wants, and whether there might be some needs and wants going unsatisfied in your area. You also need to know what your competition is doing to serve this clientele. In a competitive setting your products/services must be distinctively different from those of your competitors, and better than theirs. Otherwise customers will choose competitors who *are* different and better. Bars that simply copy other bars seldom succeed.

You should also give some thought to how you will manage to stay different and better. It means keeping your ear very close to the ground to find out what is going on in other bars and what customers think about it. If other bars copy you and do your thing even better, you no longer have your competitive edge. You also need to keep up with changing customer tastes. An innovative attitude and a means of getting customer feedback can help you to stay ahead.

The type and extent of your products and services will be limited by what you can offer profitably. This will depend in turn on the nature and size of the potential market, and that is the next thing to find out.

Analyzing the market . . .

Before you can make any final plans, you want to know where your target customers are, and how many there are, and how well they are being served by existing bar enterprises. You are interested in location in relation to clientele, numbers of potential customers, and competition, all at the same time.

If you are starting from scratch, you have a chance to pick a location—a general area of town and a specific site within that area. Where will your customers come from? If you are going to serve diners you will want to be accessible to a residential area, or to places of work, or to both. For people in search of entertainment you'll probably want an "in" location. For drop-ins you will want to be on their way to wherever they are going, while for regulars you'll want to be near the neighborhood where they live.

Once you have an area in mind, you need to know whether it contains enough potential customers to support your concept. In some cities you can get free help from city departments that can tell you about demographics (who lives where, how much money they make, education, eating and spending habits, and so on) and about neighborhoods, urban planning, traffic count on city streets. Chambers of commerce have information on business growth, tourist and convention markets, real estate development. Banks that loan money to restaurants and bars know a lot about your community; they can save you from opening the right facility in the wrong place, like a swinging singles bar in a retirement colony. Talk to real estate agents, people in restaurant and club owners' associations. Talk to local restaurant equipment firms; they usually know who is opening and where, who is closing and why.

For a small investment you can consult a demographics firm that specializes in the food-service field. It can make a computerized analysis of a specific area based on census figures, giving you data on population density, age, sex, occupation, size of household, income, ethnic makeup, money spent on dining out, or whatever demographic information you specify.[1]

For a larger investment you can hire a consultant to do a detailed study of a given area and its existing competition, to determine whether there is a market in that area for your bar concept. Or, if you already have a facility but want to revamp it, a consultant's market study can help you develop a concept adapted to the kind of market that *is* there. There is always the risk that you'll spend several thousand dollars to be told that the market you are hoping for is *not* there. But this is cheaper than finding it out the hard way.

Certain areas of cities tend to be "in" places; often bars and restaurants cluster at certain points. Competition may be heavy, but a crowded area may have advantages. Bright lights and movement of people make a festive atmosphere. People enjoy going where they've been before; it's familiar and they know how to get there. If they can't get into one place they'll try another, and the competition benefits everybody.

On the other hand, such areas do become saturated as new ventures try to imitate success. If you are thinking of such a location, two precautions are in order. One is to investigate the openings and especially the closings in the area during the past few years, to see what kinds of places have staying power and what kinds there are already too many of. The other is to study the competition carefully and be very sure of your uniqueness, your special contribution to the area. Don't copy anybody else; be different and better.

Opening in a new or isolated area may be risky, but it can be rewarding if your research uncovers a large pent-up demand. If you do make a big success in a new area, you may soon be joined by competitors, so never count on being the only place to offer something. Think long-range. Offer something really special that will build a following. And don't expand too quickly on the basis of an initial high demand.

Avoid declining areas; look for neighborhoods that are stable or growing. Consider, though, that some once-declining parts of cities are reviving, as the movement grows for restoring old buildings and rehabilitating old sections of town. Some older properties are more spacious, better built, and more generally accessible than new facilities on the outskirts of town. This could be especially important during an energy shortage.

As you are exploring areas and determining customer potential, you must also be looking at the competition. How well are the customers' needs and desires being served there already? Visit all the bars in the area that serve your type of customer, study their offerings, estimate their sales volume, and compare them to your bar concept. Then compare the total existing

[1] See *Restaurant Business*, "The Science of Site Selection," November 1979.

competition to the market potential your study has established. Is there room for you?

When you zero in on a specific site, look at it through the customer's eyes. Does it have good visibility? Adequate transportation? Plenty of parking? Is it easy to reach? Watch out for one-way streets, planned construction, heavy traffic. Keep your ear to the ground for changing circumstances; talk to other types of retailers in the neighborhood. If you are thinking of taking over an existing facility, why did the last tenant leave?

Most important of all, check out the licensing, zoning, and other restrictions. Sometimes different parts of the same city have different licensing requirements. Sometimes a nearby church or school can keep you from getting a license. Zoning restrictions can make it impossible to open the kind of facility you want, or even to open at all in your chosen location. The red tape of permits and licensing can delay your opening for months.

Finally, analyze the financial feasibility of your projected enterprise for that site and market area. You must estimate the capital needed for land and buildings, furnishings and equipment, opening expenses, and a reserve for operating at a loss in the beginning. You must work out a detailed operating plan (drink menu, staff, hours, and so on) and an operating budget (as described in Chapter 12). Estimate sales conservatively and expenses liberally. If the income does not exceed expenses by the desired profit margin, your project is not feasible for that market and location.

Atmosphere as marketing . . .

By this time you probably have a pretty good idea of the overall impression you want to create—the ***atmosphere*** or ***ambience*** of your place. These words are hard to define, but you know what they mean. They have to do with what is seen, heard, touched, and tasted—a totality of sensory impressions to which the customer adds a psychological ingredient of response. It may well be the most influential part of the customer's experience.

It is likely to make its impact immediately. One glimpse of the front of the building, one step inside the door, and the reaction is immediately one of pleasure or disappointment. If you have done your job well, this reaction will depend on the type of customer that comes to the door. What pleases those you are after may well disappoint the others. Consider the four groups of customers identified earlier.

Diners are likely to be mature men and women, alone together or with children in tow. They want to enjoy their drinks and their meal in comfort and relative quiet. They want to be able to read the menu without a flashlight, converse without shouting, watch other people like themselves, and feel well taken care of. They don't like loud music and noisy talk and don't come for fancy decor, although they don't object to it. For this group you might choose a fresh, friendly, low-key atmosphere, conservative but definitely not dull; things should certainly offer a change of pace from eating at home.

Leisure-time customers in search of entertainment are likely to want just the opposite atmosphere. They enjoy noise, action, crowds, people shouting to be heard above everyone else, loud music. They respond to the newest in decor, the latest in drinks, gimmicks, activity, something going on all the time.

A subgroup of these two groups is a younger, trendier group of diners who are really leisure-time customers for whom the eating and drinking are the evening's entertainment. The atmosphere takes its cue accordingly.

Drop-in customers are likely to be less involved with a special ambience. They want prompt service and good drinks, like to have things move along briskly, and usually enjoy a crowded, friendly bar.

Neighborhood bar customers want familiar, relaxed, comfortable surroundings and an atmosphere of goodfellowship. Never mind the hype and entertainment; they will provide their own.

There are many many exceptions and variations from these scenarios, but you get the idea: one person's meat can easily be another's poison.

As you plan the atmosphere of your bar, look at it through the eyes of the group you want to serve. Everyone sees things differently, and the way something looks to you may not be the way your customers see it at all. Their perceptions are filtered through their own attitudes, past experience, and value systems. Your awareness of their view of the world is very important: plan a decor *they* respond to, offer services *they* are looking for, train a staff that knows how to deal with *them* and is sensitive to *their* ways of perceiving things. Your customers, for example, may not share your taste in server uniforms or background music, or think your jokes are funny. They may even be offended at things it never occurred to you to question. But if you can get into their outlook on life and create an atmosphere that expresses this view, you can make them feel, *This is really my kind of place.*

Using atmosphere to attract customers is not just for the new enterprise; it can also revive one whose business is falling off. A certain amount of customer turnover is in the nature of things: job transfers to other cities, marriages, babies, inflation, shifts in buying habits, sickness, and death can eat away at your customer count at the rate of about 15% a year. Of course there are plenty of potential new customers: young people reach legal drinking age, new families move into the homes of people who moved away, new offices and shops move into the area, customers bring in friends. But not many of these people will become patrons unless you do something to attract them. Look at your premises with a critical eye and think about giving a new look and feel to your place.

Successful bar operators make frequent physical improvements, ranging from painting and papering to buying new furniture, opening up a patio, or renovating a facility completely. Customers are quick to spot neglect. Peeling paint, stained carpets, weeds in the shrubbery, and potholes in the parking lot raise doubts about whether you wash your glasses carefully and keep the roaches away. A spick-and-span look, on the other hand, invites the customer in and promises a shipshape operation with careful attention to product and service. It is something that should be budgeted right into the profit plan on a regular basis.

There are two major components to work with in creating atmosphere—physical factors and human factors.

Among the physical factors, appearance and comfort are most important. Appearance has the most immediate impact, from the entrance and the interior as a whole to the lesser details of uniforms, restrooms, table tops, glassware, matchbook covers. Restrooms deserve special attention; they can cancel out a previously favorable impression.

Using decor to create atmosphere has been treated at some length in Chapter 2. You might find it fruitful to reread that discussion in the light of all you have learned since. Think in terms of your total concept, your individuality, your image. The look of your place is one of your most potent marketing tools—it's the packaging of your product.

A brilliant example of using physical environment as a marketing vehicle is Henry Africa's Cocktail Lounge in San Francisco. Figure 14.1 shows one quarter of it. A big, bright, incredible room, it is a visual feast of collector's items, arranged to intrigue, delight, and surprise the visitor. Everything there is the real thing, from the seven honest-to-goodness Tiffany lamps (the others are "all genuine antiques but not genuine Tiffanys") to the oriental rugs under the golden-oak tables and chairs, the vintage motorcycles mounted in the huge plate-glass windows, the collector Lionel electric trains that run around the room just above the windows, and the Steinway grand piano on a platform near the ceiling for live music in the evening. Most genuine of all are the drinks—2-ounce drinks made with premium spirits, freshly squeezed juices, bottled mixers, hand-stirred or hand-shaken, the cream drinks made with Swensen's ice cream, and all at a price that many bars charge for drinks half the size and half as good. In fact, as of February 1981, you could get a 3-ounce Martini for $2 and maybe you still can.

This marketing thrust of visual theatrics plus product value was directed originally toward a young, trendy, boy-meets-girl clientele, but Henry Africa's bar quickly became a tourist attraction. Today's clientele is a cosmopolitan mix of all ages and types of sightseers, brought in entirely by image, value, and word of mouth.

Decor creates the first impression; comfort has a slower but no less important impact. Furnishings—part decor, part comfort—can be chosen to fill both needs. Lighting is also both decor and comfort, and there is sometimes a fine line of compromise to be drawn here. Temperature is not important to the customer until it is too hot or too cold; then it's very important. A ventilation system that draws smoke away and keeps the air fresh is

figure 14.1 Henry Africa's cocktail lounge, San Francisco. (Photo by Charlotte F. Speight)

an essential element of a comfortable bar environment. Noise level is still another comfort factor that you can control according to customer tastes.

Sight and sound provide the first impression; human encounter provides the second and most lasting. Customers of every group respond to the way you and your people greet them, treat them, and deliver your products and services. They like to be cordially welcomed, they like to be served promptly and efficiently, and they like to feel they are getting personal attention to their needs and desires. This single element of your atmosphere is the one most likely to bring people back and to spread the word about your place to friends of satisfied customers.

This doesn't happen by itself. It depends on selecting friendly, people-oriented staff and training them thoroughly in your products, serving routines, customer relations, and philosophy of service. It also depends on your own performance with people, both for your own personal impact and as the model you set for your employees. It depends too on whether your employees are happy in their jobs and in your enterprise—in the working atmosphere you create for *them*. An employee at odds with the management is likely to carry an edge of frustration and anger into customer relationships.

Another human ingredient of atmosphere is the customers themselves. If you focus your marketing efforts to attract a certain clientele, you will have a compatible mix and people will feel comfortable from the beginning, ready to enjoy their experience. And enjoyment itself is contagious. Today's bar is not yesterday's dark, mysterious place where people, isolated in corners and booths, drank by the light of a single candle and could not see or be seen. Everybody is visible, and people-watching and shared experience are part of the fun.

Attracting customers . . .

Let us assume you have targeted a customer group, decided on your products and services, and created an atmosphere that will favorably impress your chosen clientele. But they won't be impressed unless they come through your doors. How do you get them in?

Word of mouth. By far the most effective marketing technique is one the eager entrepreneur can approach only indirectly. This is ***word of mouth***—people telling other people about your place. You hope the comments will be favorable, but even controversy is said to be better than silence. The owner of a famous Texas steakhouse claims there is no such thing as bad publicity, but this could be one of those Texas exaggerations.

To initiate effective word-of-mouth promotion, give people something to talk about. Be different. Be better. Be special. We have touched on this subject a number of times. Here is more.

Unusual decorative items, artwork on display, arresting exhibits of any kind can often cause talk. Crazy Ed's Incredible Whiskey Machine (page 54) is a good example. For a very different clientele, Courtlandt's in Houston will give a private party in their wine cellar on table settings that have won national awards. At Tony's in the same city, wines dating back to the 1800s are displayed in a case in the lobby. Henry Africa's unexpected combination of Tiffany lamps and oriental rugs with Lionel trains and motorcycles cannot help but cause talk.

Another approach is to have the most of something. In Seattle, the bar at F. X. McRory's Steak Chop and Oyster House boasts the world's largest collection of bourbons. Bern's Steak House in Tampa has the longest wine list (500 pages, 5000 wines) and the largest cellar (400,000 bottles). Billy Bob's in Fort Worth is the world's largest night club. And guess what paying the most money ever for an old bottle of Bordeaux (page 134) did for that Memphis restaurateur.

This type of thrust for the limelight can bring free publicity, a form of word of mouth that is highly desirable. If you can manage to get your place mentioned by the local amusement or restaurant columnists, especially if the mention carries a recommendation, you have one of the best kinds of word of mouth going for you. And when you get into the guidebooks of Places To Go, you've got it made.

But you don't have to go to great lengths or spend a lot of money to cause talk. Give your place an intriguing name. A Chicago chain called Lettuce Entertain You has restaurants named R. J. Grunt's, Fritz That's It!, Jonathan's Seafood, The Great Gritzbe's Flying Food Show, and Lawrence of Oregano. When Baby Doe's Matchless Mine opened in Dallas people waited in line for two hours just to get in.

Take-home items with your name or logo are another way of spreading the word. Matchbooks, of course. Coasters, funny menus, recipes for special drinks. Anything that goes home with a customer becomes a conversation piece and an ad for your place.

You can have glassware marked with your logo, include the cost of the glass in the price of the drink, and invite the customer to take it home. In the bar it promotes the sale of the drink; at home it becomes a word-of-mouth promotion piece. Figure 14.2 (page 375) is a case in point.

Specialty drinks are good for starting word-of-mouth promotions. If you invent a new one and dress it up in a special glass, you may be lucky enough to catch customer fancy and become the talk of the town. Two historic examples are the Buena Vista's Irish Coffee and the Cock 'N Bull's Moscow Mule, drinks invented 35 and 40 years ago that are still bringing people into the bars that created them.

The classic example of using specialty drinks as a marketing technique is Victor Bergeron's drinks for Trader Vic's restaurants. He concocted a whole menuful of drinks having a polynesian aura, many of them outsize, designed a special glass for each one, and charged a price that made you know you were getting something really special. People are still talking about them.

Today many bars are experimenting with their own concoctions or are borrowing recipes from little booklets put out by liquor manufacturers. For tips on inventing your own, see Chapter 10. If you really dress them up—serve them in outsize glasses, garnish them lavishly, flame them, anything to call attention—specialty drinks will often sell themselves as you carry them through the room. Whether or not they get you word-of-mouth promotion, they are profitable. It is easier to get an extra 50¢ for a specialty drink than it is to raise the price of beer by a nickel.

You don't really need the kind of word of mouth that makes a place famous. The best kind is simply satisfied customers telling their friends that yours is *the* place to go. It helps if you can provide your own special twists that linger in the memory—your Café Diablo made with a flourish tableside, your ocean view, your fabulous hors d'oeuvres, your jukebox with everything from Bach to the Rolling Stones. Such things extend the conversation, and the detail sticks in the listener's mind better than a vague general recommendation of "We really enjoyed it." However, it is likely to be just that enjoyment—the totality of the experience you provide—that evokes the word of mouth. If you meet the customer expectations outlined earlier and your guests go home wanting to come back, they will tell their friends.

One of the major ingredients of such satisfaction is a feeling of having been well served with person-to-person attention. Personal service is a facet of marketing in which the owner/manager has a golden opportunity. The proprietor who welcomes each guest personally is a refreshing holdover from the past. In today's impersonal world of machines, numbers, computers, self-service, and canned entertainment, people are hungry for conversation and personal attention. The owner/manager, as one of the last of the species of individual entrepreneurs, has a unique chance to fill this need and to sell personality along with refreshment. The warmth of the boss, reflected in the equally warm personal attention of the staff, will give customers something really special to tell their friends about. Don't underestimate its impact; people love it.

Marketing through personal contact. But word of mouth is often capricious and always beyond your control. There is a large market out there that you can approach more directly. One way is to work through personal contact. You can broaden your own contacts, and you can get people you know to bring in people they know.

One way to broaden your contacts is to make yourself and your place known as contributing members of the community. If you support popular community causes—youth groups, church activities, fundraising for charities, and the like—and maybe even assume a leadership role, it can establish you as a respected working member of the community and broaden your acquaintance with other community leaders. Community good will is an important and continuing asset in any business, and particularly in the bar business. If in the process you offer your bar as a convenient place for meetings at quiet hours, this can also contribute directly to your business. But be careful that your community involvement is not simply for personal gain. This motive quickly becomes apparent and people resent it.

Another way of reaching new customers through personal contact is to give a party and ask your present customers to invite their friends. You print up invitations to a celebration of St. Patrick's Day or Greek Independence Day or Halloween and have your customers give them to their friends. This is likely to produce a congenial group at the party. For future use, each invitation should have a place for the guest's name and address. This gives you the nucleus of a mailing list.

Still another idea for building clientele through your customers is to organize games or contests centered around their interests. You might even form a club and give members membership cards to pass out to their friends. Such games as darts, shuffleboard, pool, and video contests are fun both to play and to watch. Interest builds if you form teams or set up tournaments—a good idea for slow evenings. Players may bring in friends to watch and cheer them on. A good crowd attracts additional business as passersby stop in to see what is going on. You do a good business that night, and some of the new people will probably come back. This sort of thing is particularly appropriate for neighborhood bars and other enterprises catering to a clientele seeking entertainment and relaxation.

Another version of the game/contest idea is to sponsor an athletic team that plays in a league of teams sponsored by other establishments. Bowling leagues are a common type; you may also find baseball, softball, touch football, and soccer leagues. A team of your own not only builds business (everybody adjourns to the bar after the game); in addition sports contests of any kind are usually good word-of-mouth generators.

Promoting through paid advertising . . .

Yet there is still a large group of potential customers out there whom you have not reached. There may also be some old customers who haven't been in lately. You need both groups; you want to keep your present customers coming and you want to add new ones. At this point you should think about spending some money on advertising.

Planning an overall program. Planning a total program of repeat advertising over a period of time is likely to give the best results. Before you do anything, however, find out what your state and local regulations allow you to do. Some areas have very specific requirements and prohibitions applying to advertising of liquor and places that sell it.

Your advertising should be targeted at several objectives:

- To create a clear, distinct, favorable image of your enterprise in the mind of the potential patron (often called ***institutional advertising***).
- To make your potential patrons continuously aware of that image (called ***reminder advertising***).
- To produce an immediate visit to your place (known in the trade as ***presell***).

The last two types are the most common and are referred to as ***competitive advertising*** because they are generally aimed at getting the customer to choose your bar over others.

We have talked a good deal about image; let us assume that you have gone to considerable pains to create this image in the bar itself—its ambience, its drinks, its staff, its uniqueness, its identity, its superiority. The object of advertising is to convey this image to potential customers in a way that makes them feel your place would satisfy their needs and desires better than any other place. Not only that; your ads should make them want to visit your place right away, and keep reminding them how much they want to come.

It takes skill and experience to turn out fine-tuned advertising that accomplishes your objectives. A skilled professional who is sensitive to your message is well worth the fee charged. In terms of the added business the fee is minimal.

And speaking of added business, be ready for it. You don't want the competitive advantage you have gained to be frittered away in frantic efforts to cope with a crowd you weren't prepared for. You must fulfill their expectations. If you do, you have a splendid chance of turning them into loyal customers.

The two advertising methods most frequently used are:

- Direct mail—reaching potential customers one by one by letter, brochure, or flier.
- Media advertising—sending out your message to a mass of people by radio, newspaper, or other medium of general communication, and hoping your potential customers will single it out.

Let's look at these methods in turn.

Direct-mail advertising. When you advertise by direct mail you can reach potential customers directly. You can send your message selectively to people living or working in a certain area, or belonging in a certain income level, or having certain interests compatible with your type of place. So the first thing you need is a selective mailing list.

The core of a good list consists of names and addresses you have collected through your own contacts, such as people who have come to your parties. You can add to this by clipping names from newspapers, people who seem to be your kind of customer. Another source is membership lists of clubs, business or social groups, lodges, fraternal orders, and so on. You can sometimes get such lists from customers who are members. Many such groups do not have their own meeting places or bar facilities and might even be interested in using your place for meetings.

You can also buy lists from other organizations. Choose them on the basis of how well they match your clientele and your geographical area. Your local phone company will rent you a directory of subscribers arranged by street address. Credit-card companies are another source of lists. American Express, for example, can sell you lists according to zip code, giving you a specific geographic area and a certain degree of income-level selectivity. Such lists cost about $50 per 1000 names (1980 figure). Your local motor vehicle bureau can supply a list of every car owner in your city for a modest price. For a higher price you can get a more selective list, such as owners of specific late-model cars.

Consider next the purpose of your mailing. You can cover your three objectives with a series of short and simple messages. *Image* will be conveyed by your logo, the look of your piece (paper, color, arrangement on the page), and the way you send it (postcard, first-class letter, hand-delivered flier). Take care that the envelope doesn't look like junk mail. The *reminder* function should be carried out through regular mailings—at least four a year; continuity is important. To get the recipient to *visit soon* (your third objective), offer a reason for coming, such as a special attraction on a specific date or a discount good for a limited time. Such specifics can also help you measure the pulling power of your mailing.

If you send a letter, make it short, informal, and personal in tone. Say something interesting about your place ("We have just added a new game room with a great bunch of video games. . . .") and end with a specific suggestion ("Join us on the 30th—we're having a beer-tasting party").

For certain types of customers a brochure may be useful. For example, you may want to tell business organizations about your meeting facilities and services. Usually a letter addressed by name to the person in charge should go along with the brochure.

Fliers or handbills delivered by hand are often effective if they are eye-catching, brief, and make an offer of some kind. They are an inexpensive and useful way to blanket a certain neighborhood or office building. "Bring in the coupon" will give you a measure of the flier's effectiveness.

Direct mail should be coordinated with the rest of your advertising plan. The style, message, and frequency should reinforce your media ads while taking advantage of the direct and selective approach. And all of it, of course, should be focused on the needs and desires of your chosen clientele.

Media advertising. You have several choices of media in which to advertise your enterprise—radio, newspapers, magazines, television, directories, outdoor advertising. All these expose your message to masses of people, in contrast to the selectivity of direct mail. Hence you must take care to place your message in a medium that your target customers are likely to see or hear, and your message must be designed to catch their specific attention.

In choosing a medium, consider all these things:

- Audience or readership: size and appropriateness.
- Media availability in relation to your needs for coverage and frequency.
- Price, discounts, and cost efficiency of each medium.
- Your ad budget.

In considering cost and budget, cost efficiency must be related to the ***cost per impression***. How many readers does a given newspaper have? How many people listen to a given radio station at a given time of day? A prime-time radio spot, though it takes the most dollars, may reach the most people. This is your cost per impression. However, in any advertising medium, you are paying for many readers, listeners, and viewers who are not potential customers, a fact that makes cost per impression meaningless unless you know how many were impressed. The only way to measure this is a resulting increase in your business—tied, if possible, to a specific offer on your part.

Radio is a medium often chosen by bars and restaurants. It is fairly reasonable, though the dollar cost of prime time is high. You would schedule your advertising to appear during the kinds of programs your target clientele might listen to, at times of day they might be likely to listen.

Newspapers are probably the most popular medium for bar and restaurant ads. A newspaper ad has visual impact, providing another dimension to your message; it is always timely and not outrageously expensive. You may have a choice of morning and afternoon papers; some people use both, so as not to risk missing half their audience. In large cities you can sometimes reach specialized audiences through foreign-language papers. Newspapers have known circulation, and by using coupons you can measure your ad's effectiveness.

The position of your ad in the paper is important. Will it most likely be seen in the entertainment pages where it will be crowded in with all the others? Will it stand out alone if you run it in the sports section, the food section, or the Sunday magazine, or won't it even be noticed there? Sometimes you can attract attention by having it placed beside a feature story.

For example, if you learn that the paper is planning an article on California wines you may be able to tie in an ad on the same page, mentioning the California wines you serve.

National magazines are seldom used unless there is a local edition. Such magazines are expensive and likely to give you limited and overselective coverage. Local magazines are a more promising medium. Each has known circulation and a predictable market, and some have columns on restaurants and night clubs. Such pages might be logical spots for your ads, especially if you are on a columnist's Recommended list. Theater programs are another ad medium worth exploring. Big-city and resort hotels often have giveaway magazines for their guests, which are great places for ads. In fast-growing areas real estate firms sometimes have giveaway magazines for their customers, good spots for your ads.

Television would be a splendid medium if it weren't so expensive. It is likely to be out of reach for all but hotels and chains. Outdoor advertising is very expensive too, and is forbidden by law in many states. Even the sign in front of your premises may be restricted. Check out this form of advertising with care.

Finally there are the Yellow Pages. Few customers will shop for a drink in the Yellow Pages, but it is the best place for making your address and phone number easy to find.

As for the cost of marketing, the crucial judgment is always the relationship between the outgo and the added business, the familiar cost/sales percentage. A more expensive medium may pay for itself several times over if it brings in a flood of new customers, while a low-cost ad in the wrong form and place may bring in no business at all and therefore be very expensive.

On-premise merchandising . . .

Inside your doors, your marketing effort continues. It has several goals: you want your guests to be favorably impressed, you want them to come back, you want them to tell their friends about you. Most of all at this point, you want them to buy your wares.

The guest's impression is a series of impacts on the mind and senses. There is the initial impression of sight and sound, followed by the human impression of staff and guests. There is another impact as the order is taken, still another as the drink is served, and yet another with the first sip. These points of impact offer possibilities for merchandising and salesmanship as unlimited as the imagination and skills of you and your staff.

Merchandising materials. You can make almost anything into a promotion piece if it is (1) eye-catching and (2) informative. Again, imagination is the key. Here are some ideas that have proven successful.

figure 14.2 Table tent featuring take-home-the-glass specialty drinks, Hyatt Regency, Dallas. (Photo by Pat Kovach Roberts)

Displays catch the eye, inform the viewer, and often provoke thirst. A wine rack in your lobby tells the arriving guest the variety and quality of your offerings and sets the mind spinning about which one to order. Labels of your pouring brands of spirits, prominently displayed, look like little coats of arms proclaiming your pedigree. Backbar displays of your call brands and reserves are the classic way of impressing guests.

Drink menus and wine lists are extremely important as merchandisers. They should attract and inform, especially if your offerings are new to the guest. People will seldom buy a drink or a wine they know nothing about—so tell them about it. Menus can also be intriguing, amusing, and appetite-provoking. Wine glasses as part of the table setting suggest wine with the meal.

Specials spark interest and can often focus a drink order. Single them out for special attention on a blackboard in your entry or at the bar, or highlight them on a table tent or a menu clip-on. Figure 14.2 shows a Hyatt Regency table tent featuring three premium-price specialty drinks. Four-color photos and a catchy description of each drink have been selling 100 dozen a month of these take-home-the-glass concoctions.

Many kinds of promotional materials are offered by manufacturers, distributors, and sales personnel, provided state laws allow such giveaways. Posters, backbar signs, ashtrays, coasters, table tents, matches, lapel buttons for servers, and booklets of recipes are some typical items.

Your own matches should carry your logo and maybe a memorable slogan for word-of-mouth advertising. They are long-term reminders of good times and good drinks.

And good drinks are of course the best promotion of all.

Personal selling. The final step on the path to profit takes place at the bar or tableside. Here the marketing effort is in the hands of your staff. You have carefully selected people who are friendly, outgoing, upbeat, alert, articulate, creative, flexible, confident, persuasive, service-oriented, sensitive to people and adept at relating to them—in short, perfect in every way. You have trained them in your routines and taught them everything you know about wines, drinks, and people. Now you are going to teach them how to sell.

The selling the staff does is twofold: they sell your beverages and they sell your place by the way they sell and serve your drinks. Their approach must be customer-oriented: it begins with the customer's needs and desires, and it fills those needs and desires in a way that maximizes both the guest's satisfaction and the profit to the enterprise. The profit goal is both short-term and long-term: maximum present sales (plus a big tip) and repeat customers (plus future tips). Sometimes short-term and long-term profits don't go together; a customer may later regret the amount drunk and the money spent and may never come back. It does not pay to oversell.

For personal selling to be effective it should be:

- *Clear, informative, factual.* The staff should be able to explain clearly the ingredients in mixed drinks, the house specialties, the tastes and pedigrees of wines. Servers should stick to the facts and present them objectively. Predictions are risky: "Try it, you'll like it" puts you out on a limb. If the customer doesn't like it you've lost your credibility and probably another sale. "It's one of our most popular drinks" is factual and makes the same point.
- *Persuasive.* The best persuasion uses suggestion and reasoning, never pressure or manipulation. You can offer, inform, suggest, explain, excite interest. Usually customers react positively if their needs and desires have been correctly perceived, but they resent being pushed.

Good staff members develop their own successful sales techniques. But sometimes even the best techniques meet sales resistance. There are two kinds of resistance—logical and psychological.

Logical resistance objects to a suggestion on its merits. Maybe the price is high, the brand is unfamiliar, or the customer doesn't like the drink ingredients. Be prepared with alternatives.

Psychological resistance is far more complex. One has to look beyond what the customer is saying to behavior and feelings. Resistance to change, for example, may cause a customer to turn down a new imported beer. It isn't the beer the customer is resisting, it's the change from the familiar. Another kind of resistance is the satisfaction some people get out of being wise guys and tough customers. All these forms of resistance should be respected. It is far more important for the customer to get what is wanted than to sell a high-profit item or a house special. You want the customer to feel taken care of, so take care.

The taking of an order and the serving of the drink are the culmination of this final phase of marketing. The sum of all these individual encounters adds up to the success of your enterprise. Only if each sale is a mutually beneficial experience will the success be lasting. The talents and skills of your staff are the final critical link in your marketing.

Good sales personnel are also good sources of ideas for specialty drinks, promotions, selling techniques. They can tell you about trends in customer tastes and sometimes what your competition is doing. A good staff is an invaluable resource, deserving praise, encouragement, and even occasional pampering.

Pricing as a merchandising tool . . .

We have already seen pricing in several different roles. Earlier in this chapter we noted that the general price level will be very influential in determining the nature and numbers of potential customers and the image of an enterprise. In general, the higher the prices the smaller the group, the higher their income level, and the more exclusive the bar's image. In Chapter 12 we examined the most basic role of pricing, the financial role of producing income when prices are multiplied by volume. We also noted there the effect of price on demand, and herein lies its merchandising role: reduce a price and you increase demand.

You can use the lowering of prices as a short-range tactic for several purposes. One is to draw customers away from other bars or to hold the line against other bars' efforts to draw customers away from you. Another is to attract new customers with special advertising offers. A third way is to lower prices as a promotional effort in place of spending money on ads or other forms of promotion. All these can be effective merchandising tools in even the most prestigious of bars. On the other hand, if you offer specials all the time, customers may think your regular prices are inflated.

The most familiar and most common use of pricing as a merchandising tool is the ***happy hour***. Its usual format is a time period—generally in the late afternoons or early evenings—when the drinks are reduced in price or two drinks are offered for the price of one. There are many variations in hours—even late-night or after-midnight happy hours—and many variations in pricing. One such variation used successfully for a time at Daddy's Money featured a time clock: the earlier the customer clocked in, the lower the price of the drink.

The happy hour as an institution has enjoyed a long but not always happy life. On the plus side it can enable an enterprise to attract new customers. If they are impressed with the overall ambience, they may come back during regular hours, or they may stay on and pay the regular prices. Another stroke will have to go on the plus side for hanging onto one's clientele in the face of happy hours in other bars all over one's neighborhood. There are places that cannot afford to be without a happy hour.

In many instances the happy hour itself is not a profitable period; one simply breaks even. One way to improve this is to drop prices only on certain drinks and to discreetly encourage the buying of regular-price drinks. Another way is to serve larger-size drinks for the usual price during happy hour; you usually come out with a better profit margin. The whole happy-hour price structure deserves careful attention and analysis.

Because prices are directly related to image, the overall level of prices should be kept stable. When increased costs force you to raise prices, set them high enough so that you won't have to raise them again soon. Customers can take an occasional change in stride, but several raises in a row can raise temperatures as well.

Above all else, customers must feel that the value they receive from you is equal to or greater than the prices they pay—a fair deal, their money's worth. It does not matter what the actual costs and money values are. If they think they are paying too much for what they are getting, you have set your prices too high, because they won't be back.

The marketing role of pricing must not interfere with its financial role. The two must continue together in a coordinated relationship; both are important to profitability.

Summing up . . .

Since customers are the marketing goal, they must also be the starting point of any marketing effort. Once you have identified a clientele you know are out there, your entire enterprise should be shaped to their needs, desires, and expectations. Create an ambience they respond to, serve drinks to their liking, and make their total experience memorable, so they'll come back and bring their friends.

Advertising and promotion are ongoing parts of marketing. An overall plan, with the media and the message directed to your chosen clientele, should project your distinctive image and create a desire to visit your place. On premise, the marketing focus shifts to sales. Displays, special drinks, and special prices are good merchandising tools, but the real merchandisers are your serving staff, with personal attention and interpersonal communication in first place. To sell the product, serve the customer!

FEASIBILITY STUDY FOR A HYPOTHETICAL BAR

1. Amount of investment planned	$160,000
2. Return on investment desired	25% ($160,000 × 0.25 = $40,000)
3. Profit goal	20%
4. Sales needed at that profit to generate desired ROI	$200,000 ($40,000 ÷ 0.20 = $200,000)
5. Average drink price	$2.50
6. Number of drinks needed to generate dollar sales needed	80,000 ($200,000 ÷ $2.50 = 80,000)
7. Average drinks per customer	2
8. Customers needed per year	40,000 (80,000 ÷ 2 = 40,000)
9. Days open per year	360
10. Customers needed per day	112 (40,000 ÷ 360 = 112)

Conclusion: With 112 customers per day consuming an average of 2 drinks each, the operation is feasible because:

112 × 2 = 224 drinks per day

224 × 360 days = 80,640 drinks sold

chapter 15 . . .

REGULATIONS

There are many many federal, state, and local laws that apply to the sale of alcoholic beverages to be consumed on the premises. Some of these laws you are surely aware of; others might never cross your mind. Many can have serious consequences if you break them. It is crucially important for you to know and observe all laws that apply to you.

The purpose of this chapter is to inform you of federal regulations and to make you aware of typical state and local beverage laws. Since these vary greatly, it is up to you to find out what laws are in force in your locale.

It is often pointed out that the alcoholic beverage industry is the most regulated industry in the United States. Regulations of one sort or another can affect where you locate, what you serve, whom you serve, how many you serve, the days and hours you stay open, where you buy your supplies, what records you keep, when you pay your bills, whether you can serve liquor with food or food with liquor, your advertising, your reputation, what you do with your empty bottles, and your ability to do any business whatsoever.

Why so? Two blanket reasons: public revenues and public interest.

There is nothing new about either reason. Riotous drinking was the subject of legislation in Babylonia 4000 years ago, and it was the tavernkeeper who was punished for allowing it by losing a limb or even life itself. In colonial America laws restricted hours of sale and prohibited selling to minors, drunks, Indians, servants, and slaves. The tavernkeeper was selected and approved by the town fathers, and the tavern was hedged about with rules for keeping it a sober and respectable part of the community.

As for revenue, licensing fees brought income to early colonial towns and cities. British taxes on Dutch rum helped to bring on the American Revolution. A tax on whisky caused the Whisky Rebellion in 1794. When Washington crushed the rebellion he established firmly the right and tradition of federal taxation of liquor for general revenue. In 1980 taxes of $10.50 per proof gallon on beverage alcohol brought nearly $4 billion to the federal government, while state and local governments collected nearly $3 billion. The retail price of a typical bottle of spirits is more than 50% taxes and fees paid into federal, state, and local treasuries.

Today's public-interest regulations date from the repeal of Prohibition, and they reflect the earnest desire of citizens and governments to prevent both the excesses of the pre-Prohibition era and the abuses by the illegal industry during the Prohibition years. Some regulations also represent that "cultural hangover" from the temperance movement, reinforced by genuine concern over the rise of alcoholism and accidents from drunken driving. Some are considered unreasonable by the beverage industry, since they sometimes act as a straitjacket to development and a deterrent to profits. But whether they help you or hem you in, you are urgently advised to know and observe every regulation—federal, state, and local—that applies to you.

What you must do before opening your doors . . .

If you own a bar or restaurant serving alcoholic beverages, you are defined in legal terms as an ***on-premise retail dealer.*** In this role you are subject to federal laws, state laws, county laws, and municipal laws, as well as zoning ordinances, building codes, health codes, and fire codes. State and local laws vary widely and there is no way in the world this book can cover them all. But we can indicate what kinds of restrictions are typical and what you ought to check out before you take action.

Both you and your premises must meet several requirements in order to open at all. Some of them involve a good deal of time and legwork. You may have to attend a city council meeting, go before a judge, post a notice in a newspaper, be fingerprinted—whatever the state and local laws require. The first thing to do, even before you sign a lease or buy a property, is to check out all the state and local regulations. Your local licensing agency will be able to provide them.

Regulations that affect your choice of location. In Oklahoma, state law does not allow the sale of mixed drinks at all. In Kansas and West Virginia only private clubs may serve mixed drinks. Oklahoma and Colorado limit retail beer sales to 3.2% beer. Many states limit the alcohol content of malt liquor.

Most states have ***local option*** laws—that is, the state allows the people of local communities to choose whether or not they will allow the sale of alcoholic beverages, and if so how. Thus, although state law may allow liquor to be sold by the drink, some communities may not—even different parts of the same city or county. If your city or county is one of these, it behooves you to check carefully where the lines of demarcation are drawn. You must also find out what beverages are permitted in your locale, because there are often different regulations for spirits, beers, and wines.

Even in areas that allow liquor by the drink there may be zoning laws that rule out operating a bar in your chosen location. For example, if you want to remodel an old house, the area may turn out to be zoned only for residential use. In other instances an area zoned for commercial use may specifically prohibit retail sale of liquor.

Many city ordinances have parking-lot requirements—a parking space for every so-and-so many square feet of customer floor space. Is there room on your chosen premises to meet the requirement?

In many areas you cannot locate a bar within a certain distance of a church, school, or hospital—say 300 feet or a city block. The way the distance is measured—lot line to lot line or door to door—may disqualify the premises you are planning to use.

Some local authorities limit the number of licenses they will grant. Other local regulations may so restrict the way you can operate that you may decide to move somewhere else. Some areas restrict hours of sale more than others. In Oregon local communities may prohibit entertainment and dancing. In some places you *must* serve food with drinks; in others you *can't.*

Licensing and registration requirements. Once you have determined that your chosen location meets both your requirements and the law's, you must secure the required state and local licenses and permits and register your place of business with the federal government before you open your doors.

A ***license*** and a ***permit*** for on-premise retail sales are essentially the same thing—a document granting permission to sell specific beverage types at a specific location by a specific business entity provided certain conditions are carried out. In some jurisdictions the two terms are used to distinguish between various types of permission—for example, a permit to sell wholesale and a license to sell retail, or a permit to sell mixed beverages and a license to sell beer (or vice versa).

The federal government does not issue licenses or permits for retail sales. Jurisdiction over retail sales is reserved for the state. The state may in turn pass on the right to local bodies. The intricacies of state and local laws are so complex and vary so widely there is no way to indicate here what licenses and permits you will need in any given location in order to sell what you plan to sell. The licenses vary; so do the requirements you must meet, the procedures you must go through, and the fees you must pay.

In some places you must obtain a state license, in others a local license, and in still others both. In some states a state license must have local approval; in other states it is the other way around. In some areas you must have separate licenses or permits for spirits, beers, wine, and mixed drinks. In some areas—notably the states of West Virginia and Kansas—you can get a license to sell spirits only if you operate a private club. In a few places you must sell your drinks by the miniature bottle. In Utah you cannot get a license to sell spirits at all; you must arrange to have a state store on your premises, which will sell liquor to your customers in miniature bottles. You sell the setups and the service.

Licenses and permits typically run for a year, requiring annual renewal and payment of an annual fee. Such fees vary widely. Some are flat fees. Others may be tied to sales volume, or number of rooms in a hotel or seats in a restaurant, or even to the population of the issuing community. Amounts vary from as little as, say, $25 for a beer permit to $2500 or more for a mixed-beverage license, Honolulu County, using a percent-of-sales system, sets its ceiling at $7500.

In some areas a separate operator's license may be required in addition to the license for the business or premises. This may have such requirements as age (over 21), character ("good moral character"), U.S. citizenship, or residency of a certain length of time in the city or state.

The penalties for selling alcoholic beverages without a license are severe. In fact, you could put yourself out of the bar business forever.

When you have your state and local permits and licenses in hand or are sure you are going to receive them, it is time to turn your attention to some federal requirements.

Federal law requires a retail dealer in alcoholic beverages to register each place of business owned and to pay a ***special occupational tax*** for each place of business each year. In return the government issues a special ***tax***

stamp for each place of business. If you as a retail dealer are serving wine or spirits or both, your tax for each location is an annual $54; if you serve only beer it is $24 per location per year. If your business is an airline, a passenger train or boat, a circus, a carnival, or some other enterprise traveling from place to place, you are considered a ***dealer at large*** and your stamp and tax ($54) cover all your activities throughout the United States.

The vehicle for registering, paying the special taxes, and receiving your stamps is IRS Form 11, which you may get from any Internal Revenue Service office. You must file this form and pay your tax before you begin selling any drinks, and each year thereafter. You must keep your special tax stamp and your tax receipt available for inspection by officers of the federal Bureau of Alcohol, Tobacco and Firearms (BATF). If you deliberately sell alcoholic beverages without paying the special tax you may be fined up to $5000, or imprisoned for up to two years, or both.

If you move your place of business you must file an amended return (another Form 11) within 30 days, and send in your tax stamp to be amended accordingly. You cannot transfer your registration or tax stamp to anyone except your spouse. If you sell your business the new owner must start over with the IRS and new stamps. If you buy someone else's business you must start fresh. State and local license regulations have similar provisions for change of ownership. It is government's way of providing accountability.

The federal registration and tax payment do not permit you to sell liquor; they just prohibit you from selling liquor without them. Registration is mainly a mechanism for keeping track of retail liquor dealers in order to enforce some of the public-interest laws having to do with the liquor itself. We will discuss that a little later.

Other local regulations. Your premises must meet the standards required by various other local codes.

If you build or renovate a structure the finished work must pass the building-code inspection. Electrical work and plumbing must meet electrical and plumbing codes. The entire facility must meet the code of the fire district. This will include the requisite number and placement of exits, lighted exit signs, fire extinguishers, fire doors with panic bars that must open outward and be kept unlocked during business hours, sprinkler systems in kitchens and high-rise buildings. There is also usually an occupancy requirement limiting the number of persons allowed inside the premises at one time. There will also be specifications about employee training and fire drills.

When you are all set up ready to open, your facility and personnel must pass inspection by the local health department. Usually you must have certain specified equipment such as a triple or quadruple sink or mechanical dishwasher and hand sink. All equipment must be NSF-approved and must be installed according to health department rules. For example, your ice machine and other equipment with drains must either be sealed to the floor or have a minimum of 6 inches of space below allowing for adequate cleaning. Other requirements pertain to floors (tile or concrete), counters, general cleanliness, storing food properly, proper temperatures for refrigerators and freezers, freedom from pests (rodents and roaches), sanitary practices such as keeping everything but ice out of the ice bin, air-drying glasses, keeping bathrooms clean and stocked with soap, towels, and tissue. You may not open for business until you have a certificate of occupancy.

Health departments typically require that employees pass a health examination and that you keep their health certificates available for inspection. They also require personal cleanliness. Generally smoking behind the bar is a violation of the health code. Health inspectors usually make periodic visits to see that you are continuing to meet their requirements.

What, when, and to whom you may sell . . .

Generally the regulations governing drinks, hours, and customers are state or local and there is considerable variation. The only federal requirement is that you may not sell any type of beverage for which you have not paid the special occupational tax, nor any spirit not carrying a portion of the federal strip stamp or other approved seal on the bottle.

What you may sell. What you may sell is governed everywhere by the type of licenses you hold (mixed drinks, wine, beer) and special restrictions of the kind we have already mentioned, such as minibottles in place of custom-mixed drinks or beers of no more than a certain strength. In some states you cannot buy a spirit stronger than 100 proof, which means you cannot make a bona fide Zombie or serve a straight shot of Wild Turkey 101-proof bourbon.

Some state and local laws say you may serve liquor *only* with food; others specify that you must *not* serve liquor with food. Thus Colorado stipulates that hotels, restaurants, and taverns may serve drinks only with food, and in Virginia your food sales must exceed your beverage sales. In Kansas, on the other hand, liquor may be sold only by the package in private clubs that are physically separated from food-service areas.

When you may sell. Nearly all states have regulations governing the hours an on-premise retail dealer may be open. Generally these hours are fairly generous. Most states require closing for at least a few early-morning hours, and most require Sunday closing for part or all of the day. Many states prohibit the sale of liquor on Election Day, at least while the polls are open, and many specify closing on specific national holidays, Christmas being most commonly mentioned. Many states allow local bodies to further curtail hours and days of service.

To whom you may sell. All states and the District of Columbia prohibit sale of spirits to minors; most prohibit sale of beer and wine to minors as well. (In Wisconsin you can sell beer to any minor accompanied by a parent or guardian.) The age identifying a person as a minor varies from 18 to 21 in different states. Some specify a lower age for beer (usually 18) than for spirits, and sometimes for wine as well. Recently several states have passed laws raising the drinking age.

The term ***minor*** used in the context of selling alcohol refers specifically to beverage laws and not to any other legal meaning of the term. An individual may be of legal age to marry without parental consent, sign contracts, rent an apartment, and run a business, yet still be defined as a minor when it comes to buying a drink.

It is up to you as the seller and all your personnel who serve drinks (and by the way, they too *must* be of legal age) to judge whether each prospective customer is of legal age. If you serve a minor, it is generally you and not the minor who is held responsible. You may be subject to such penalties as fines, loss of license, or a short jail term.

If there is any doubt at all about a customer's age, the server should ask for proof of age, such as a driver's license, voter registration card if the legal age in your locality is 18, passport, birth certificate, or special ID card issued by state or local government authorities. Some proof-of-age documents are easy to fake and even easier to borrow from a friend, so they must be examined closely. Many laws specify a document with a signature, description, or picture. If the picture, description, and signature fit the individual, you as the seller are usually (though not everywhere) presumed innocent in case of trouble. The law sometimes contains a key word for your protection: you may not "knowingly" sell to a minor. However, you are entitled to refuse to serve anyone you may suspect.

The other category of persons you may not sell to is the person who is clearly intoxicated. Many if not most states and communities have this law on their books. It protects the individual concerned from being taken advantage of by an unscrupulous seller, and it may incidentally save him from a horrendous hangover. It also protects your customers from the unpleasant antics of the overimbiber. Most important, it protects those innocent bystanders who might be killed or injured by a drunken driver headed for home or for another bar. Some laws include the habitual drunkard (even if sober) and the insane in their ban.

It is a delicate matter to refuse to serve a customer; you risk a scene and the embarrassment other customers may feel. But it must be done. Some laws require you to remove an intoxicated person from your premises entirely, including your parking lot.

It is, of course, against the law to refuse to serve a customer on the basis of race, creed, color, sex, or national origin. If you are dealing with an intoxicated person who happens to be the subject of such laws, you will be wise to make sure you have witnesses to the intoxication in case the person concerned charges discrimination.

In recent years the growing numbers of drunk-driving accidents have added another dimension to the legal consequences of serving an intoxicated person. Because of widespread public concern over the menace to innocent victims, a number of states have laws known as Dram Shop Acts. Under these laws a person who sells or gives liquor to an intoxicated person or a minor can be held liable for damages caused by that person's drunkenness. This concept is known as ***third-party liability.*** It is a reversal of the old common-law concept which held that the drinker, not the seller of the drink, is responsible for damages the drinker does to others. That view of the seller stemmed from the early days when taverns were required by law to provide food, lodging, and refreshment to travelers—days when alcoholic beverages were considered a necessary part of a meal.

Dram Shop Acts not only prohibit the sale of liquor to minors and all visibly intoxicated persons; they encourage large damage suits against the seller of the liquor. The Stacy case in California is probably the most famous of these suits. In this episode actor James Stacy, while driving a motorcycle, was hit by a drunken driver. His passenger was killed and he himself lost an arm and a leg. He sued the Chopping Block bar, which had served the driver the offending drinks, and was awarded $1.9 million in damages.

Following this award the California Restaurant Association lobbied successfully to have the California law modified. But the moral is clear: never serve a customer too many drinks. The trend of the law is to hold you responsible for what that customer does.

Regardless of the law, it is a basic principle of responsible bar management to allow no one to drink more than he or she can handle. But it is difficult to translate that into numbers of drinks. The National Safety Council measures intoxication by levels of alcohol in the bloodstream. Here are the standards they use in drunk driving cases:

0.00 to 0.05% (1–2 beers or 1–2 ounces whisky): not under the influence.

0.05 to 0.15% (3–8 beers or 3–8 ounces whisky): possibly to probably intoxicated.

Over 0.15%: under the influence of alcohol.

Over 0.20%: definitely intoxicated.

Over 0.30%: seriously intoxicated.

Over 0.40%: fatally intoxicated.

These measures are necessarily arbitrary, since individuals and conditions vary, and it would be hard to translate them into the drinking any one person might do at your bar. But consider them flags of caution, and make rules for yourself and your staff, such as:

- Refuse to serve another drink to anyone you or your employees think has had enough. Offer a cup of coffee on the house.
- Train your bar personnel how to anticipate problems and how to handle troublesome customers.
- See that an intoxicated person has someone else to do the driving.
- If all else fails, call for help—the police if necessary.

There are other things related to drunken customers for which you may be liable under various laws. Among them are allowing injuries to take place on your premises, even when inflicted by other customers (stop the fight), allowing an intoxicated customer to drive away from your place (call a cab), and using more than "reasonable" force to eject an intoxicated person. It is a good idea to examine your liability insurance policies carefully to see that they do not exclude or limit coverage relating to the sale of alcoholic beverages.

Regulations that affect purchasing . . .

Federal, state, and local regulations may affect your purchasing both directly and indirectly.

Where you can buy. As we mentioned in Chapter 11, in 18 states the sale of distilled spirits is a state monopoly, and bar operators must buy from state-owned stores. Prices are the same in all state stores, since there is a markup formula mandated by the state. The markup plus state taxes ranges from 33.9% in Mississippi to 94% in Oregon. Discounts to on-premise retail dealers range from nothing at all to 17% (Michigan). Thus, although manufacturers' prices may be the same everywhere, the price to the bar operator varies from one state to another. Nevertheless you must buy within the state.

In the remaining states, bar operators can buy only from suppliers who have the required licenses and permits—state, local, and federal. The federal document is known as a ***wholesaler's basic permit*** and is free; the wholesaler must also have paid a federal special occupational tax. In some states you can also buy directly from state-licensed manufacturers or distributors. In Texas you can buy only from retail package stores that also hold a federal wholesaler's permit. This came about because the package stores opposed legalizing liquor-by-the-drink because it would cut heavily into their business; before that the customer took a bottle of liquor in a brown bag and the bar host provided the setups and ice.

In most license states the state does not control prices or markups. They may vary from one supplier to another, though the manufacturer or distributor price must be the same to all wholesalers. Most license states require manufacturers and wholesalers to file brand and price information with the state beverage-control agency and to post prices and discounts. Publication of suppliers' prices is usually required. The overall objective is to avoid price wars and monopoly-building of the kind that caused the beverage industry to degenerate in the years before Prohibition. Many state laws specify that manufacturers' prices must be "no higher than the lowest price" offered outside the state. This gives the producers' price structure a certain homogeneity countrywide, varied somewhat by the amount of state and local taxes and fees.

Bottle sizes available. Federal regulations and some state laws specify the size containers that may be sold. State laws follow federal specifications, but in a few states certain bottle sizes are not available at the large and small ends of the scale (gallons, miniatures, tenths, half pints). Malt-beverage container sizes for sale (kegs, half-kegs, and so on) must also meet federal and state requirements. Federal regulations have recently been liberalized to allow 7-, 8-, and 10-ounce cans and bottles in addition to the 12-ounce size.

Credit restrictions. Nearly all states restrict credit given by the supplier to the on-premise retailer, requiring either cash or payment within a specific time period, usually 30 days or less. Extensions of credit are prohibited in many states, and federal laws prohibit extensions of credit beyond 30 days on goods figuring in interstate commerce.

Relationships with suppliers. State laws regulate or prohibit certain interrelationships between on-premise dealers on the one hand and wholesalers, manufacturers, importers, and distributors on the other. The purpose of these laws is to maintain fair competition and prevent retailers from being controlled by other segments of the industry as the saloons were controlled by brewers in the old days.

Here are some of the most important prohibitions typically found in state beverage codes:

- No ***tied-house*** relationships. No supplier may have a financial or legal interest in your business, premises, or equipment.

- Suppliers may not furnish equipment or fixtures.
- A supplier may not pay your debts or guarantee their payment.
- A supplier may not sell to you on consignment—that is, postpone payment until goods are sold, with return privileges for goods unsold. This is also a federal law.
- Suppliers may not give you special discounts that are not available to all on-premise dealers.
- Suppliers may not give you gifts, premiums, prizes, or anything of value except specified items of limited dollar value for advertising or promotional purposes, such as table tents or a beer sign for your window, or consumer giveaways such as matches, recipe booklets, napkins, coasters.
- A supplier may not induce you to purchase all or a certain quota of your beverage supplies from his or her enterprise.
- Bribery for control of your purchasing carries severe penalties.

In sum, suppliers may not own a piece of your business nor *induce* you to give them your trade. You too are breaking the law if you accept such inducements. On the other hand, there is no law against giving most of your business to one supplier if you make the choice without inducement.

If you are tempted to accept deals and favors—and who doesn't love to get something for less or for nothing?—check out your state and local laws to be sure the deal or the favor is legal and has no strings. Laws that maintain a competitive beverage market, though framed primarily in the public interest, protect you as an entrepreneur as well.

Regulations that affect operations . . .

In addition to regulating your buying and selling, other regulations—both federal and state—touch some of the smallest details of your beverage operation.

Policing the product. During Prohibition the illegal production and sale of alcoholic beverages was so widespread, and so much of the stuff was harmful and even deadly, that it was a major reason why Prohibition was finally repealed. It had been impossible to control illicit production. The only answer was for the government to work hand in hand with a legal beverage industry and to set up controls to protect the public from unscrupulous producers and bad products.

Accordingly the federal government established Standards of Identity for each product, plus systems of inspection and control at licensed distilleries, breweries, and wineries to see that the Standards were met. Today federal regulations assure you, the customer, that the product inside the bottle is exactly what the label says. Labels must correctly state the product class and type as defined in the Standards of Identity, its alcoholic content (except for beer), the net contents of the bottle, and the name of the manufacturer, bottler, or importer. Periodic on-premise inspections by federal agents of the BATF assure that contents and label agree. Bonded warehouses under lock and key are further controls against substandard merchandise. Imports are controlled in similar fashion through customs regulation, inspection, and labeling requirements.

Federal control of the product extends right into your storeroom, speed rack, and backbar via the strip stamp and certain other regulations.

The ***strip stamp***—that red or green stamp spanning the top of each bottle of distilled spirits—is a key control in three federal government activities:

- Collecting revenue from distillers.
- Maintaining product quality.
- Preventing the sale of illegal spirits.

The strip stamp goes on the bottle at the time of bottling, when taxes are paid and the bottle, its contents approved, leaves the warehouse. The stamp—green for 100-proof Bottled in Bond straight whisky, red for all other spirits—is affixed in such a way that the bottle cannot be opened without breaking the stamp. That stamp must remain attached and unbroken until you open the bottle at your bar. It is your assurance that no one has tampered with its contents; they are exactly as they were when they left the distillery.

If a delivery from a supplier includes one or more bottles having broken or mutilated stamps or no stamps at all, you should not accept such bottles. It may be an accident, but it can also mean that the original contents have been altered and you are not getting what you ordered. Besides, possession of bottles without stamps is against federal law. Substitute stamps are available from BATF, but let your supplier cope with getting new stamps. Send the bottle back.

When you open a bottle you must leave a portion of its stamp attached to the bottle, as in Figure 15.1. You must leave the liquor in the stamped bottle and not transfer it to any other bottle. Make this very clear to all your personnel. If you have unstamped liquors on your premises you are subject to a fine of up to $10,000 and/or imprisonment for up to five years. Again, a bottle without a stamp may be an accident, but if an inspecting federal agent should find it, it puts you under suspicion of having bought illicit goods.

The federal government has recently approved some alternatives to the strip stamp, such as certain pull tabs and other devices that make clear

figure 15.1 Strip stamps before and after opening bottle. (Photo by Pat Kovach Roberts)

whether a bottle has been opened. One of them is a screw top that leaves a metal ring around the bottle neck when you open it, similar to the closure on many brands of scotch. The ring must remain on the bottle, like the portion of the strip stamp. On domestically bottled liquors this type of closure plus a code number on the bottle will make a strip stamp unnecessary.

For you, a broken or missing stamp on a full bottle may be a clue that someone having access to your liquor supply has substituted something else and gone off with your good liquor. So the strip stamp can work for you as well as for the government.

It is also against federal law for you—for anyone—to reuse an empty spirits bottle for any purpose whatever. The primary thrust of this law is to prevent illegal distillers from filling such bottles with their product and passing them off as the real thing. But it is just as illegal for you to use an empty liquor bottle for your simple syrup or your sweet-and-sour mix.

Federal law does allow you to return empties to a bottler or importer who has federal permission to reuse them, or to destroy them on your premises, or to send them elsewhere for destruction. But many state or local laws require spirits bottles to be broken as soon as they are empty. Moonshining and bootlegging are far from dead, and both federal and state beverage-control authorities are actively engaged in rooting it out. A recent federal-state crackdown in the Virginia hills destroyed 97 stills capable of producing 8000 gallons of moonshine every four or five days.

There is also a federal law against adding anything—other spirits, water, any substance—to the original contents in a bottle. This forbidden practice is known as ***marrying***. You may not even combine the contents of two nearly used bottles of an identical product.

In some states your liquor bottles may carry an additional state stamp. It may show that state taxes have been paid, or it may have some other identification purpose. These state stamps too should be left on the bottles.

Records and inspections. The federal government requires you to keep records of all spirits, wines, and beers received—quantities, names of sellers, and dates received. You may keep this in book form, or in the form of all your invoices and bills. State and local governments may require you to keep daily records of gross sales, especially if you must pay sales taxes or collect them from the customer. You must keep all your records for at least three years.

Your place of business, your stock of liquors, your records, and your special tax stamp and receipt are subject to inspection at any time by BATF officers. This is not likely to happen unless there is some reason to suspect you of breaking federal law, or unless they think you might have some evidence that would help them in tracing someone who has broken the law.

Your licenses and permits, your records, and your entire operation are also subject to inspection by state and local officials if your state and local codes so state—and they probably do.

All such officials carry identification which they must show to you. Once you have verified their identity you must show them anything they want to see.

Other state/local regulations. If there are state or local sales taxes, you will undoubtedly be required to collect them from your customers. For this you must keep daily records of sales of taxable items, with separate records for each category of taxables (beer, wine, spirits, mixed drinks) if the tax rates distinguish between them. The simplest way to keep these records is to have a cash register with a separate key for each category. Each category will be totaled separately, and you will have a printed record on the tape.

You will be required to make sales-tax reports and payments at regular intervals, and your records will be subject to audit. You must keep your daily records for however long state/local laws specify.

Another frequent provision of state beverage codes has to do with the credit you may extend to customers. Some states forbid it entirely, requiring that all drinks be paid for in cash. Others permit customers to charge drinks to their hotel bills or club membership accounts or to pay by credit card. Still others have no credit restrictions. In Wisconsin consumer liquor debts are uncollectable under law.

Nearly all codes set limits on advertising involving alcoholic beverages, and the federal government has stringent regulations designed to prevent false, misleading, or offensive product advertising. Most states follow suit. Code provisions that might affect a beverage-service enterprise concern billboards advertising your bar or restaurant, content of your magazine and newspaper ads, signs identifying your place or your products, window displays. Many beverage codes forbid ads associating your products with provocative women, biblical characters, or Santa Claus.

A final category of regulation has to do with keeping your enterprise an acceptable member of your community. Here are some kinds of things for which, under some state or local laws, you may be liable for fines, imprisonment, or loss of license:

- Possessing or selling an illicit beverage.
- Permitting lewd, immoral, or indecent conduct or entertainment on your premises (exposure of person, obscene language, disturbing display of a deadly weapon, prostitution, narcotics possession or use).
- Disturbing the peace.

Some final words of wisdom . . .

In the foregoing discussion we have focused on the beverage laws that apply to you as a dispenser of refreshment to thirsty patrons. Regulation, of course, encompasses the entire industry: no aspect of manufacturing, importing, selling, or advertising escapes the scrutiny of at least one level of government. But that side of the story is beyond the scope of this book.

If you find the laws bewildering, contradictory, frustrating, and time-consuming, you are not alone. It helps to see them in the perspective of history and to realize that the seeming chaos simply grew that way in response to differing local needs and desires and a good bit of emotion and politics along the way. It also helps to realize that frustration and contradiction are not forever but can be amended through education, the political lobby, and the voting machine.

It helps, too, to realize that many of the regulations benefit and protect *you.* Thanks to regulations you have a product of guaranteed consistency and an environment of free and fair competition. Within the framework of regulation the beverage industry is taking its place as an essential and respected element in the economic and social fabric of America. When it comes to profitability, that is worth a lot in goodwill.

So inform yourself of your local laws, mind your Ps and Qs, pay your beverage taxes (budgeting them firmly into your profit plan), and comfort yourself with the thought that beverage laws are necessary precisely because the demand for your product is inexhaustible.

Summing up . . .

The sale of alcoholic beverages is heavily regulated in many places, and the beverages themselves are heavily taxed. There might be even more regulation if alcohol wasn't such a good source of revenue for governments on all levels.

The observance of existing laws, however absurd some of them may seem, is important for two reasons. One is for your own immediate good, and the other is for the good of the industry as a whole. If it is to maintain a respected place in American life, we must all be good, law-abiding citizens.

The Dram Shop laws are a reminder that the dual nature of alcohol is always with us and that we operate within the shadow of its dark side. Whatever the laws and the prevailing legal philosophy, let us not make our profit at anyone's expense.

dictionary of drinks . . .

This dictionary is intended as a springboard for developing your own custom-made recipe file. The list includes what we consider to be today's popular drinks. It also includes the drinks mentioned in the text, some historic drinks anyone associated with a bar should know, and a few of our local favorites. It is not intended to be an authoritative list, and certainly not an exhaustive one.

The recipes follow the format used in the text. The ingredients are listed in order of use. The mixing method is given in a word or two at the beginning: Build, Shake (meaning by hand or with shake mixer), Stir, Blend, Shake/Build; for expanded definitions see Chapters 9 and 10. The glass specified is what we have found to be the current (1982) mode but, in an industry that thrives on creativity, this is only a guideline. It does, however, indicate a quantity relationship with the drink ingredients that you should consider when substituting another glass.

In adapting these recipes to your clientele, we recommend you taste-test our proportions and modify them for your standardized version to suit your patrons' tastes, while retaining the basic character of the drink. In most cases we give you a choice of fresh ingredients or mixes. For 1 teaspoon of sugar you can use ¼ ounce of simple syrup. For a teaspoon use a barspoon. Ingredients in parentheses are optional.

In developing your own recipe file keep your eye out for enticing recipes that often appear in such trade magazines as *Restaurants & Institutions, Hospitality*, and *Restaurant Business*. Another source is the free recipe booklets put out by distillers, available through your wholesaler or by writing to the manufacturer. For creating your own special drinks, Chapter 10 contains some suggestions for success. It is our hope that every bar manager or owner will create his or her own specialties. These are fast becoming today's hallmark of a place that is pouring for profit.

Acapulco — Shake or Blend

Cocktail glass, chilled, or
Rocks glass filled with cube ice
1 jigger light rum or tequila
½ jigger lime juice
½ jigger triple sec
½ tsp sugar or ¾ tsp simple syrup
(Fresh mint leaves), (dash pineapple juice)

Aggravation — Build

Highball glass
¾ glass cube ice
1 jigger scotch
½ jigger Kahlúa
Milk to fill

Alexander Shake or Blend

Cocktail glass, chilled, or
Rocks glass filled with cube ice
¾ jigger brandy or other liquor of choice
¾ jigger crème de cacao (light with gin, vodka, tequila, rum; dark with brandy, whisky)
1 jigger cream

Alice in Wonderland (*Dallas Alice*) Build (*pousse-café*)

Straight-sided pony or cordial glass
⅓ Tia Maria
⅓ gold tequila
⅓ Grand Marnier

Americano Build

Rocks glass
Full glass cube ice
1 jigger Campari
1 jigger sweet vermouth
Lemon twist or orange slice

Americano Highball Build

Highball glass
¾ glass cube ice
1 jigger Campari
1 jigger sweet vermouth
Soda to fill
Lemon twist or orange slice

Angel's Kiss Build (*pousse-café*)

Straight-sided pony or cordial glass
¼ dark crème de cacao
¼ crème d'Yvette
¼ brandy
¼ cream

Angel's Lips Build (*pousse-café*)

Straight-sided pony or cordial glass
⅔ B & B
⅓ heavy cream

Bacardi Shake or Blend

Cocktail glass, chilled, or
Rocks glass filled with cube ice
1 jigger Bacardi rum
1 jigger lime juice
¼ oz grenadine

Banana Daiquiri Blend

Large cocktail glass, chilled
1 jigger light rum
1 jigger lime juice
1 tsp sugar or ¼ oz simple syrup
½ ripe banana, crushed
Banana slice

Banshee (*White Monkey*) Shake or Blend

Cocktail glass, chilled, or
Rocks glass filled with cube ice
¾ jigger light crème de cacao
¾ jigger crème de banana
1 jigger cream

Between the Sheets Shake or Blend

Cocktail glass, chilled, or
Rocks glass filled with cube ice
½ jigger rum
½ jigger brandy
½ jigger lemon juice
½ jigger triple sec

Blackjack Stir

Cocktail glass filled with crushed ice (frappé style), or
Rocks glass filled with cube ice
1 jigger Kirschwasser
1 jigger coffee, iced
¼ oz brandy

Black Magic Build

Rocks glass
Full glass cube ice
1 jigger vodka
½ jigger Kahlúa
Few drops lemon juice
Twist

Black Russian Build

Rocks glass
Full glass cube ice
1 jigger vodka
½ jigger Kahlúa

Black Velvet Build

Collins, cooler, or zombie glass
1 part Guinness, chilled
1 part champagne
Pour simultaneously into glass; do not stir.

Black Watch Build

Rocks glass
Full glass cube ice
1 jigger scotch
½ jigger Kahlúa
Twist

Bleeding Clam, Clamdigger Build

Highball or collins glass
½ glass cube ice
1 jigger vodka
1 jigger clam juice
3 oz tomato juice
½ oz fresh lemon juice
2–3 dashes Worcestershire
2 drops Tabasco
Salt, pepper
(Or Bloody Mary mix to fill)
Lemon wheel

Bloody Bull Build

Highball or collins glass (rimmed with salt)
½ glass cube ice
1 jigger vodka
1½ oz tomato juice
1½ oz beef bouillon
½ oz fresh lemon juice
2–3 dashes Worcestershire
2 drops Tabasco
Salt, pepper, (celery salt)
(Or equal parts beef bouillon and Bloody Mary mix to fill)
Celery stick or lemon wheel

Bloody Maria Build

Highball or collins glass (rimmed with salt)
½ glass cube ice
1 jigger tequila
3 oz tomato juice
½ oz fresh lemon juice
2–3 dashes Worcestershire
2 drops Tabasco
Salt, pepper, (celery salt)
(Or Bloody Mary mix to fill)
Celery stick or lemon or lime wedge

Bloody Mary Build

Highball or collins glass (rimmed with salt)
½ glass cube ice
1 jigger vodka
3 oz tomato juice
½ oz fresh lemon juice
2–3 dashes Worcestershire
2 drops Tabasco
Salt, pepper, (celery salt)
(Or Bloody Mary mix to fill)
Celery stick, lemon wheel, or lime wedge

Blue Blazer
(*Jerry Thomas's recipe, word for word*)

2 large, silver-plated mugs with handles
1 wineglass scotch in one mug
1 wineglass boiling water in the other mug
Ignite the whisky with fire, and while blazing mix both ingredients by pouring them four or five times from one mug to the other. If well done this will have the appearance of a continued stream of liquid fire.
Sweeten with 1 teaspoonful of powdered white sugar, and serve in a small bar tumbler, with a piece of lemon peel.

Blue-Tailed Fly Shake or Blend

Cocktail glass, chilled, or
Rocks glass filled with cube ice
¾ jigger light crème de cacao
¾ jigger blue curaçao
1 jigger cream

Boccie Ball Build

Highball glass
½ glass cube ice
1 jigger amaretto
Orange juice to fill, or
Equal parts orange juice/soda to fill

The Boss Build

Rocks glass
Full glass cube ice
1 jigger bourbon
½ jigger amaretto

Brandy Alexander Shake or Blend

Cocktail glass, chilled, or
Rocks glass filled with cube ice
¾ jigger brandy
¾ jigger dark crème de cacao
1 jigger cream

Brandy Milk Punch. *See* Milk Punch

Brave Bull Build

Rocks glass
Full glass cube ice
1 jigger tequila
½ jigger Kahlúa

Brown Squirrel Shake or Blend

Cocktail glass, chilled, or
Rocks glass filled with cube ice
¾ jigger dark crème de cacao
¾ jigger amaretto
1 jigger cream

Buck Build

Highball glass
¾ glass cube ice
Lemon wedge, squeezed, shell in glass
1 jigger liquor of choice
Ginger ale to fill

Bullshot Build

Highball or collins glass
¾ glass cube ice
1 jigger vodka
¼ oz lemon juice
2–3 dashes Worcestershire
2 drops Tabasco
(Salt), (celery salt)
Beef bouillon to fill
Lemon or lime wedge or celery stick

Café Calypso Build

Cup, mug, or wine glass
1 jigger rum
½ jigger dark crème de cacao or Tia Maria or 1 tsp brown sugar
Coffee to fill, hot
Top with whipped cream

Café Diablo Build

Cup, mug, or wine glass
1 jigger cognac
¾ jigger Grand Marnier or Cointreau
½ jigger Sambuca
¼ tsp sugar or ⅜ tsp simple syrup
Grated orange rind
Sprinkle cinnamon, cloves, allspice
Coffee to fill, hot
Float ½ oz cognac, flame

Café Pucci Build

Cup, mug, or wine glass
¾ jigger Trinidad rum
¾ jigger amaretto
Coffee to fill, hot
Top with whipped cream

Café Royale I Build

Cup, mug, or wine glass
1 jigger bourbon or brandy
1 tsp sugar or ¼ oz simple syrup
Coffee to fill, hot
Top with whipped cream

Café Royale II Build

Cup, mug, or wine glass
½ jigger Metaxa
½ jigger Galliano
Coffee to fill, hot
Top with whipped cream

Campari and Soda Build

Highball glass
¾ glass cube ice
1 jigger Campari
Soda to fill
Lime wedge

Canadian Driver Build

Highball glass
½ glass cube ice
1 jigger Canadian whisky
Orange juice to fill

Cape Codder Build

Highball glass
½ glass cube ice
1 jigger vodka
Cranberry juice to fill
Lime wedge

Cappucchino Build

Cup, mug, or wine glass
½ oz dark crème de cacao
½ oz Galliano
¼ oz rum
¼ oz brandy
¼ oz cream
Espresso to fill, hot
Top with whipped cream

Champagne Cocktail Build

Champagne glass, chilled
1 sugar cube or 1 tsp sugar or ¼ oz simple syrup
1 dash Angostura bitters
Fill with champagne
Twist

Charley Goodlay, Charley Goodleg Build

Highball glass
½ glass cube ice
1 jigger tequila
Orange juice to fill
Float ½ jigger Galliano

Chi Chi — Blend

12-oz glass
¾ glass ice, cube or crushed
1½ jiggers vodka
1–2 jiggers coconut milk or cream of coconut
1–2 jiggers pineapple juice or crushed pineapple
(Or Build with Piña Colada mix to fill)
Cherry, pineapple spear, lime

Clover Club — Shake or Blend

Cocktail glass, chilled, or
Rocks glass filled with cube ice
1 jigger gin
½ jigger lemon juice
¼ jigger grenadine
1 tsp egg white

Cobra (*Hammer, Sloe Screw*) — Build

Highball glass
½ glass cube ice
1 jigger sloe gin
Orange juice to fill

Colorado Bulldog — Shake/Build or Blend/Build

Highball glass
¾ glass cube ice
1 jigger vodka or tequila
½ jigger Kahlúa
1 jigger cream
Cola to fill

Comfortable Screw — Build

Highball glass
½ glass cube ice
1 jigger Southern Comfort
Orange juice to fill

Cossack (*White Spider*) — Build

Rocks glass
Full glass cube ice
1 jigger vodka
½ jigger white crème de menthe

Cuba Libre — Build

Highball glass
¾ glass cube ice
1 jigger rum
Cola to fill
Lime wedge, squeezed

Cuban Screwdriver — Build

Highball glass
½ glass cube ice
1 jigger rum
Orange juice to fill

Cucumber — Shake or Blend

Cocktail glass, chilled, or
Rocks glass filled with cube ice
1 jigger green crème de menthe
(½ jigger brandy or gin)
1 jigger cream

Daiquiri — Shake or Blend

1 jigger light rum
1 jigger lime juice
1 tsp sugar or ¼ oz simple syrup
(Or 1 jigger sweet-sour mix)

Daisy — Build

Highball or stem glass or silver mug
Full glass crushed ice
1 jigger liquor of choice
½ jigger lemon juice
½ tsp sugar or ¾ tsp simple syrup
(Or ½ jigger sweet-sour mix)
1 tsp grenadine
Stir until glass frosts
Cherry, orange slice

Danish Mary Build

Highball or collins glass (rimmed with salt)
½ glass cube ice
1 jigger aquavit
3 oz tomato juice
½ oz fresh lemon juice
2–3 dashes Worchestershire
2 drops Tabasco
Salt, pepper, (celery salt)
} Or Bloody Mary mix to fill
Celery stick or lemon or lime wedge

Diablo Stir

Cocktail glass, chilled, or
Rocks glass filled with cube ice
½ jigger brandy
½ jigger dry vermouth
2 dashes bitters
3 dashes orange bitters or ½ oz orange curaçao or triple sec
Cherry or twist

Dirty Mother Build

Rocks glass
Full glass cube ice
1 jigger brandy
½ jigger Kahlúa
(Cream float)

Dreamcicle Shake/Build or Blend/Build

Highball glass
¾ glass cube ice
1 jigger amaretto
2 oz orange juice
1 jigger heavy cream
½ tsp sugar or ¾ tsp simple syrup
Float ½ jigger Galliano

Dry Manhattan Stir

Cocktail glass, chilled, or
Rocks glass filled with cube ice
6 parts whisky
1 part dry vermouth
Lemon twist

Dutch Coffee Build

Cup, mug, or wine glass
1 jigger Vandermint
Coffee to fill, hot
Top with whipped cream

Easy Rider Build

Highball glass
¾ glass cube ice
1 jigger amaretto
Soda to fill
Cherry

Eggnog Shake or Blend

Collins glass, chilled
1 egg
1 jigger brandy
1 tsp sugar or ¼ oz simple syrup
4 oz milk
Sprinkle nutmeg

Eggnog, hot Blend/Build

Highball glass
1 egg
½ jigger rum
½ jigger brandy
1 tsp sugar or ¼ oz simple syrup
Hot milk to fill
Sprinkle nutmeg

El Presidente — Shake or Blend

Cocktail glass, chilled, or
Rocks glass filled with cube ice
6 parts gold rum
1 part orange curaçao
1 part dry vermouth
Dash grenadine
Dash bitters or lime juice

Flip — Shake or Blend

Stem glass, 5 to 8 oz, chilled
1 egg
1 jigger liquor or fortified wine of choice
1 tsp sugar or ¼ oz simple syrup
Sprinkle nutmeg

Fog Cutter — Shake or Blend

Collins glass
½ glass cube ice
½ jigger light rum
½ jigger brandy
½ jigger gin
1 jigger orange juice
1 jigger Mai Tai mix
Cherry, pineapple spear

Fog Horn — Build

Highball glass
¾ glass cube ice
Juice of half lime; drop shell in glass
1 jigger gin
Ginger ale to fill

Frappé — Build

Cocktail glass, chilled
Crushed ice to fill
1 jigger liqueur of choice poured evenly over ice
Short straws

Freddy Fudpucker — Build

Highball glass
½ glass cube ice
1 jigger tequila
Orange juice to fill
Float ½ jigger Galliano

French Connection — Build

Rocks glass
Full glass cube ice
1 jigger brandy
½ jigger amaretto

French Screwdriver — Build

Highball glass
½ glass cube ice
1 jigger cognac or armagnac
Orange juice to fill

French 75 — Shake/Build or Blend/Build

Collins glass
¾ glass ice, cube or crushed
1 jigger gin
1 jigger lemon juice
1 tsp sugar or ¼ oz simple syrup
Or
1 jigger sweet-sour mix
Champagne to fill
Lemon wedge, cherry, orange slice
Long straws

French 95 — Shake/Build or Blend/Build

Collins glass
¾ glass ice, cube or crushed
1 jigger bourbon
1 jigger lemon juice
1 tsp sugar or ¼ oz simple syrup
Or
1 jigger sweet-sour mix
Champagne to fill
Lemon wedge, cherry, orange slice
Long straws

French 125 — Shake/Build or Blend/Build

Collins glass
¾ glass ice, cube or crushed
1 jigger brandy
1 jigger lemon juice } Or
1 tsp sugar or ¼ oz simple syrup } 1 jigger sweet-sour mix
Champagne to fill
Lemon wedge, cherry, orange slice
Long straws

Galliano Stinger — Build

Rocks glass
Full glass cube ice
1 jigger brandy
½ jigger Galliano

Gibson — Stir

Cocktail glass, chilled, or
Rocks glass filled with cube ice
6 parts gin
1 part dry vermouth
Cocktail onion

Gimlet — Shake or Blend

Cocktail glass, chilled, or
Rocks glass filled with cube ice
1 jigger gin
½ jigger Rose's lime juice
Squeeze of lime

Gin and Tonic — Build

Highball glass
¾ glass cube ice
1 jigger gin
Tonic to fill
Lime wedge, squeezed

Gin Fizz — Shake/Build or Blend/Build

Highball glass
¾ glass cube ice
1 jigger gin
1 jigger lemon juice } Or
1 tsp sugar or ¼ oz simple syrup } 1 jigger sweet-sour mix
Soda to fill

Godfather — Build

Rocks glass
Full glass cube ice
1 jigger scotch or bourbon
½ jigger amaretto

Godmother — Build

Rocks glass
Full glass cube ice
1 jigger vodka
½ jigger amaretto

Golden Cadillac — Shake or Blend

Cocktail glass, chilled, or
Rocks glass filled with cube ice
¾ jigger light crème de cacao
¾ jigger Galliano
1 jigger cream

Golden Dream — Shake or Blend

Cocktail glass, chilled, or
Rocks glass filled with cube ice
1 jigger Galliano
½ jigger Cointreau or triple sec
½ jigger orange juice
1 jigger cream

Golden Fizz — Shake/Build or Blend/Build

Highball or collins glass
¾ glass cube ice
1 jigger gin
1 jigger lemon juice } Or
1 tsp sugar or ¼ oz simple syrup } 1 jigger sweet-sour mix
1 egg yolk
Soda to fill

Golden Screw — Build

Highball glass
½ glass cube ice
1 jigger Galliano
Orange juice to fill

Good and Plenty — Build

Rocks glass
Full glass cube ice
1 jigger Kahlúa or Tia Maria
1 jigger ouzo or anisette

Grasshopper — Shake or Blend

Cocktail glass, chilled, or
Rocks glass filled with cube ice
¾ jigger green crème de menthe
¾ jigger light crème de cacao
1 jigger cream

Grasshopper (*ice cream version*) — Blend or Shake-Mix

8-oz stem glass, chilled
1 scoop vanilla ice cream
¾ jigger green crème de menthe
¾ jigger white crème de cacao

Greek Coffee — Build

Cup, mug, or wine glass
1 jigger Metaxa
⅓ jigger ouzo
Coffee to fill, hot
Top with whipped cream

Greek Screwdriver — Build

Highball glass
½ glass cube ice
1 jigger ouzo
Orange juice to fill

Greek Stinger — Build

Rocks glass
Full glass cube ice
1 jigger Metaxa
½ jigger white crème de menthe

Greyhound — Build

Highball glass
½ glass cube ice
1 jigger vodka
Grapefruit juice to fill

Hammer (*Cobra, Sloe Screw*) — Build

Highball glass
½ glass cube ice
1 jigger sloe gin
Orange juice to fill

Harbor Lights I — Build (*pousse-café*)

Straight-sided pony or cordial glass
2 parts tequila
1 part Kahlúa
Float 151° rum
Ignite rum and serve flaming

Harbor Lights II Build (*pousse-café*)

Straight-sided pony or cordial glass
2 parts vodka
1 part green crème de menthe
Orange wheel across top of glass
1 cube sugar soaked in 151° rum on orange wheel
Ignite sugar and serve flaming

Harvey Wallbanger Build

Highball glass
½ glass cube ice
1 jigger vodka
Orange juice to fill
Float ½ jigger Galliano

Henrietta Wallbanger Build

Highball glass
½ glass cube ice
1 jigger vodka
Grapefruit juice to fill
Float ½ jigger Galliano

Hot Buttered Rum Build

Cup, mug, or old-fashioned glass
1 jigger dark rum
1 tsp butter-sugar mixture (equal parts butter/sugar premixed with cinnamon, nutmeg, cloves, salt)
Boiling water to fill
Stick cinnamon as stir stick

Hot Lemonade Build

Highball glass
1 jigger whisky, rum, or brandy
1 jigger lemon juice } Or 1 jigger sweet-sour mix
1 tsp sugar or ¼ oz simple syrup }
Hot water to fill

Hot Toddy Build

Cup, mug, or old-fashioned glass
1 jigger liquor of choice
1 tsp sugar or ¼ oz simple syrup
(Mixed spices: cloves, cinnamon, nutmeg)
Hot water to fill
Sprinkle nutmeg
Cinnamon stick stirrer

Hurricane Shake or Blend

Hurricane glass
¾ glass crushed or cube ice
1 jigger light rum
1 jigger dark rum
1 jigger lime juice } Or 2–3 jiggers passion fruit syrup
1 jigger pineapple juice }
1 jigger orange juice }
1 tsp sugar or ¼ oz simple syrup
¼ jigger grenadine
(Lime wheel, flagged orange, pineapple spear, mint sprig)

Il Magnifico Shake or Blend

Cocktail glass, chilled, or
Rocks glass filled with cube ice
½ jigger Tuaca
½ jigger triple sec
1 jigger cream

International Stinger Build

Rocks glass
Full glass cube ice
1 jigger brandy
1 jigger white crème de menthe

Irish Coffee Build

Cup, mug, or wine glass
1 jigger Irish whiskey
1 tsp sugar or ¼ oz simple syrup
Coffee to fill, hot
Top with whipped cream

Irish Mafia (*a St. Patrick's Day Special*) Blend

8-oz wine glass, chilled
1 jigger green crème de menthe
1 jigger amaretto
1 jigger heavy cream
Ice to just above liquor level in blender cup
Blend until cream is whipped

Irish Screwdriver Build

Highball glass
½ glass cube ice
1 jigger Irish whiskey
Orange juice to fill

Irish Stinger Build

Rocks glass
Full glass cube ice
1 jigger brandy
½ jigger green crème de menthe

Italian Screwdriver Build

Highball glass
½ glass cube ice
1 jigger Galliano
Orange juice to fill

Jack Rose Shake or Blend

Cocktail glass, chilled, or
Rocks glass filled with cube ice
1 jigger applejack or apple brandy
½ jigger lemon or lime juice
½ jigger grenadine
(1 tsp egg white)

Jamaican Coffee Build

Cup, mug, or wine glass
1 jigger rum
½ jigger Tia Maria
Coffee to fill, hot
Top with whipped cream

Jenny Wallbanger Build

Highball glass
½ glass cube ice
1 jigger vodka
Equal parts orange juice/heavy cream to fill
Float ½ jigger Galliano

Joe Canoe (*Jonkanov*) Build

Highball glass
½ glass cube ice
1 jigger rum
Orange juice to fill
Float ½ jigger Galliano

John Collins Shake/Build or Blend/Build

Collins glass
¾ glass ice, cube or crushed
1 jigger bourbon
1 jigger lemon juice } Or 1 jigger sweet-sour mix
2 tsp sugar or ½ oz simple syrup }
Soda to fill
Cherry, (orange slice)

Kentucky Screwdriver Build

Highball glass
½ glass cube ice
1 jigger bourbon
Orange juice to fill

Kioki (*Keoke*) Coffee Build

Cup, mug, or wine glass
1 jigger brandy
½ jigger dark crème de cacao
Coffee to fill, hot
(1 tsp brown sugar)
Top with whipped cream

Kir Build

8-oz wine glass
½ glass cube ice
4 oz chilled white wine
½ oz crème de cassis
(Twist)

Knucklebuster (*Knucklehead*) Build

Rocks glass
Full glass cube ice
1 jigger scotch
½ jigger Drambuie

Left-Handed Screwdriver Build

Highball glass
½ glass cube ice
1 jigger gin
Orange juice to fill

Little Princess Stir

Cocktail glass, chilled, or
Rocks glass filled with cube ice
1 jigger rum
1 jigger sweet vermouth

Long Island Tea Build

10- to 12-oz collins or wine glass
Full glass ice
⅓ oz each vodka, rum, triple sec
¼ oz each gin, tequila
1½ oz sweet-sour mix
1½ oz 7-Up
½ oz cola or
to the color of iced tea
Lime wedge, squeezed

Madras Build

Highball glass
½ glass cube ice
1 jigger vodka
Equal parts orange juice/cranberry juice
to fill

Mai Tai Shake or Blend

Collins glass
¾ glass cube or crushed ice
1 jigger light rum
(1 jigger dark rum)
1 lime (juice and shell)
½ jigger orange curaçao or triple sec
½ oz simple syrup or 2 tsp sugar
½ jigger orgeat
(Pineapple juice)
(Orange juice)
} Or Mai Tai mix to fill
Pineapple stick, cherry, mint sprig

Mamie's Sister Build

Collins glass
¾ glass cube ice
Lime half, squeezed, shell in glass
1 jigger gin
Ginger ale to fill

Mamie's Southern Sister Build

Highball glass
¾ glass cube ice
1 jigger bourbon
Ginger ale to fill

Mamie Taylor Build

Highball glass
¾ glass cube ice
1 jigger scotch
Ginger ale to fill
Lemon twist or squeeze of lime

Manhattan Stir

Cocktail glass, chilled, or
Rocks glass filled with cube ice
6 parts whisky (bourbon or blend)
1 part sweet (Italian) vermouth
(Dash bitters)
Stemmed cherry

Margarita Shake or Blend

Cocktail or Margarita glass, chilled, rimmed with salt
1 jigger tequila
⅓ jigger lime juice
⅓ jigger triple sec

Margarita, Frozen Blend

8- to 10-oz stem glass, chilled, rimmed with salt
1 jigger tequila
½ jigger lime juice
½ jigger triple sec
Crushed ice in blender to just above liquor level

Martini Stir

Cocktail glass, chilled, or
Rocks glass filled with cube ice
6 parts gin
1 part dry (French) vermouth
Olive or lemon twist

Mexican Coffee Build

Cup, mug, or wine glass
1 jigger tequila
½ jigger Kahlúa
Coffee to fill, hot
Top with whipped cream

Mexican Screwdriver Build

Highball glass
½ glass cube ice
1 jigger tequila
Orange juice to fill

Milk Punch Build

Collins glass
¾ glass cube ice
1 jigger liquor of choice
1 tsp sugar or ¼ oz simple syrup
4 oz milk
Sprinkle nutmeg

Milk Punch, Hot Build

Mug or collins glass
¾ jigger rum
¾ jigger brandy
1 tsp sugar or ¼ oz simple syrup
Hot milk to fill
Sprinkle nutmeg

Mimosa Build

Wine or champagne glass, chilled
Equal parts champagne and orange juice to fill

Mint Julep Build

16-oz glass, chilled
10–12 fresh mint leaves, bruised gently
1 tsp sugar with splash of soda, or ¼ oz simple syrup
Muddle sugar and mint
Add ½ glass finely crushed ice
1 jigger bourbon
Stir up and down until well mixed
Add crushed ice to fill
Add another jigger bourbon
Stir up and down until glass frosts
Long straws
Mint sprigs dipped in sugar, (fruit)

Mist Build

Rocks glass
Full glass cracked or crushed ice
1 jigger liquor of choice
(Twist)

Moscow Mule Build

Copper mug, highball size
¾ mug cube ice
1 jigger vodka
Ginger beer to fill
Lime half, squeezed

Mount Olympus Coffee Build

Cup, mug, or wine glass
1 jigger Metaxa
Coffee to fill, hot
Top with whipped cream
Lace whipped cream with amaretto

Muddy River (*Sombrero*) Shake or Blend

Cocktail glass, chilled, or
Rocks glass filled with cube ice
1½ jiggers Kahlúa
1 jigger cream

Mud Slide Build

Cocktail glass, chilled, or
Rocks glass filled with cube ice
¾ jigger Bailey's Irish Cream
¾ jigger dark crème de cacao

Negroni Stir

Cocktail glass, chilled, or
Rocks glass filled with cube ice
1 jigger gin
½ jigger sweet vermouth
½ jigger Campari
Lemon or orange twist

Negroni Highball Stir/Build

Highball glass
¾ glass cube ice
1 jigger gin
½ jigger sweet vermouth
½ jigger Campari
Soda to fill
Lemon twist or orange slice

New Orleans Fizz Shake/Build or Blend/Build

Collins or 10-oz stem glass, chilled
1 jigger gin
1 jigger lemon juice
½ jigger lime juice
1 tsp sugar or ¼ oz simple syrup
1 egg white
(½ jigger cream)
3 dashes orange flower water (or float)
Soda to fill

Old-Fashioned Build

Old-fashioned glass
1–2 tsp sugar or ¼–½ oz simple syrup or
  1 sugar cube dissolved in splash of
  soda or water
1–3 dashes Angostura bitters
Full glass cube ice
Up to 2 oz bourbon or blended whisky
(Soda or water to fill)
Cherry, orange slice, (lemon twist)

Orange Blossom Stir (or Blend)

Cocktail glass, chilled, rimmed with sugar
1 jigger gin
1 jigger orange juice
(½ tsp sugar)
(Half orange slice)

Paddy — Stir

Cocktail glass, chilled, or
Rocks glass filled with cube ice
1 jigger Irish whiskey
1 jigger sweet vermouth
1 dash bitters

Panama (*Rum Alexander*) — Shake or Blend

Cocktail glass, chilled, or
Rocks glass filled with cube ice
¾ jigger rum
¾ jigger light crème de cacao
1 jigger cream

Peach Margarita, Frozen — Blend

8-oz or 10-oz stem glass, chilled, (rimmed with sugar)
1 jigger tequila
½ jigger peach liqueur
1½ jiggers pureed peaches
Crushed ice to just above ingredients
Fresh peach wedge in season

Perfect Manhattan — Stir

Cocktail glass, chilled, or
Rocks glass filled with cube ice
6 parts whisky
½ part dry vermouth
½ part sweet vermouth
Lemon twist

Perfect Martini — Stir

Cocktail glass, chilled, or
Rocks glass filled with cube ice
6 parts gin
½ part dry vermouth
½ part sweet vermouth
(Dash bitters)
(Lemon twist)

Persuader — Build

Highball glass
½ glass cube ice
½ jigger amaretto
½ jigger brandy
Orange juice to fill

Piña Colada — Blend

12-oz collins or stem glass
¾ glass crushed ice
1–1½ jiggers light rum
1–2 jiggers coconut milk or cream of coconut
1–2 jiggers pineapple juice or crushed pineapple
} Or Build with Piña Colada mix to fill
Cherry, pineapple spear, lime

Pineapple Rum Royal — Shake or Blend

Whole fresh pineapple, hollowed out
1 scoop cube ice in pineapple
1 jigger Jamaica rum
1 jigger light rum
½ jigger cherry liqueur
½ jigger lemon juice
½ jigger orange juice
½ jigger pineapple juice
1 jigger cream
Pineapple spear, red and green cherries, long straws

Pineapple Sunrise — Build

Highball glass
½ glass cube ice
1 jigger tequila
Pineapple juice to fill
2 dashes grenadine

Pink Lady (*current version*) — Shake or Blend

Cocktail glass, chilled, or
Rocks glass filled with cube ice
1 jigger gin
½ jigger grenadine
1 jigger cream

Pink Lady (*yesterday's version*) — Shake or Blend

Cocktail glass, chilled
1 jigger gin
1 jigger apple brandy
½ jigger lemon or lime juice
½ jigger grenadine
1 tsp egg white

Pink Squirrel — Shake or Blend

Cocktail glass, chilled, or
Rocks glass filled with cube ice
¾ jigger light crème de cacao
¾ jigger crème de noyaux
1 jigger cream

Planter's Punch — Shake or Blend

Collins glass
¾ glass ice, cube or crushed
1 jigger light rum
1 jigger Jamaica rum
1 oz lemon or lime juice
1 tsp sugar or ¼ oz simple syrup
} Or 1 oz sweet-sour mix
1 oz orange juice
(1–2 oz pineapple juice)
½ oz grenadine
(Soda to fill)
Pineapple, (orange, cherry, mint)

Portuguese Driver — Build

Highball glass
½ glass cube ice
1 jigger port wine
Orange juice to fill

Pousse-café — Build

Straight-sided cordial glass
⅙ grenadine (red)
⅙ dark crème de cacao (brown)
⅙ white crème de menthe (white)
⅙ apricot-flavored brandy (orange)
⅙ green Chartreuse (green)
⅙ brandy (amber)

Prairie Oyster (*a historic hangover "cure"*) — Build

Cocktail glass
(1 jigger brandy)
Egg yolk, whole
2 dashes vinegar
1 tsp Worcestershire
1 dash Tabasco
(1 tsp catsup)
Sprinkle salt
Swallow in one gulp

Presbyterian — Build

Highball glass
¾ glass cube ice
1 jigger any liquor
Equal parts ginger ale/soda to fill, or
Equal parts water/soda (Northeast), or
Equal parts 7-Up/soda (South)

Purple Bunny — Shake or Blend

Cocktail glass, chilled, or
Rocks glass filled with cube ice
¾ jigger cherry-flavored brandy
1 jigger cream

Ramos Fizz — Shake/Build or Blend/Build

Collins glass, chilled (sugared rim)
1 jigger gin
½ jigger lemon juice
½ jigger lime juice
1 tsp sugar or ¼ oz simple syrup
1 egg white
(Or 1 jigger sweet-sour mix with frothee)
1 jigger cream
3–4 dashes orange flower water (or float)
Soda to fill

Ramos Fizz *(ice cream version)* — Blend/Build or Shake-Mix/Build

12-oz glass, chilled
3 scoops vanilla ice cream (3¾ oz)
1 jigger gin
½ jigger lemon juice
½ jigger lime juice
1 tsp sugar or ¼ oz simple syrup
1 egg white
(Or 1 jigger sweet-sour mix with frothee)
3–4 dashes orange flower water
Soda to fill

Red Snapper (*Gin Bloody Mary*) — Build

Highball or collins glass (rimmed with salt)
½ glass cube ice
1 jigger gin
3 oz tomato juice
½ oz fresh lemon juice
2–3 dashes Worcestershire
2 drops Tabasco
Salt, pepper, (celery salt)
(Or Bloody Mary Mix to fill)
Celery stick or lemon wheel or lime wedge

Rickey — Build

Highball glass
¾ glass cube ice
Half lime, squeezed, shell in glass
1 jigger liquor of choice
Soda to fill

Rob Roy — Stir

Cocktail glass, chilled, or
Rocks glass filled with cube ice
4 parts scotch
1 part sweet vermouth
1 dash bitters (Angostura or orange)
Stemmed cherry

Rob Roy, Dry — Stir

Cocktail glass, chilled, or
Rocks glass filled with cube ice
4 parts scotch
1 part dry vermouth
Olive or lemon twist

Roman Coffee — Build

Cup, mug, or wine glass
1 jigger Galliano
Coffee to fill, hot
Top with whipped cream

Royal Fizz — Shake/Build or Blend/Build

Collins glass, chilled
1 jigger gin
1 jigger lemon juice
1 tsp sugar or ¼ oz simple syrup
(Or 1 jigger sweet-sour mix)
Whole egg
(½ jigger cream)
Soda to fill

Royal Street Coffee — Build

Cup, mug, or wine glass
1 jigger amaretto
½ jigger Kahlúa
Coffee to fill, hot
Top with whipped cream
Sprinkle nutmeg

Rum Martini Stir

Cocktail glass, chilled, or
Rocks glass filled with cube ice
4 parts rum
1 part dry vermouth
(1 dash orange bitters)
Olive or onion or lemon twist

Rum Milk Punch. ***See* Milk Punch**

Russian Bear (*Vodka Alexander*) Shake or Blend

Cocktail glass, chilled, or
Rocks glass filled with cube ice
¾ jigger vodka
¾ jigger light crème de cacao
1 jigger cream

Russian Sunrise Build

Highball glass
½ glass cube ice
1 jigger vodka
Orange juice to fill
½ jigger grenadine

Rusty Nail Build

Rocks glass
Full glass cube ice
1 jigger scotch
½ jigger Drambuie

Rye and Ginger Build

Highball glass
¾ glass cube ice
1 jigger blended whisky
Ginger ale to fill

Salty Dog Build

Collins glass, salt-rimmed
½ glass cube ice
1 jigger vodka
Grapefruit juice to fill

Sazarac Build

Old-fashioned glass, chilled
Splash of pernod rolled in glass to coat inside; discard excess
2 dashes Peychaud's bitters
1 tsp simple syrup
Up to 2 oz straight rye (or bourbon)
1 or 2 large ice cubes
Lemon twist or cherry

Scarlett O'Hara Shake or Blend

Cocktail glass, chilled, or
Rocks glass filled with cube ice
1 jigger Southern Comfort
1 jigger cranberry juice
¼ jigger lime juice

Scorpion Shake or Blend

Collins glass
¾ glass ice, cube or crushed
1 jigger rum
1 jigger brandy
2 jiggers orange juice
1 jigger lemon juice
1 jigger orgeat
Orange, pineapple, mint, or 151° rum float

Scotch and Milk Build

Highball glass
¾ glass cube ice
1 jigger scotch
Milk to fill

Scotch and Soda Build

Highball glass
¾ glass cube ice
1 jigger scotch
Soda to fill

Scotch Driver — Build

Highball glass
½ glass cube ice
1 jigger scotch
Orange juice to fill

Scotch Plaid (Rusty Nail) — Build

Rocks glass
Full glass cube ice
1 jigger scotch
½ jigger Drambuie

Screwdriver — Build

Highball glass
½ glass cube ice
1 jigger vodka
Orange juice to fill

Seven and Seven — Build

Highball glass
¾ glass cube ice
1 jigger 7-Crown whisky
7-Up to fill

Siberian — Build

Rocks glass
Full glass cube ice
1 jigger vodka
½ jigger Kahlúa
Float ½ jigger brandy

Side Car — Shake or Blend

Cocktail glass, chilled, (rimmed with sugar)
1 jigger brandy
1 jigger lemon juice
1 jigger triple sec or Cointreau

Silver Bullet I — Stir

Cocktail glass, chilled, or
Rocks glass filled with cube ice
6 parts gin
1 part scotch
Lemon twist

Silver Bullet II — Stir/Build

Cocktail glass, chilled, or
Rocks glass filled with cube ice
6 parts gin
1 part dry vermouth
Float 1 part scotch

Silver Fizz — Shake/Build or Blend/Build

Highball or collins glass
¾ glass cube ice
1 jigger gin
1 jigger lemon juice
1 tsp sugar or ¼ oz simple syrup
1 egg white
} Or 1 jigger sweet-sour mix with frothee
Soda to fill

Singapore Sling — Shake/Build or Blend/Build

Collins glass
¾ glass cube ice
1 jigger gin
¼ jigger cherry-flavored brandy
1 jigger lemon juice
1 tsp sugar or ¼ oz simple syrup
} Or 1 jigger sweet-sour mix
Soda to fill
(4 drops Bénédictine)
(4 drops brandy)

Skip and Go Naked — Build

Collins glass
¾ glass ice, cube or crushed
1 jigger gin or vodka
Beer to fill or equal parts beer and soda

Sloe Gin Fizz — Shake/Build or Blend/Build

Highball glass
¾ glass cube ice
1 jigger sloe gin
1 jigger lemon juice
1 tsp sugar or ¼ oz simple syrup
} Or 1 jigger sweet-sour mix
Soda to fill

Sloe Screw (Cobra, Hammer) — Build

Highball glass
½ glass cube ice
1 jigger sloe gin
Orange juice to fill

Smash — Build

Old-fashioned or rocks glass
1 cube sugar in splash of soda or water or ¼ oz simple syrup, muddled with
4 sprigs fresh mint
¾ glass cube ice
Up to 2 oz liquor of choice
(Orange slice, cherry)

Smith & Kerns — Build

Highball glass
¾ glass cube ice
1 jigger dark crème de cacao
Milk or cream to ¾ full
Soda to fill

Smoothy — Build

Rocks glass
Full glass cube ice
1 jigger bourbon
½ jigger white crème de menthe

Snowshoe — Build

Rocks glass
Full glass cube ice
1 jigger tequila or brandy
½ jigger peppermint schnapps

Sombrero I — Shake or Blend

Cocktail glass, chilled, or
Rocks glass filled with cube ice
1½ jiggers Kahlúa
1 jigger cream

Sombrero II — Build

Highball glass
¾ glass cube ice
1 jigger any coffee-flavored liqueur
Milk or half and half to fill

Sour — Shake or Blend

Sour glass, chilled, or
Rocks glass filled with cube ice
1 jigger liquor of choice
1 jigger lemon juice
1 tsp sugar or ¼ oz simple syrup
} Or 1 jigger sweet-sour mix
Cherry and orange slice

Southern Screwdriver — Build

Highball glass
½ glass cube ice
1 jigger Southern Comfort
Orange juice to fill

Spanish Fly — Build

Rocks glass
Full glass cube ice
1 jigger tequila
½ jigger amaretto

Spritzer Build

Highball or collins or wine glass
¾ glass cube or crushed ice
Half fill with white wine
Soda to fill
Twist, lemon slice, or lime wedge

Starboard Lights Build (*pousse-café*)

Straight-sided cocktail or wine glass, 3½ oz
1½ oz vodka
1 oz green crème de menthe
Float 1 tsp 151° rum
Ignite rum and serve flaming

Stars and Stripes Build (*pousse-café*)

Straight-sided cordial glass
⅓ grenadine
⅓ milk or cream
⅓ crème d'Yvette

Stinger Build

Rocks glass
Full glass cube ice
1 jigger brandy
½ jigger white crème de menthe

Strawberries Diana Blend

10-oz wine glass, chilled
¼ cup fresh strawberries
1½ oz white wine
1 tsp simple syrup
Crushed ice to just above ingredients in cup

Strawberry Daiquiri Blend

Large cocktail glass, chilled
1 jigger light rum
½ jigger lime juice
1 tsp sugar or ¼ oz simple syrup
1½ jiggers fresh or frozen strawberries, crushed
Whole fresh strawberry

Strawberry Daiquiri, frozen Blend

8- to 10-oz stem glass, chilled
1 jigger rum
½ jigger lime juice
1 tsp sugar or ¼ oz simple syrup
1 jigger pureed strawberries or 6 fresh berries cut up
Crushed ice to just above ingredients in cup
Whole berry garnish

Susie Taylor Build

Collins glass
¾ glass cube ice
½ lime squeezed, hull in glass
1 jigger rum
Ginger ale to fill

Swizzle Build

Collins glass
1 tsp sugar dissolved in 2 oz soda
1 jigger lime juice
2 dashes bitters
Crushed ice to fill
1 jigger liquor of choice
Soda to fill
Stir until glass frosts

Tango Stir

Cocktail glass, chilled, or
Rocks glass filled with cube ice
1 jigger gin
¼ jigger sweet vermouth
4 dashes triple sec or curaçao

Tequila Fresa Shake or Blend

Rocks glass
Full glass crushed ice
1 jigger tequila
¼ cup fresh strawberries } Or ½ jigger strawberry liqueur
1 tsp sugar or ¼ oz simple syrup }
¼ jigger lime juice
Dash of orange bitters
Lime slice, (fresh strawberry)

Tequila Rose Shake or Blend

Cocktail glass, chilled, or
Rocks glass filled with cube ice
1 jigger tequila
⅓ jigger lime juice
⅓ jigger grenadine

Tequila Sunrise Build

Highball glass
½ glass cube ice
1 jigger tequila
Orange juice to fill
½ oz grenadine; do not stir

Tequini (*Tequila Martini*) Stir

Cocktail glass, chilled, or
Rocks glass filled with cube ice
4 parts tequila
1 part dry vermouth
Lemon twist or olive

Tijuana Sunrise Build

Highball glass
½ glass cube ice
1 jigger tequila
Orange juice to fill
1 dash Angostura bitters

Tom and Jerry Build

Mug or cup
1 heaping tablespoon Tom & Jerry batter (below)
1 jigger rum
1 jigger whisky or brandy
Hot milk or water to fill
Sprinkle nutmeg

Tom and Jerry batter

12 egg yolks }
6 Tbs sugar } Beat together
¼ tsp cinnamon }
¼ tsp cloves }
12 egg whites beaten until stiff
Fold whites into yolk mixture

Tom Collins Shake/Build or Blend/Build

Collins glass
¾ glass ice, cube or crushed
1 jigger gin
1 jigger lemon juice } Or 1 jigger sweet-sour mix } Or Build with collins mix to fill
2 tsp sugar or ½ oz simple syrup }
Soda to fill
Cherry, orange slice

Tootsie Roll Build

Highball glass
½ glass cube ice
1 jigger dark crème de cacao
Orange juice to fill

Velvet Hammer Shake or Blend

Cocktail glass, chilled, or
Rocks glass filled with cube ice
¾ jigger vodka or Cointreau
¾ jigger light crème de cacao
1 jigger cream

Vermouth Cassis Build

Highball or wine glass
½ glass cube ice
2 oz dry vermouth
½ oz crème de cassis
Soda to fill
(Lemon twist)
(May be served on the rocks without soda)

Virgin Mary Build

Highball glass
½ glass cube ice
4 oz tomato juice
½ oz lemon juice
2–3 dashes Worcestershire
2 drops Tabasco
Salt, pepper, (celery salt)
} Or 4 oz Bloody Mary Mix

Vodkatini (*Vodka Martini*) Stir

Cocktail glass, chilled, or
Rocks glass filled with cube ice
6 parts vodka
1 part dry vermouth
Lemon twist

Ward 8 Shake or Blend

Sour glass, chilled, or
Rocks glass filled with cube ice
1 jigger bourbon or blended whisky
1 jigger lemon juice
1 tsp sugar or ¼ oz simple syrup
} Or 1 jigger sweet-sour mix
¼ jigger grenadine

Watermelon Shake or Blend

Cocktail glass, chilled, or
Rocks glass filled with cube ice
¾ jigger Southern Comfort
¾ jigger orange juice
¾ jigger amaretto

White Cadillac Shake or Blend

Cocktail glass, chilled, or
Rocks glass filled with cube ice
¾ jigger light crème de cacao
¾ jigger Cointreau or triple sec
1 jigger cream

White Elephant (*Vodka Alexander*) Shake or Blend

Cocktail glass, chilled, or
Rocks glass filled with cube ice
¾ jigger vodka
¾ jigger light crème de cacao
1 jigger cream

White Monkey (*Banshee*) Shake or Blend

Cocktail glass, chilled, or
Rocks glass filled with cube ice
¾ jigger light crème de cacao
¾ jigger crème de banana
1 jigger cream

White Russian Shake or Blend

Cocktail glass, chilled, or
Rocks glass filled with cube ice
1 jigger vodka
¾ jigger Kahlúa
1 jigger cream

White Spider (*Cossack*) Build

Rocks glass
Full glass cube ice
1 jigger vodka
½ jigger white crème de menthe

White Way Build

Rocks glass
Full glass cube ice
1 jigger gin
½ jigger white crème de menthe

Wine Cooler Build

8-oz or 10-oz glass

¾ glass cube or crushed ice

Half fill with red wine

7-Up to fill

(Flagged orange/cherry), (mint sprigs), (twist)

Yellow Jacket Build

Highball glass

½ glass cube ice

1 jigger bourbon

Orange juice to fill

Zombie Shake or Blend

Zombie glass, 12–14 oz

¾ glass ice, cube or crushed

1 jigger light rum

1 jigger dark rum

1 jigger pineapple juice

1 jigger orange juice

½ jigger lime juice

(½ jigger papaya juice)

½ jigger apricot-flavored brandy

½ jigger passion fruit syrup or orange curaçao

Float ½ jigger 151° rum

Pineapple spear, orange wheel, (mint)

index . . .